Near-Rings and Near-Fields

Mathematics and Its Applications

Managing Editor:

M. HAZEWINKEL

Centre for Mathematics and Computer Science, Amsterdam, The Netherlands

Volume 336

Near-Rings and Near-Fields

Proceedings of the Conference on Near-Rings and Near-Fields,
Fredericton, New Brunswick, Canada, July 18–24, 1993

edited by

Yuen Fong
Department of Mathematics,
National Cheng Kung University,
Tainan, Taiwan, Republic of China

Howard E. Bell
Department of Mathematics,
Brock University,
St. Catherines, Ontario, Canada

Wen-Fong Ke
Department of Mathematics,
National Cheng Kung University,
Tainan, Taiwan, Republic of China

Gordon Mason
Department of Mathematics and Statistics,
University of New Brunswick,
Fredericton, New Brunswick, Canada

and

Günter Pilz
Institute for Mathematics,
Johannes Kepler Universität,
Linz, Austria

SPRINGER-SCIENCE+BUSINESS MEDIA, B.V.

A C.I.P. Catalogue record for this book is available from the Library of Congress.

ISBN 978-94-010-4160-7   ISBN 978-94-011-0359-6 (eBook)
DOI 10.1007/978-94-011-0359-6

Printed on acid free paper

Table of Contents

FOREWORD

This volume contains three invited lectures and 26 other papers presented at the 12th International Conference on Near-rings and Near-fields held at Fredericton, New Brunswick, Canada, July 18-24, 1993.

All the papers in this volume have been refereed. Thanks are due to all the referees, who were helpful and prompt in submitting their reports.

I am pleased to acknowledge financial assistance for the Conference provided by the Natural Sciences and Engineering Research Council of Canada, The New Brunswick Department of Advanced Education, Brock University (St. Catharines, Ontario) and the University of New Brunswick.

I wish to record my thanks to the Editorial Board: Howard Bell, who was also co-organizer of the Conference; Gunter Pilz, whose advice on conferences and their proceedings was most helpful; Yuen Fong, who so ably made the arrangements with the publisher; and the Managing Editor, Wen-Fong Ke, whose mastery of TEX has been of great help. Finally, I extend my sincerest appreciation to Mrs. Linda Guthrie of the UNB Deparmtent of Mathematics and Statistics, whose secretarial work for both the Conference and its Proceedings has been invaluable.

The Fredericton Conference was held 25 years after the first conference devoted entirely to near-rings and near-fields which took place at the Mathematisches Forschungsinstitut, Oberwolfach, Germany in 1968. To mark the occasion, Betsch's invited lecture gives a historical overview of the development of the subject while the other invited talks by Heatherly and Malone combine a retrospective look at two of the main themes of near-ring research (distributivity conditions and endomorphism near-rings, respectively) with up-to-the minute results.

In fact the development of near-rings and near-fields has matured to the point where the theory is substantial, the applications are numerous and both are amply documented. There are now four text books ([3], [5], [6], [9]) and five volumes of conference proceedings ([1], [2], [4], [7], [8]). Moreover, thanks to the foresight of the early workers in the field, a comprehensive classified bibliography was established and is updated regularly in the near-ring newsletter.

The papers in this volume reflect some of the diversity of the subject:

- Connections to geometry are discussed by Kiechle and Ke, Kolb, Kreuzer, and Ney.
- Centralizer near-rings are investigated by Bagley, Cannon, Kabza, Maxson, Smith, and van der Merwe.
- Ideals and radicals are considered in the invited lecture by Heatherly, in the papers of Birkenmeier and Botha and the paper by Birkenmeier, Heatherly, and Lee.
- Various near-ring modules are examined by Ahsan, Niewieczerzal, Peterson, and Scott.

- Topological considerations are found in Magill and in Magill and Misra.
- Applications to error-correcting codes are considered by Eggetsberger, and Wagner.
- Endomorphism near-rings are the subject of the paper by Saad, Syskin and Thomsen as well as the invited lecture by Malone.

The remaining papers touch on a variety of other topics: Connections to probability (Blackett), syntactic near-rings (Fong, Ke and Wang), minimal varieties (Fong, Kaarli and Ke) and isomorphism questions (Fray).

References

[1] G. Betsch (ed.), Near-rings and near-fields, Proceedings of the 1985 Tübingen Conference, North-Holland, Amsterdam, 1987.

[2] G. Betsch (ed.), Near-rings and near-fields, Proceedings of the 1989 Oberwolfach Conference, to appear.

[3] J. Clay, Nearrings: Geneses and Applications, Oxford Univ. Press, Oxford, 1992.

[4] G. Ferrero and C. Ferrero-Cotti (eds.), Proceedings of the 1981 San Benedetto Conference, 1982.

[5] J. Meldrum, Near-rings and their links with groups, Pitman, London, 1985.

[6] G. Pilz, Near-rings (2nd Ed.), North-Holland, Amsterdam, 1983.

[7] G. Pilz, (ed.)', Proceedings of the 1991 Linz Conference, Contributions to General Algebra 8, Hölder-Pichler-Tempsky, Wien, 1992.

[8] Proceedings of the 1978 Edinburgh Conference, Proc. Edin. Math. Soc. 23 (1980), 1-140.

[9] H. Wähling, Theorie der Fastkorper, Thales Verlag, Essen, 1987.

Gordon Mason
Fredericton, July 1994

ACKNOWLEDGMENT

On behalf of all the nearringers, I would like to express my deepest thanks to Mr. Fong Ka-Wai and Prof. C. S. Wang for their generous financial support in publishing this new proceedings. Without their kindest understanding and enthusiasm, this new volume is surely still in the air. One very special remark should go to Prof. W. F. Ke, my dearest former student, who has spent over hundreds of hours for all the computer works which were compulsorily needed for the birth of this new creature. To him, I must say that "Thanks Ke!"

Last but not the least, thanks are also due to Prof. A. V. Mikhalev, the head of the Mathematics Institute of Moscow State University, who has built up the bridge between Dr. Paul Roos, the representative of Kluwer Academic Publishers and myself. Without his help, I am quite sure that we can't find such a good place for our new-born baby! Of course, one should finally say that "Thanks ◎!"

Yuen Fong
Spring, 1995

Remark for fun:

◯ = Yuen (in Chinese 源, circle)

▢ = Fong (in Chinese 方, square)

$$\bigcirc + \square = \begin{cases} \text{◎} & \text{(ancient Chinese coin, or simply money)} \\ \text{Mathematics} & \text{(square-circle problem)} \end{cases}$$

ON THE BEGINNINGS AND DEVELOPMENT OF NEAR-RING THEORY

Gerhard Betsch

Mathematical Institute, University of Tübingen, Auf der Morgenstelle 10
72076 Tübingen, GERMANY

Introduction

Twenty-five years ago, late in 1968, the first conference on near-rings and near-fields took place in the Mathematische Forschungsinstitut Oberwolfach (Germany). Hence it seems appropriate to present a (mainly) historical survey "on the beginnings and development of near-ring theory". I hope this will provide some welcome background information on our subject. And sometimes returning to important work of the past stimulates new ideas how to proceed further.

I shall put emphasis on the early history and on facts which might be less well known. Also, I shall stress the great lines of tradition and development, and have to avoid many interesting details.

For the convenience of the audience the talk at Fredericton was supported by lists, indexes, diagrams, etc. I decided to keep this factual and sketchy approach for the written version, rather than rewriting the material into a "beautiful" essay. The lists, tables etc. are not intended as replacement for bibliographies, or biographical lists, but it should be easy to complete e. g. bibliographical data. (The Near-ring Bibliography will help.) I use fairly common abbreviations, like TAMS. "1893 WEBER, (On Galois Theory) Math. Ann. **43**" e. g. means, that the 1893 paper by WEBER in Math. Ann. **43** was on Galois Theory, while I refrain from giving the exact title. Papers (and authors) having an impact on the theory are mentioned in the text. A few more sources, mainly for historical information, are listed at the end of the paper.

The long section 1.1 is a response to the discussions at Fredericton.

List of Contents

Y. Fong et al. (eds.), Near-Rings and Near-Fields, 1–11.
© 1995 *Kluwer Academic Publishers.*

(c) H. NEUMANN, A. FRÖHLICH.
(d) Dissertations by LAXTON 1961, MALONE 1962, BETSCH 1963, BEIDLE-
MAN 1964 and others.

3. *International near-ring meetings:* 1968 Oberwolfach–1993 Fredericton N. B.

1. Early History of near-field theory

1.1 The context: HILBERT's program of axiomatisation

My claim is: The discovery and early investigation of near-fields fits very
well into the mainstream development of mathematics in general, and algebra in
particular, around 1900. This development is dominated to a large extent by D.
HILBERT's program of axiomatisation. Let me give a chronology of pertinent
papers and events (Table 1). Also, it might be useful to visualize the "Relations
of biographical data" of people who played a central role in the development of
abstract algebra and getting near-field and near-ring theory started (Table 2).

Table 1: The Early Papers and Books

1882 PASCH, Vorlesung über neuere Geometrie.
1893 WEBER, (On Galois Theory) Math. Ann. **43**.
    1895 HILBERT appointed Professor at Göttingen.
1899 HILBERT, (Foundations of Geometry/Festschrift).
1899 HILBERT, Jahresber. Dt. Math.-Ver.
1900 HILBERT, 23 Problems.
1901 DICKSON, Linear Groups.
1903 DICKSON, Linear associative algebras. TAMS 4.
1905 WEDDERBURN, A theorem on finite algebras. TAMS **6**.
1905 HESSENBERG, (Pappos/Pascal $\Rightarrow$ Desargues) Math. Ann. **61**.
1905 DICKSON, Definition of a group and a field by independent postulates.
    TAMS **6**.
1905/1907 VEBLEN-WEDDERBURN, Non-desarguesian and non-pascalian ge-
    ometries. TAMS **8**.
1908 WEDDERBURN, On hypercomplex numbers. PLMS(2), **6**.
1910 STEINITZ, (Algebraic Theory of Fields) Crelle **137**.
1914 DICKSON, Linear Algebras.

To deduce mathematical theorems from a suitable system of axioms is a
very old procedure. EUCLID's "Elements" are an early triumph of this axiomatic
method. But up to HILBERT's time this method was confined to geometry, and
the axioms had to be "evident". Take for instance PASCH, a contemporary of
HILBERT's. He did consider geometry as a system of logical deductions, but he
still believed that its fundamental concepts are based on "unmittelbare Beobach-
tungen" [immediate observations]. In the second half of the 19th century the old
question of the *truth* of geometry, or more general, of the applicability of mathematics to the "real" world, was discussed again. This was partly a result of the

Table 2: Relations of Biographical Data

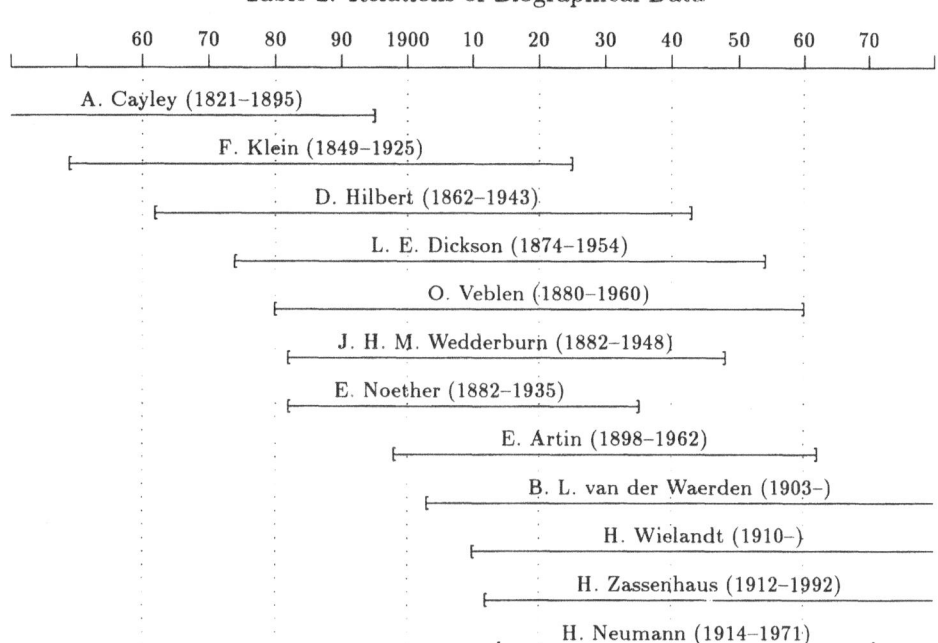

discovery of the non-euclidean geometries. And this discussion contributed to a change in attitude towards axioms and fundamental concepts. Axiomatic mathematics, as we know it today, was brought about mainly by David HILBERT. His contributions in this context may be summarized as follows:

1. In his "Grundlagen der Geometrie" (published first as "Festschrift" to celebrate the inauguration of the Gauß-Weber monument at Göttingen) and in other publications on related subjects, he put an end to speculations on the evidence of axioms. Instead of evidence he stressed the consistency of systems of axioms.

2. HILBERT extended the axiomatic method to other branches of mathematics, e. g., to algebra or the construction of number systems. In fact, in his note "Über den Zahlenbegriff" (JDMV **8**, 1899) he introduced first the concept of a genetic method (start from natural numbers to construct real numbers) as opposed to the axiomatic method. Then he goes on to state: "Trotz des hohen pädagogischen und heuristischen Wertes der genetischen Methode verdient doch zur endgültigen Darstellung und völligen logischen Sicherung des Inhalts unserer Erkenntnis die axiomatische Methode den Vorzug." [Despite of the high pedagogical and heuristic value of the genetic method, it is the axiomatic method which deserves preference for a final presentation and complete logical stabilisation of the substance of our (mathematical) knowledge.]

3. By his famous "Zahlbericht" (on algebraic number theory), by his "Grund-

lagen der Geometrie", and by his 1900 lecture on the 23 main problems, he
established his new style of doing mathematics, and so helped to pave the
way for the axiomatic method e. g., in the field of algebra. The "program
of axiomatization" of algebra was actually carried out by DICKSON, WED-
DERBURN, STEINITZ, Emmy NOETHER and her students, and others,
Emmy NOETHER being a predominant figure.

However, HILBERT should not be misunderstood as an advocate of "playing with axioms". In his 1919 course on "Natur und mathematisches Erkennen"
[Nature and mathematical insight, published Göttingen as edited reprint] he said
(p. 19): "Es bilden also die verschiedenen vorliegenden mathematischen Diszi-
plinen notwendige Glieder im Aufbau einer systematischen Gedankenentwicklung,
welche von einfachen, naturgemäß sich bietenden Fragen anhebend, auf einem
durch den Zwang innerer Gründe im wesentlichen vorgezeichneten Wege fortschrei-
tet. Von Willkür ist hier keine Rede. Die Mathematik ist nicht wie ein Spiel, bei
dem die Aufgaben durch willkürlich erdachte Regeln bestimmt werden, sondern
ein begriffliches System von innerer Notwendigkeit, das nur so und nicht anders
sein kann." [Therefore, the different existing mathematical disciplines form necessary links in building up a systematic development of thought, which, starting
from simple, natural questions proceeds on a way essentially determined by the
force of inherent reasons. There can be no question of arbitrariness. Mathematics
is not like a game, the problems of which are determined by arbitrarily conceived
rules, but it is a conceptual system of inner necessity, which can only be in this
state, and not in a different one.]

The advantages of the new axiomatic method may be marked by the key
words: Clarity, simplicity, maximal generality. Since no special properties of a
model are involved, axiomatic considerations are most economical. A very important point: According to HILBERT the objects of a theory are not *defined* by
the axioms; the axioms only regulate how to deal with these objects. All these
statements seem rather obvious to the 1993 mathematician. But they were by
no means obvious around 1900. HILBERT's contemporaries duly esteemed his
achievements. For example HURWITZ wrote to HILBERT in 1903: 'Sie haben da
ein unermeßliches Feld mathematischer Forschung erschlossen, welches als 'Mathematik der Axiome' bezeichnet werden könnte und weit über das Gebiet der Geometrie hinausreicht" [You have opened up an immeasurable field of mathematical
research, which might be called 'Mathematics of Axioms', which extends far be-
yond the area of geometry].

The goal of an axiomatic theory, which may be very difficult to achieve, is
a complete classification of all structures satisfying a certain set of axioms, or at
least a complete list of possible types. A quotation from STEINITZ' famous 1910
paper may clarify this point: "Eine Übersicht über alle möglichen Körpertypen
zu geben und ihre Beziehungen untereinander in ihren Grundzügen festzustellen,
kann als Programm dieser Arbeit gelten" [To give a survey on all possible types
of fields, and to establish their mutual relationships, may be considered to be the
purpose of this paper].

A recent example of classification: The "list" of finite simple groups.

1.2. DICKSON: Impact of his research. VEBLEN-WEDDERBURN. REIDE-MEISTER

Commutativity of addition is not independent, and may be proved from other field axioms (HANKEL 1867). In spring 1905, WEDDERBURN presented his famous result, that a finite skew-field is a commutative field. This means: In the finite case, commutativity of multiplication is not independent, and may be proved from the other usual axioms of a field. Almost simultaneously, in 1905 L. E. DICKSON published two papers on the independence of postulates for groups and fields, in particular in the finite case. He gave the first examples of what we now call finite left near-fields, thus proving that commutativity of multiplication and the left distributive law can not be proved from the other axioms of a skew-field, not even in the finite case. Here is DICKSON 's very first example of a near-field from his paper in TAMS **6** (1905).

"Nine elements $a + bj$, $a, b \equiv 0, 1, -1 \pmod 3$;

$$(a + bj) \oplus (c + dj) = a + c + (b + d)j, \tag{1}$$
$$(a + bj) \otimes (c + dj) = ac - (-1)^{ab}bd + (bc + (-1)^{ab}ad)j, \tag{2}$$

where the exponent ab is taken in the form 0, 1, or -1. Then $j \otimes j = -1, j \otimes (1+j) = -1 + j, (1 + j) \otimes j = 1 - j$, so that $6^\times$ [i. e., the commutativity of multiplication] and the right-hand distributive law fail. For $4^\times$ [i. e., existence of a multiplicative inverse],

$$(a + bj) \otimes \left(\frac{a - (-1)^{ab}bj}{a^2 + b^2} \right) = 1 = i_x \tag{3}$$

[= multiplicative identity], since $a^2 + b^2 \equiv 0$ only when $a \equiv b \equiv 0 \pmod 3$. Computation shows that $2^\times$ [i. e., the associative law of multiplication] holds."

In a second paper "On finite algebras", submitted in the same year 1905 (Nachr.Ges.Wiss.Göttingen, Math.-phys.Klasse), DICKSON attacked the difficult problem of determining all finite near-fields. He did not achieve this goal, but obtained large classes of finite near-fields. And he proved: Any finite near-field has prime power order and commutative addition. DICKSON's method of constructing finite near-fields: Take the additive group $(K, +)$ of a finite field K, and define multiplication $\circ$ by

$$a \circ b := a \cdot \varrho_a(b) \quad \text{for } a \neq 0, \tag{4}$$
$$0 \circ b := b \circ 0 := 0. \tag{5}$$

Here ϱ_a is an automorphism of the field K depending on a and $a \cdot \varrho_a(b)$ means ordinary multiplication in K.

Today, a near-field constructed in this way from a (not necessarily finite) field K is called "Dickson near-field" or "coupled with K". DICKSON 's idea to define a multiplication is a very fruitful one, cf. for instance the definition of multiplication when constructing a planar near-ring from a "Ferrero pair". Even as early as

1905 it could be expected, that a projective plane over a near-field ("field with ex-
tenuating circumstances") should have peculiar geometric properties. In fact, VE-
BLEN and MACLAGAN-WEDDERBURN, in their paper on "Non-desarguesian
and non-pascalian geometries" (TAMS **8** (1907)) applied DICKSON's new number
systems (finite near-fields) and the example given above in particular, to construct
a finite projective plane which is non-desarguesian and non-pascalian (today we
would say "Pappian" instead of "pascalian"). And from HESSENBERG's theorem
of 1905 (!) (Beweis des Desarguesschen Satzes aus dem Pascalschen. Math. Ann.
61) it follows that a non-desarguesian plane has to be non-pascalian anyway.

It is worth mentioning, that VEBLEN and WEDDERBURN submitted the
original version of their paper in the same year 1905, at almost the same time as
DICKSON finished his two papers quoted above. The first example of an infinite
affine plane over a near-field was given by REIDEMEISTER in his "Vorlesungen
über Grundlagen der Geometrie" (Berlin 1930; 2nd ed.1968). He also established
a connection between (right) near-fields and the theory of "webs". A web satisfies
REIDEMEISTER's "Schließungsbedingung" (closing condition) $\sum 4$ if and only if
the coordinate domain of the web satisfies both distributive laws.

The VEBLEN-WEDDERBURN paper of 1907 is an early example of a prominent line of research in the foundations of geometry, investigating the correspondence between *geometric* properties of a projective or affine plane, and *algebraic*
properties of its coordinate domain. This line of research was intensified, after
M. HALL 1943 and L. A. SKORNYAKOW 1951 introduced ternary fields to "coordinatize" arbitrary projective and affine planes. Let me give you a brief indication
of the correspondence I mentioned. We have the following list:

Affine plane	Coordinate domain of A
general	ternary field
translation plane	quasi-field, VEBLEN-WEDDERBURN system
special translation plane	planar near-field (note, any finite near-field is planar)
desarguesian plane	skew field (division ring)
pappian (pascalian) plane	commutative field

*1.3 Sharply 2-transitive permutation groups and near-fields. CARMICHAEL,
ZASSENHAUS*

Another line of research opened up and new impetus was given to near-field
theory when CARMICHAEL in 1931 discovered that the finite sharply 2-transitive
permutation groups are precisely the groups of affine transformations

$$x \mapsto ax + b \quad (a \neq 0)$$

of a finite near-field into itself. (Amer. J. Math. **53** (1931), 631–644).

The correspondence between finite 2-transitive permutation groups and finite near-fields is rather simple and instructive. Therefore I take a "detour" to
sketch this correspondence, following the presentation in CARMICHAEL's book

(Introduction to the theory of groups of finite order, pp. 396–401. 1937, reprint 1956)

Let G be a sharply 2-transitive permutation proup on a set F. Some facts from the general theory:

- $|F| = p^n$, where p is a prime, and $|G| = p^n(p^n - 1)$.
- $H := \{g \in G | g$ has no fixed point$\} \cup \{identity\}$ is an abelian normal subgroup of G, and $|H| = p^n$ (Carmichael, p. 145)
- For any $a \in F$, the stabilizer of a has order $p^n - 1$ and acts sharply 1-transitive (regularly) on $F \setminus \{a\}$.

Now we select two arbitrary elements from F, which we denote by 0 and 1. Let M be the stabilizer of 0. Then for any $a \in F$ there exists exactly one element $\eta_a \in H$ such that $\eta_a(0) = a$; for any $0 \neq a \in F$ there exists exactly one element $\mu_a \in M$ such that $\mu_a(1) = a$.

And now we define addition and multiplication on F by

$$a + b := \eta_b(a),$$
$$a \cdot b := \mu_a(b) \text{ if } a \neq 0,$$
$$0 \cdot b := b \cdot 0 := 0.$$

$(F, +, \cdot)$ turns out to be a left near-field of order p^n. The commutativity of addition follows again from the commutativity of H.

Conversely, let $(F, +, \cdot)$ be a finite near-field. Then the affine transformations

$$x \mapsto ax + b \quad (a \neq 0)$$

of F into itself form a group Γ acting sharply 2-transitive on F. (Hence we have again $|F| = p^n$, p a prime.)

We may think of F as being constructed in the way described above. We only have to consider the maps:

$$\mu_a : x \mapsto ax \quad (a \neq 0),$$
$$\eta_b : x \mapsto x + b.$$

Obviously, the μ_a's $(a \neq 0)$ and the η_b's generate the group Γ.

Now returning to CARMICHAEL's paper of 1931: He also noticed, that some near-fields lead to 3-transitive groups. ZASSENHAUS in his 1934 dissertation (publ. in Abh. Math. Sem. Univ. Hamburg **11** (1936), 17–40) took up this observation. He proved: Any finite sharply 3-transitive permutation group is isomorphic to a group of permutations (projective transformations)

$$x \mapsto \frac{a + b \circ x}{c + d \circ x},$$

where addition and division is from a suitable Galois field, and the multiplication ∘ is either the multiplication of this Galois field, or the multiplication of a proper

near-field. Now DICKSON's problem of determining *all* finite near-fields was
even more important. This problem was solved by ZASSENHAUS in 1936 (Über
endliche Fastkörper, Abh. Math. Sem. Univ. Hamburg **11** (1936), 132–145). He
proved: Up to seven exceptional near-fields which he described precisely, any finite
near-field may be derived from a finite field by DICKSON's method of changing the
multiplication. In other words: Any finite near-field is either a Dickson near-field,
or it is one of seven well known exceptional near-fields. A detailed classification of
the isomorphism types involved was given later, after 1960.

1.4. Remarks on incidence groups and the modern theory of near-fields

Modern near-field theory started around 1960, where the main motivation
came from geometry. It turned out that near-fields play an essential role in the "al-
gebraization" of geometric structures, like the weak affine spaces of E. SPERNER,
the projective incidence groups of H. KARZEL, the parallel structures of J. AN-
DRÉ, and circle geometries. TITS, GRÄTZER, and KARZEL extended the correspondence with sharply 2-transitive and 3-transitive permutation groups to the
infinite case, in which case a near-field has to be replaced by the more general
structure of a near-domain. The additive structure of a near-domain is a not necessarily commutative loop, and some other, technical conditions have to be satisfied. A finite near-domain is a near-field. So far, no proper near-domain (which
is *not* a near-field) is known. Of course, studying infinite sharply 2-transitive (3-
transitive) permutation groups stressed the importance of and encouraged interest
in infinite near-fields. The first systematic attempt to construct infinite near-fields
by "coupling" on fields is due to KARZEL 1965. KARZEL also gave an axiomatic
treatment of DICKSON's construction method. For details I refer to the pertinent publications of KARZEL and his school, and to WÄHLING's book "Theorie
der Fastkörper" 1987. The introduction to this book gives a very good historical
account.

Also, I mention only in passing the large area of research on topological near-
fields.

2. Origins of near-ring theory in particular

This chapter can be rather brief, because the basic facts are well known to the
participants. Also, in earlier survey lectures, special aspects of near-ring history
have been presented.

To the best of my knowledge, Helmut WIELANDT was the first to investigate
near-rings. His starting points were FITTING's "endomorphism domains" of non-
abelian groups (1932) and ZASSENHAUS' work on finite near-fields. His aim
was to obtain a tool for group theory. In his own words (translated from the
abstract of a lecture delivered at the 1980 Oberwolfach Near-ring Conference):
He developed the "basic facts of the structure theory [of near-rings] up to the
analog of Wedderburn's theorem on simple rings". He presented his results in
a lecture delivered on 20 May 1937 at a group theory conference in Hamburg.

It was WIELANDT's first public lecture. A brief abstract of this lecture, as part of a conference report formulated by ZASSENHAUS, appeared in Deutsche Mathematik **3** (1938), 9–10.

In the same year 1937 the monograph "Lehrbuch der Gruppentheorie" by ZASSENHAUS appeared in print. On pages 71–73 simple properties of near-rings are discussed.

Around 1950 WIELANDT resumed studying the subject of near-rings. From earlier results he developed a density theorem (in the sense of N. JACOBSON) for 2-primitive near-rings with identity (modern terminology). He presented this theorem, and other results in this context during the fifties in several colloquium lectures. For a more detailed discussion of WIELANDT's work on near-rings and its impact on the theory I refer to my essay in H. WIELANDT, Mathematical Works, Vol. 1.

Also, in the fifties, the investigation of general near-rings was started by D. W. BLACKETT and W. E. DESKINS, while H. NEUMANN and A. FRÖH-LICH initiated the theory of distributively generated near-rings (cf. table 4). In 1950, D. W. BLACKETT completed his thesis "Simple and semisimple near-rings" under the supervision of E. ARTIN. It was the very first thesis on near-rings, and contained a Wedderburn-Artin theory of near-rings with DCC for N-subgroups. The bulk of this thesis was published in Proc AMS **4** (1953), 772–785.

Table 3: Important publications on algebra, and on near-rings/near-fields in particular 1926–1953

1926	E. NOETHER, (Abstract ideal theory) Math. Ann. **96**.
1927	ARTIN, (Rings with minimum condition) Abh. Hamb. **5**.
1929	E. NOETHER, Hyperkomplexe Größen und Darstellungstheorie. Math. Z. **30**.
1930/31	van der WAERDEN, Moderne Algebra I, II.
1930	REIDEMEISTER, (Foundations of Geometry).
1936	ZASSENHAUS, (Sharply 3-transitive finite permutation groups) Abh. Hamb. **11**.
	ZASSENHAUS, (On finite near-fields) Abh. Hamb. **11**.
1937	CARMICHAEL, Introd. to the theory of groups of finite order.
1936	TAUSSKY and F.-TAUSSKY, (Commutativity of addition).
1937, 20 May:	WIELANDT,(On domains of group mappings) Deutsche Math. **3** (1938).
1937	ZASSENHAUS, Lehrbuch der Gruppentheorie.
1943	JACOBSON, Theory of rings.
1945	JACOBSON, The radical and semisimplicity for arbitrary rings. AJM **67**.
1950	BLACKETT, Dissertation (PAMS **4** (1953)).
1952	ARTIN-NESBITT-THRALL, Rings with Minimum Condition.
1950-1960	WIELANDT, Colloquium lectures (unpublished; Density theorem).
1954	DESKINS, A radical for near-rings. PAMS **5**.
1952/54	AMITSUR, (A general theory of radicals.) AJM.
1953	KUROSH, Radicals of rings and algebras. Math. Sb.

Table 4. Early publications on d. g. near-rings

1954 H. NEUMANN, Near-rings connected with free groups.
Proc. Intern. Congress of Math. Amsterdam II, 46–47.
H. NEUMANN, On varieties of groups and their assoc. near-rings.
MZ **65**, 36–69.
1958 A. FRÖHLICH, D. G. Near-rings.
Proc. LMS (3) **8**, 76–94; 95–108 JLMS **33**, 95–107.

Between 1961 and 1970 several young mathematicians wrote their doctoral dissertations on near-rings, thus joining the small group of near-ring specialists. See the following table.

Table 5. Early dissertations on near-rings

1961	LAXTON
1962	MALONE, ROTH
1962/63	Van der WALT
1963	BETSCH
1964	BEIDLEMAN
1966	CLAY
1967	MAXSON, PILZ
1968	HEATHERLY, RAMAKOTAIAH
1969	LIGH
1970	HOLCOMBE, SCOTT

The first monograph on the subject was

1977 G. PILZ, Near-rings.

3. International Near-ring Meetings

Table 6. List of International Meetings on Near-rings and Near-fields

1968	5. 12.–8. 12.	Oberwolfach
1972	30. 1.–3. 2.	Oberwolfach
1976	27. 6.–3. 7.	Oberwolfach
1978	6. 8.–12. 8.	Edinburgh
1980	12. 4.–19. 4.	Oberwolfach
1981	13. 9.–19. 9.	San Benetto del Tronto
1983	1. 8.–6. 8.	Harrisonburg/Va.
1985	7. 1.–11. 1.	Nagarjuna University
	4. 8.–10. 8.	Tübingen
1987	2. 8.–8. 8.	Middlesbrough
1989	5. 11.–11. 11.	Oberwolfach
1991	14. 7.–20. 7.	Linz
1993	18. 7.–24. 7.	Fredericton N. B.

Many more to come!

So far, I attended all international meetings listed above, with one exception, which I regret. My impression is, that all meetings were quite successful:

- they all had a rather broad scope, covering the current fields of research on near-rings, but at the same time they all had a few "focuses";
- they stimulated active research;
- they all took place in a very friendly and cooperative atmosphere, which I am very grateful for.

These claims may be easily proved from the pertinent conference reports and proceedings, and from personal recollections.

As a concluding remark, I come back to a statement in my introduction. I am convinced that returning to important work of the past may, and frequently will, inspire new ideas how to proceed further. But if this is true, then it is worthwhile to study "classical" papers on our subject carefully again and again, and it might also be worthwhile to look at "the beginnings and development of near-ring theory."

4. Bibliographical Note

This survey of the history of near-ring theory, including near-field theory, is based mainly upon the important original papers. For the historical context and other background information, I used the following sources.

G. Betsch: Fastringe. Zulassungsarbeit. Tübingen 1959.

G. Betsch: Wielandt's work on near-rings. In: Helmut Wielandt, Mathematical Works, Volume 1, pp. 741–745. Berlin-New York, de Gruyter 1994.

J. J. Gray, H. Kaiser, E. Scholz: Ausblick auf Entwicklungen im 20. Jahrhundert. Chapter 15 of E. Scholz (Ed.), Geschichte der Algebra. Mannheim 1990.

H. Wähling: Introduction to his book: Theorie der Fastkörper. Essen: Thales 1987.

LOCALIZED DISTRIBUTIVITY CONDITIONS

Henry E. Heatherly

Department of Mathematics, University of Southwestern Louisiana
Lafayette, Louisiana 70504, U. S. A.

ABSTRACT

Various conditions have been used in lieu of full two-sided distributivity to obtain structure theory for near-rings, most successful and most widely known being the distributively generated condition, but also including a plethora of other conditions, some being quite exotic. Imperative in justifying the use of such a condition is not only that it gives satisfory results, but that also there is a wide class of natural examples which satisfy the condition. *Definition:* Let K, S, and T be subsets of a (left) near-ring R, with K and T non-empty. We say "K is (S,T)-distributive" if $s\,(k_1 + k_2)\,t = sk_1t + sk_2t$, for each $k_1, k_2 \in K$, $s \in S$, $t \in T$. We say "K is (S,T)-d.g. on X" if K is (S, X)-distributive and T is contained in the subgroup of R which is generated additively by X. Generically we call such conditions "localized distributivity conditions". First we give a historically oriented survey of such conditions and how they were used. This serves as one source of motivating examples. Then certain localized distributivity conditions on special R-subgroups or ideals are discussed with applications to the study of minimal ideals, various types of prime ideals and their associated radicals, and primitivity and the J_ν radicals. Examples of near-rings of mappings or homomorphisms on additive groups are given to illustrate how these particular conditions discussed arise naturally, thus satisfying both justification criteria.

0 Introduction

All near-rings enjoy at least a little "localized distributivity". By this we mean there are subsets K, S and T of a near-ring, with K and T non-empty, such that $s\,(k_1 + k_2)\,t = sk_1t + sk_2t$, for each $k_1, k_2 \in K$, $s \in S$, $t \in T$. If S is empty, then delete the "s" from the expression. At one extreme there are the constant near-rings, where zero is the only element which distributes over the sum of a given pair of elements. At the other extreme we have distributive near-rings (which includes rings). This paper discusses the effects of various types of localized distributivity, giving examples of where such arise. Commensurate with the origin of the paper as coming from one of the hour addresses at the Fredericton conference and following the suggestion from the conference directors that those addresses give an historical perspective to the theory, attention will be given to how the theory developed and where it fits in the context of the unfolding panorama of near-rings. After a survey of earlier uses of localized distributivity conditions and a cornucopia of motivating examples, the paper zeroes in on three diverse areas where localized distributivity

13

Y. Fong et al. (eds.), Near Rings and Near Fields, 13–29.
© 1995 *Kluwer Academic Publishers.*

has been successfully utilized by the author and his collaborators: minimal ideals, various types of prime ideals and their associated radicals, and primitivity and the J_ν radicals. The section on minimal ideals is almost entirely new results developed by the author since the Fredericton conference. The subsequent sections primarily report on recent results (since 1989) by Birkenmeier, Heatherly, and Lee.

Herein all near-rings are left near-rings. The symbol R will denote a near-ring throughout. Unless otherwise specified the terminology and notation will follow that in the book of Meldrum [47].

Definition 0.1. Let K, S, T be subsets of R, with T and K non-empty. Then:
(0.1.1) K is (S,T)-distributive if for each $s \in S$, $k_1, k_2 \in K$, $t \in T$, we have
  $s(k_1 + k_2)t = sk_1t + sk_2t$;
(0.1.2) K is (S,T)-d.g. on X if K is (S,X)-distributive and $T \subseteq \mathbf{gp}(X)$. (Herein
  $\mathbf{gp}(X)$ denotes the subgroup of $(R,+)$ which is generated additively by X).

In the above if S is empty, then delete s in the equation. If the generating set X plays no explicit role in a discussion , then we may use the terminology "K is (S,T)-d.g.". Informally 0.1.1 and 0.1.2 will be called "localized distributivity conditions". Note that if Λ is the set of all subsets V of R such that K is (S,V)-disributive, then K is (S,X)-distributive, where $X = \cup\Lambda$, and K is (S,T)-d.g. on X, where $T = \mathbf{gp}(X)$.

In the sequel we use the following notation. If Y is a non-empty subset of R, then $\langle Y \rangle_R$ and $\mathbf{ngp}_R(Y)$ are the ideal generated by Y in R and the normal subgroup generated by Y in $(R,+)$, respectively. If no ambiguity will arise we use $\langle Y \rangle$ and $\mathbf{ngp}(Y)$ for convenience. The *Lie Commutator* ideal of R is defined as: $\langle R, R \rangle = \langle\{ab - ba : a, b \in R\}\rangle$. We use $\Delta(R) = \{(a + b)c - bc - ac\}$ for the distributor set of R. Observe that $R/\langle R, R \rangle$ is a commutative near-ring and $R/\langle \Delta(R) \rangle$ is a distributive near-ring; so $\langle \Delta(R) \rangle \subseteq \langle R, R \rangle$. The set of all distributive elements of R is denoted by $\mathcal{D}(R)$. The n-th commutator subgroup of $(R,+)$ is denoted by $R^{(n)}$ and $R' = R^{(1)}$.

1 Motivating Examples

In this section we give a wide selection of types of near-rings each satisfying some localized distributivity condition. With each type one or more representative results are given which involve the localized distributivity condition, as well as comments putting the topic in historical perspective. Some natural examples arising from functions and endomorphisms on groups conclude the section.

1.1.   If R is distributive, then R is $(\emptyset, R)$-distributive. This was the first kind of localized distributivity condition studied, actually predating the report by Zassenhaus (in 1938) of Wielandt's seminal 1937 lecture on the theory of abstract near-rings [61]. (I am indebted to Professor G. Betsch for this information concerning the early work of Wielandt on near-rings). In 1937 Olga Taussky [57] published the first paper on what are now called "distributive near-rings". (She used the term "rings with non-commutative addition", which was still in use as late as 1975 in Weinert's [59, 60] comprehensive two part study). Ferrero [20] took up the subject of distributive near-rings in the early 1960's, obtaining, *inter alia*, Sylow-like theorems, representative of which is the following: if G is a distributive near-ring of order $p^{\alpha}m$, where p is a prime, α and m are positive integers, and p does not divide m, then G has a subnear-ring of order p^{α}. The subject of distributive near-rings has been addressed by several others in addition to those already mentioned, but it has yet to become central to theory of near-rings. (However, see Section 26.1 in the recent monograph by Clay [16]).

1.2.   The class of d.g. near-rings is exactly the class of all R such that R is $(\emptyset, R)$-d.g. on $\mathcal{D}(R)$. Recall [47, Corollary 9.49], that if R is d.g. and $(R, +)$ is solvable of length $n \geq 2$, then the commutator subgroup, R', is a nilpotent ideal and $(R')^n = 0$. So R' is $\left(\emptyset, (R')^{n-1}\right)$-distributive. Thus if $(G, +)$ is a solvable group of length $n \geq 2$ and R is anyone of $I(G)$, $A(G)$, or $E(G)$, then R' is $\left(\emptyset, (R')^{n-1}\right)$-distributive.

1.3.   If I is an ideal of any near-ring R and $I^q = 0$, $q \geq 2$, then R is (I^m, I^n)-distributive, where $m + n = q$. Nilpotent ideals are relatively abundant in many near-rings.

1.4.   In one of the early attempts to get ordinary matrix multiplication to work right for near-rings, the concept of a pseudo-distributive near-rings arose: R is *pseudo-distributive* if (1) $ab + cd = cd + ab$, for $a, b, c, d \in R$ and (2) $(\sum a_i b_i) r = \sum a_i b_i r$, for each $r, a_i, b_i \in R$ and $i = 1, \cdots, n$, for any n. Observe that if R is pseudo-distributive, then $\mathbf{gp}(R^2)$ is $(\emptyset, R)$-distributive (and conversely). Ligh [40] showed that for a near-ring R, the following are equivalent:

(i) for each n the set of all n-by-n matrices with entries from R is a near-ring under ordinary matrix addition and multiplication;

(ii) R is pseudo-distributive.

(Caveat: These are <u>not</u> matrix near-rings in the sense of Meldrum and Van der Walt [49]). Pseudo-distributive near-rings have also been used in the contexts of group or semigroup near-rings and of polynomials and formal power series over near-rings. (See for example Heatherly and Ligh [28], and Mason [43]; the term "pseudo-distributive" was introduced in the former).

1.5   A non-empty subset S of R is said to satisfy a *permutation identity* if there is some non-identity permutation σ on n symbols such that $\Pi s_j = \Pi s_{\sigma(j)}$, $j = 1, \cdots, n$, for all $s_1, \cdots, s_n \in S$. If $S = R$ we say R is a *permutation identity near-ring* or R "satisfies a permutation identity". Permutation identity near-rings were considered by Scapellato [54, 55], and studied in depth by Birkenmeier and Heatherly [7, 4, 6]. In section 4 certain localized distributivity conditions are used in permutation identity near-rings to obtain results concerning the nilpotence of various ideals and the import of this for prime radicals.

If R satisfies a permutation identity σ of n symbols and if $\sigma(n) = k < n$, then R is $\left(R^{k-1}, R^{n-k}\right)$-distributive, see [7, Proposition 2.5]. If R is such a near-ring and R satisfies either a.c.c. or d.c.c. on ideals, then every nil subnear-ring of R is nilpotent; see [7, Theorem 2.7].

We say R is *right permutable* if $abc = acb$ identically. Such a near-ring is a permutation identity near-ring with permutation $\sigma = \begin{pmatrix} 1 & 2 & 3 \\ 1 & 3 & 2 \end{pmatrix}$ and having $\sigma(3) = 2$. These were considered at length by Birkenmeier and Heatherly [4], where - *inter alia* - examples and example classes are given. One such example arises from a construction on a free group as follows. Let $(G, +)$ be a free group on the free generators r, s. Let $h : G \to G/(r)$. (Note $G/(r) \approx (s) \approx Z^+$). Define $x \cdot y = h(y)x$, for each $x, y \in G$. Then $(G, +, \cdot)$ is a d.g. near-ring which is not distributive and is right permutable. (For a more general construction see [4, Example 4.9]).

1.6. Let A be an ideal of R. Define

$$\Delta_R(A) = \{(x + y)r - yr - xr : x, y \in A, r \in R\}$$

Recursively define $D^n(A)$ via:

$$D^1(A) = \mathbf{ngp}_R(\Delta_R(A)) \, and D^{j+1}(A) = \mathbf{ngp}_R \Delta_R\left(D^j(A)\right).$$

Each $D^n(A)$ is an ideal of R; see [53, Lemma 5.8]. If $D^m(R) = 0$ for some m, then R is said to be *weakly distributive* of class m. For d.g. near-rings this is equivalent to the concept of "weakly distributive" introduced by Fröhlich [21, p.92] and discussed at length in [47, Chapter 9]. Esch [19] extended Fröhlich's weakly distributive concept to zero symmetric near-rings and Roberts [53] carried this investigation much further. (Also see Meldrum [48]). Observe that R is weakly distributive of class $n + 1$ if and only if $D^n(R)$ is $(\emptyset, R)$-distributive. Thus one immediately has that all weakly distributive near-rings are zero symmetric. (This was first proved in [53, Lemma 5.15] and also is proved in [48, Théoréme 1.6]; however the proof from the $D^n(R)$ is $(\emptyset, R)$-distributive viewpoint becomes a triviality). There is a rich interconnection between the series $D^n(R)$, the series $R^{(n)}$, and multiplicative nilpotence of certain ideals of R. Exemplary of this are the following results:

(1.6.1) let R be d.g. with $R^2 = R$. Then R is weakly distributive if and only if $(R, +)$ is solvable; see [21, Theorem 4.4.7];

(1.6.2) if for some n, $(D^n(R), +)$ is solvable, then $(R, +)$ is solvable; see Esch [19];

(1.6.3) it R is weakly distributive with a right unity, then $(R, +)$ is solvable; see [53, Corollary 6.16];

(1.6.4) if R is d.g. and R is weakly distributive, then $D^1(R)$ is multiplicatively nilpotent; see [47, Corollary 9.55].

Roberts [53] investigated several other generalizations of distributivity, e. g., "feeble distributivity" which is more general than weakly distributive. (Also see Meldrum [48] for generalizations of distributivity along similar lines).

We next consider some situations were just a tiny bit of distributivity plays a decisive role.

1.7. Let R be a non-trivial integral near-ring. (See [51, Chapter 9]). Let $\Delta(x) = \{a \in R : (nx)a = n(xa), for each natural number n\}$, for $x \in R$, $x \neq 0$.

(1.7.1) If there exists an $x \in R$ with finite additive order such that $\Delta(x) \neq 0$, then there exists a prime p such that all elements of finite order in R have order p; see [27, Theorem 3.3]. Thus if $\mathcal{D}(R) \neq 0$, then R has characteristic zero or a prime.

(1.7.2) If R has d.c.c. on monogenic right R-subgroups and if $\mathcal{D}(R) \neq 0$, then R is a near-field; see [39, Corollary 5].

(1.7.3) Let A be reduced (no non-zero nilpotent elements) and zero symmetric. Using 9.39 and 9.41 of Pilz [51], we have that if A has d.c.c. on right A-subgroups and $\mathcal{D}(A) \neq 0$, then A is a finite direct sum of near-fields.

This section concludes with some examples of localized distributivity in certain near-rings of functions.

1.8. Let $(G, +)$ be a non-abelian group with subgroups H and K such that $0 \neq K \subseteq H$ and K is abelian and is fully invariant in $(G, +)$. (Such a K will always exist if $(H, +)$ is solvable). Let $R = \mathbf{gp}(Hom((G, +), (H, +)))$. Then $(R, +, \circ)$ is a d.g. near-ring with $T = \{\sigma \in R : G\sigma \subseteq K\}$ is an ideal. Observe that R is (T, T)-distributive and T is a ring. In general, $T \nsubseteq \mathcal{D}(R)$.

1.9. Let H be a subgroup of a (not necessarily abelian) group $(G, +)$. Let $R = \{\alpha \in M(G) : H\alpha \subseteq H\}$ and let $S = \{\beta \in M(G) : G\beta \subseteq H\}$. Then R is a subnear-ring of $(M(G), +, \circ)$ and S is a two-sided R-subgroup of R. Take $T = \{\delta \in R : \delta\,restricted\,to\,H\,is\,an\,endomorphism\,on\,(H, +)\}$. Then T is the largest subset of R such that R is (S, T)-distributive, and T is a subsemigroup of $(R, \circ)$. Define A to be the set of all finite sums of those elements of T which are automorphisms on $(H, +)$. Then R is (S, A)-d.g. on T.

Other such examples can be found in [7, 9, 10, 38].

2 The Minimal Ideal Problem

The problem is to characterize those near-rings which are minimal ideals of some near-ring. The corresponding problem for rings is not difficult, althrough it was not resolved until 1958 by V. Andrunakievic [1]: if I is a minimal ideal of a ring, then either $I^2 = 0$ or I is a simple ring. (This removes the problem to one of classifying simple rings, which is a formidable open problem). Work on minimal ideals in near-rings goes back (at least) to Blackett [13]. The first paper devoted entirely to characterizing them is due to Scott [56]. Kaarli [32, 33, 34] has obtained significant results on the problem, beginning in the mid-1970's and continuing to the present. From Kaarli [33] we know that an analogous result to that for rings does not hold in general for near-rings, even for finite, zero symmetric, abelian near-rings. In early 1991 Birkenmeier and Heatherly [8, 9] obtained some new results on the problem. In this section is given some recent results obtained by the author which extend those by Birkenmeier and Heatherly [9]. To avoid pathology all near-rings in this section are zero symmetric.

It is of interest to see why the technique used by Andrunakievic for rings does not work for near-rings. Crucial in his development is the following lemma, which is useful in other ways as well. (For example, see [18, p.107]).

Andrunakievic Lemma (cf. Andrunkievic [1], and [18, Lemma 61]). Let I be an ideal of a ring A and let B be an ideal of the ring I. Then $(\langle B \rangle_A)^3 \subseteq B$.

The near-ring analog of this does not hold, as can be seen using Kaarli's example mentioned above, where the heart of that subdirectly irreducible near-ring contains a proper non-zero ideal which is square zero. A version of the Andrunakievic Lemma for near-rings with a certain localized distributivity condition was given by Birkenmeier and Heatherly [9].

Proposition 2.1. Let I be an ideal of a near-ring R such that I^k is (I^m, I^n)-distributive for some $k, n \geq 1$, $m \geq 0$. If B is an ideal of the near-ring I, then $(\langle B \rangle_R)^j \subseteq B$, where $j = k + m + n + 1$.

However, for near-rings this does not open the door to use a proof technique like that of rings to characterize minimal ideals. The crux of the problem is that $\mathbf{ngp}_R((\langle B \rangle_R)^j$ need not be contained in B. So other techniques are needed and a near-ring version of the Andrunakievic Lemma so far has played no part in them. (Note: Developments after this paper was submitted have shown that the previous remark is too pessimistic. In early 1994, Birkenmeier, Heatherly, and Lee obtained significant results on the structure of minimal ideals in near-rings which satisfy a certain type of Andrunakievic condition and have also shown that wide classes of near-rings satisfy this condition, including d.g. near-rings.)

We begin the development with three useful lemmas.

Lemma 2.2. Let S and A be right R-subgroups of R and let T be a non-empty subset of R such that $RT \subseteq T$. If there exists $k > 0$, $m \geq 0$ such that S^k is (A^m, T)-distributive, then $A^m S^k \langle X \rangle \langle T \rangle = 0$, whenever $A^m S^k XT = 0$.

Proof. We give the proof explicitly for $m > 0$, but for $m = 0$ one need only delete the appropriate terms. Let $x, x_1, x_2 \in X$, $r, r_1, r_2 \in R$, $s \in S^k$, $a \in A^m$. Observe that $as(-r)t + asrt = a[s(-r) + sr]t = 0$; so $as(-r)t = -(asrt)$. Thus $A^m S^k(-X)T = 0$. Next,

$$as(x_1 + x_2 t) = a(sx_1 + sx_2)t = asx_1 t + asx_2 t = 0;$$

$$as(r + x - r)t = a[s(r + x) + s(-r)]t = as(r + x)t + as(-r)t$$
$$= a(sr + sx)t - asrt = asrt + asxt - asrt = 0;$$

$$as[(x + r_1)r_2 - r_1 r_2]t = a[s(x + r_1)r_2 + s(-r_1 r_2)]t = as(x + r_1)r_2 t$$
$$+as(-r_1 r_2)t = a(sx + sr_1)r_2 t - asr_1 r_2 t = asxr_2 t + asr_1 r_2 t - asr_1 r_2 t;$$

and $asrxt \subseteq A^m S^k XT = 0$. Repeatedly applying these results yields $A^m S^k \langle X \rangle T = 0$.

Observe that if M is a product of right R-subgroups of R and Y is a non-empty subset of R such that $MY = 0$, then $M \langle Y \rangle = 0$. Consequently $A^m S^k \langle X \rangle \langle T \rangle = 0$.

Lemma 2.3 ([8, Proposition 2]). Let I be a minimal ideal of R with $I^2 \neq 0$. Then $(0 : I)$ is a prime ideal of R and either:
(1) $(0 : I) = 0$ and R is subdirectly irreducible with heart I, or
(2) $(0 : I) \neq 0$, $\bar{R} = R/(0 : I)$ is subdirectly irreducible with heart $\bar{I}$, and $\bar{I}$ is isomorphic to I.

Lemma 2.4. Let R be a d.g. near-ring and let I be a minimal ideal of R with $I^2 \neq 0$. Then either:
(1) I is a simple ring and $I \subseteq \mathcal{D}(R)$; or
(2) $I \subseteq R^{(n)}$ for each n.

Proof. If there exists n such that $I \not\subseteq R^{(n)}$, then $I \cap R^{(n)} = 0$ and hence $R^{(n)} \subseteq (0 : I)$. So $\bar{R} = R/(0 : I)$ is a d.g. near-ring whose additive group is solvable. Consequently, $\bar{R}'$ is multiplicatively nilpotent, (see [47, Corollary 9.49]). Since $\bar{R}$ is also prime, $\bar{R}^+$ must be abelian and $\bar{R}$ is a ring. Then $\bar{I}$ is a simple ring and so is I. Since $\bar{R}$ is a ring, $(a + b)c - bc - ac$ is in $(0 : I)$ for each $a, b, c \in R$. Using $c \in I$ we have $(a + b)c - bc - ac$ is in $(0 : I) \cap I = 0$ and hence $(a + b)c = ac + bc$. So $I \subseteq \mathcal{D}(R)$.

One immediate consequence of this lemma is that in a d.g. near-ring with solvable additive group, each minimal ideal is square zero or is a simple ring. The Feit-Thompson Theorem and the Artin-Wedderburn Theorem then allow one to classify minimal ideals of d.g. near-rings of odd order as being either square zero or a full matrix ring over a field.

We next give two localized distributivity conditions which are central in the development which follows. For each of these I is an ideal of a near-ring R.

Condition (1). There exists a non-empty subset T of R such that $IT \neq 0$, $RT \subseteq T$, and I^k is (I^m, T)-distributive for some $m \geq 0$, $k > 0$.

Condition (2). R is d.g. and there exists $u \in R$ with $Iu \neq 0$ such that I^k is $(I^m, \{u, u + u\})$-distributive for some $m \geq 0$, $k > 0$.

Theorem 2.5. Let I be a minimal ideal of R, with $I^2 \neq 0$, which satisfies either Condition (1) or Condition (2). Then I is a simple ring and $I \subseteq \mathcal{D}(R)$. If R also satisfies the d.c.c. on right (left) ideals, then $(0 : I)$ is a maximal ideal and $R = I \oplus (0 : I)$.

Proof. First assume Condition (1) holds. Let $c, b, r \in R$, $s \in I^k$, $a \in I^m$, $t \in T$. (If $m = 0$, then delete the appropriate terms). From $as(-c)t + asct = a(s(-c) + sc)t = 0$, we have $as(-c)t = -(asct)$. Then

$$as((b + r)c - rc - bc)t = a(s(b + r)c + s(-rc - bc))t$$
$$= as(b + r)ct + as(-rc - bc)t = a(sb + sr)ct - a(sbc + src)t$$
$$= asbct + asrct - (asbct + asrct) = 0.$$

So $I^m I^k \cdot \Delta(R) \cdot T = 0$ and hence $I^m \cdot I^k \langle \Delta(R) \rangle \cdot \langle T \rangle = 0$. Consider the image in $\bar{R} = R/(0 : I)$ of this product of ideals. Since $\bar{R}$ is prime and since neither I nor $\langle T \rangle$ are in $(0 : I)$, we have $\langle \Delta(R) \rangle \subseteq (0 : I)$ and hence $\bar{R}$ is distributive. However, a prime distributive near-ring must be a ring. Consequently $\bar{I}$ and hence

I are simple rings. Proceed as in the last part of the proof of Lemma 2.4 to get $I \subseteq \mathcal{D}(R)$.

Next assume Condition (2) holds. Let c, b, r, s and a be as above. Similar to the previous calculation, $as\,(-c)\,u = -\,(ascu)$. Expanding $as\,(c + b)\,(u + u)$ in two different ways and utilizing the localized distributivity condition of the hypothesis we have $ascu + ascu + asbu + asbu = ascu + asbu + ascu + asbu$, which yields $ascu + asbu = asbu + ascu$. So $as\,(c + b - c - b)\,u = ascu + asbu - ascu - asbu = 0$. Consequently $I^m I^k \cdot (\mathbf{gp}\,\{x + y - x - y : x, y \in R\}) \cdot u = 0$ and hence $I^m I^k R' \cdot \langle u \rangle = 0$. Since $u \notin (0 : I)$ and hence $I \subseteq \langle u \rangle_R$, the assumption that $I \subseteq R'$ yields I is nilpotent, a contradiction. So $I \nsubseteq R'$, which by Lemma 2.4 yields I is a simple ring; also $R' \subseteq (0 : I)$ and hence $\bar{R}$ is a ring.

So under either condition $\bar{R}$ is a prime ring. Invoke the d.c.c., which is inherited by $\bar{R}$, and obtain $\bar{R}$ is a non-nilpotent simple ring. So $(0 : I)$ is a maximal ideal and hence $R = I \oplus (0 : I)$.

Note that the d.c.c. on left ideals is both unusual and comparatively weak for a left near-ring. The crux of the matter here is that the action takes place in the ring $\bar{R}$.

The next goal is to transfer the results of Theorem 2.5 to all the non-nilpotent minimal ideals. To do this we introduce global analogs of Conditions (1) and (2).

Condition (3).   There exists an ideal A and a non-empty subset T such that $RT \subseteq T$, neither A nor T is in the right annihilator of any non-nilpotent minimal ideal of R, and A^k is (A^m, T)-distributive for some $k > 0$, $m \geq 0$.

Condition (4). R is d.g. and there exist $u \in R$ and positive integers m, n such that u is not in the right annihilator of any non-nilpotent minimal ideal of R and such that A^m is $(A^m, \{u, u + u\})$-distributive, where $A = R^{(n)}$.

Corollary 2.6.   If R satisfies either Condition (3) or Condition (4), then each minimal ideal of R is either square zero or is a simple ring. If R also has d.c.c. on right (left) ideals, then $R = H \oplus B$, where H is a finite direct sum of full matrix rings over skewfields, $H \subseteq \mathcal{D}(R)$, and each minimal ideal of B is square zero. (This includes the possibility that H or B may be zero).

Proof.   Under Condition (3), each non-nilpotent minimal ideal of R is contained in A and hence for each such ideal Condition (1) is satisfied. If Condition (4) holds, then for a non-nilpotent minimal ideal I, either $I \subseteq R^{(n)}$, and hence Condition (2) holds for I, or by Lemma 2.4 (1) we also obtain I is a simple ring with $I \subseteq \mathcal{D}(R)$.

Invoke the d.c.c. on right (left) ideals. Then $\bar{R} = R/(0 : I)$ is a subdirectly irreducible prime ring which is Artinian semisimple, for each minimal ideal I of R, with $I^2 \neq 0$. So $\bar{R}$ is an Artinian simple ring and consequently $\bar{R} = \bar{I}$ and $\bar{I}$ is a full matrix ring over a skewfield, which then transfers up to I as well. Also, as seen in Theorem 2.5, each such I is a direct summand of R. Write $R = I_1 \oplus R_1$, where I_1 is such a minimal ideal and $R_1 = (0 : I_1)$. Note that each minimal ideal of R_1 is a minimial ideal of R. Repeat the process with each such minimal ideal to obtain after a finite number of steps that $R = H \oplus B$, where H is a finite direct sum of non-nilpotent minimal ideals of R, and B has no non-nilpotent minimal ideals. Since each of the summands constituting H are in $\mathcal{D}(R)$, then so is H.

Recall that for $(G, +)$ a finite, simple, non-abelian group, the near-ring $M_0(G)$ is d.g. and simple [22]. This gives us a minimal ideal in a "nice and natural" d.g. near-ring which does not satisfy Condition (2). It also serves to indicate the intricacies of the problem of classifying minimal ideals, even in finite, d.g. near-rings.

We end this section with statements of the two main open questions concerning minimal ideals.

Open Question I. Is a minimal ideal of a near-ring always either subdirectly irreducible or square zero?

Open Question II. Is a minimal ideal of a d.g. near-ring always either a simple near-ring or square zero?

3 Prime Ideals and Prime Radicals

The study of prime ideals in near-rings began in the early 1960's with independent investigations by Van der Walt [58] and Laxton [37]. While Laxton's work was on d.g. near-rings, Van der Walt handled the general near-ring case. Comments in each of these papers indicate they were motivated by McCoy's work on prime ideals in rings [45].

As is noted in many introductory books on ring theory, there are many equivalent conditions for an ideal of a ring to be prime. (For example, see [46, Section 18]). Most of these conditions are not equivalent for the class of near-rings. One such is the following:

(3.1) an ideal P of a near-ring R is a *type one prime ideal* of R if $xRy \subseteq P$ implies $x \in P$ or $y \in P$.

We say R is a *type one prime near-ring* if (0) is a type one prime ideal of R.

Type one prime ideals have been studied under several guises, often with somewhat stronger conditions being required. Beidleman [2] defined a "strictly prime d.g. near-ring", which in the d.g. setting is equivalent to a type one prime near-ring. This equivalence was noted by Groenewald [25, p.339]. Oswald [50] defined "strictly prime" near-rings in the general zero symmetric setting; when such a near-ring has a right unity, then it is type one prime. Definition (3.1) was given explicitly by Ramakotaiah and Rao [52] in 1979; however, this was done in the zero symmetric setting and almost all of their results are for and about IFP near-rings. In a paper written in 1991, Birkenmeier, Heatherly, and Lee [10] gave an extensive treatment of type one prime and - *inter alia* - used localized distributivity conditions to establish when prime ideals are type one prime. This section will give some of those results and some from the recent dissertation by Enoch Lee [38]. Important results on various types of prime ideals and their relation to primitive ideals are developed by Groenewald [26]. He uses the terminology "3-prime ideal" for what herein is called a "one-prime ideal" and he uses "1-prime ideal" for an allied but different type of primeness.

It is easy to see that a type one prime ideal is a prime ideal. The converse is not true, even in the class of zero symmetric, finite near-rings with unity, as was shown in [10, Examples 1.1, 1.2]. Denote the class of all near-rings for which every prime ideal is type one prime by $\mathfrak{R}^1$ and let $\mathfrak{R}^1_0 = \{R \in \mathrm{R}^1 : R \text{ is zerosymmetric}\}$. Some examples of near-rings in $\mathfrak{R}^1_0$ are:

(3.2.1) $M_0(G)$, where $(G, +)$ is any group;
(3.2.2) $I(G)$, $A(G)$, or $E(G)$ for $(G, +)$ a solvable group;
(3.2.3) $E(G)$ for $(G, +)$ a symmetric group of degree $n \geq 5$;
(3.2.4) all rings; all distributive near-rings;
(3.2.5) all finite, non-trivial, integral near-rings.

Two questions arise naturally.
Question 1. When is a prime ideal a type one prime?
Question 2. When is a near-ring in $\mathfrak{R}^1$?

These questions were addressed by Birkenmeier, Heatherly and Lee [10]. The following includes a sampling of those results therein that pertain to localized distributivity conditions. Here R is zero symmetric.

Theorem 3.3 ([10, Theorem 4.7]). Let K, S, T be ideals of R such that K^j is (S^m, T^n)-d.g. on X, for some $j, n \geq 1$, $m \geq 0$. If P is a prime ideal of R such that none of K, S, or T is a subset of P and if there exists $x \in X$ such that $2x \in X$ and $x \notin P$, then R/P is a ring and P is a type one prime ideal.

Corollary 3.4 ([10, Corollary 4.9]). If R is d.g. and P is a prime ideal of R such that for some $d \in \mathcal{D}(R)$, both $d \notin P$ and $2d \in \mathcal{D}(R)$, then P is a type one prime ideal.

These two results can be put in a global setting. First some terminology and notation. Herein $\mathbf{P}_0(R)$ denotes the intersection of all the prime ideals of R, the usual prime radical. (See for example, [47, Chapter 6] or [51, Chapter 5]). Also $\mathbf{P}_1(R)$ is the intersection of all the type one prime ideals of R. (This uses the notation and terminology of Birkenmeier, Heatherly and Lee [10] and Ramakotaiah and Rao [52]; Groenewald [26] uses $\mathbf{P}_3(R)$).

Corollary 3.5. Let R^j be (R^m, R^n)-d.g. on X, for some $j, n \geq 1$, $m \geq 0$. If $2x \in X$ for each $x \subset X \backslash \mathbf{P}_0(R)$, then:
(1) $R \in \mathfrak{R}^1$ and $R/\mathbf{P}_0(R)$ is a ring; and
(2) if R has d.c.c. on right (left) ideals, then each proper prime ideal of R is maximal.

Proof. Let P be a proper prime ideal of R. Then $X \nsubseteq P$, so there exists $x \in X$, $x \notin P$, and hence $2x \in X$. The hypotheses of Theorem 3.3 are satisfied and hence P is a type one prime ideal, and R/P is a ring. So $R \in \mathfrak{R}^1$ and $R/\mathbf{P}_0(R)$ is a subdirect product of prime rings, yielding $R/\mathbf{P}_0(R)$ is a ring. Invoke the given d.c.c. to get that each R/P is a prime ring with d.c.c., and hence is a simple ring. So P is a maximal ideal.

Using a different approach in the proofs than that above, some allied results were obtained by Birkenmeier, Heatherly and Lee [10].

Proposition 3.6 ([10, Proposition 4.11]). If for each $x \in R$, there exists j, m, n with $j, n \geq 1$, $m \geq 0$, such that $\langle x \rangle^j$ is $(\langle x \rangle^m, \langle x \rangle^n)$-distributive, then $R \in \mathfrak{R}^1$ and $R/\mathbf{P}_0(R)$ is a ring. (Note: Here j, m, or n may depend on x).

Proposition 3.7 ([10, Proposition 4.12]). Let R be a d.g. near-ring generated by a multiplicative semigroup X of distributive elements. If for each $x \in X$ we have $2x \in \mathcal{D}(R)$, then $R \in \mathfrak{R}^1$ and $R/\mathbf{P}_0(R)$ is a ring.

Observe that if R^j is (R^m, R^n)-distributive, for some $j, n \geq 1$, $m \geq 0$, then R satisfies the hypothesis of Corollary 3.5. It was shown in [10, Theorem 5.4] that for such near-rings $\mathbf{P}_1(I) = I \cap \mathbf{P}_1(R)$, for each ideal I of R. Lee [38, Propostion 3.1.20] has improved on that with the following.

Proposition 3.8. Let S be a two-sided R-subgroup of R. If S^j is (S^m, S^n)-distributive, for some $j, n \geq 1$, $m \geq 0$, then $\mathbf{P}_1(S) = S \cap \mathbf{P}_1(R)$.

Now for some examples illustrating the above results.

Example 3.9. Let $(G, +)$ be a Redei group (a non-abelian group such that each of its proper subgroups is abelian), and let $S = \{\alpha \in End(G, +) : \alpha \text{ is not surjective}\}$. Let R be the subnear-ring of $E(G)$ generated by S. Then R is d.g. near-ring with distributive generating semigroup S. Observe that $2\alpha \in S$ for each $\alpha \in S$ and thus R satisfies Proposition 3.7. Using $G = S_3$ and this process yields near-ring number 36 in Clay's tables [15].

Example 3.10. Let $(G, +)$ be group and H a normal subgroup of G. Let:

$$R = \{\alpha_c \in M_0(G) : (H)\alpha_c \subseteq H \text{ and } (G \backslash H)\alpha_c \subseteq c + H, \text{ where } c \in G\}.$$

Then R is a subnear-ring of $M_0(G)$. Let $S = \{\sigma \in R : (G)\sigma \subseteq H\}$. Then S is an ideal of R. Take $T = \{\tau \in R : \tau \in End(H, +)\}$. Then R is (S, T)-distributive and T is a multiplicative semigroup of $(R, \circ)$. Furthermore T is the largest subset of R such that R is (S, T)-distributive; hence $\mathcal{D}(R) \subseteq T$. Observe that T contains the ideal $I = \{\beta \in R : (H)\beta = 0\}$; I is the largest ideal of R such that R is (S, T)-distributive. Now assume H is a finite, simple, non-abelian group. Then $R = \mathbf{gp}(T)$ and R is (S, R)-d.g. on T. Note that I and S are type one prime ideals of R. If H has index two in R, then R has unity. In the particular case where $G = S_3$ and $H = A_3$, then R is not d.g.

4 Completely Prime Ideals and Their Associated Radical

Of rising interest is the concept of a completely prime ideal. One reason for this is certain algebras connected to mathematical physics (e.g. Weyl algebra, certain enveloping algebras of solvable Lie groups, and their non-linear analogs) have been shown to enjoy the property that each prime ideal is completely prime. (For more on this for rings see, for example: Birkenmeier, Heatherly and Lee [11], Borceux and Van den Bossche [14], and McConnell and Robson [44]). Recall that an ideal I in a near-ring N is a *completely prime ideal* of N if $ab \in I$ implies $a \in I$ or $b \in I$. The study of completely prime ideals in near-rings goes back at least to 1979 by Ramakotaiah and Rao [52], where such an ideal is called a "type 2 prime ideal". The terminology "completely prime" is now standard in ring theory and becoming so for near-rings. (See Birkenmeier, Heatherly and Lee [11] and Groenewald [23, 24]). The intersection of all the completely prime ideals of R is called the *completely prime radical* of R, and is denoted herein by $\mathbf{P}_2(R)$. (This notation was introduced by Ramakotaiah and Rao [52]. Different notation was used by Groenewald [23, 24], wherein he carried out the first major investigation of this radical for near-rings). Localized distributivity conditions have been used to obtain some results [7, 38] on completely prime ideals and the completely prime radical. A sampling of these results are given below. First some terminology and notation need be discussed. (Note: Here R is used to denote an arbitrary near-ring, not necessarily zero symmetric).

The class of all near-rings for which each prime ideal is completely prime is denoted by $\mathfrak{R}^2$, and the subclass of it which consists of zero symmetric near-rings is denoted by $\mathfrak{R}_0^2$.

Lemma 4.1. Let R satisfy a permutation identity. If B^k is (B^m, B^n)-d.g., for some $n, k \geq 1$, $m \geq 0$, where B is one of $\langle R, R \rangle$, $\langle \Delta(R) \rangle$, or $\langle R' \rangle$, with the addendum that $\langle R' \rangle^k R \subseteq \langle R' \rangle$ in the latter case, then R is zero symmetric and B is nilpotent.

For a proof of this lemma see [7, Theorem 1.14]. Note that for R in the above lemma we have $B \subseteq \mathbf{P}_0(R)$. A moments reflection establishes that if $\theta : R \to A$ is any surjection near-ring homomorphism with $\mathbf{P}_0(R) \subseteq Ker\theta$, then $A \in \mathfrak{R}^2$ implies $R \in \mathfrak{R}^2$. This will be useful in the sequel.

Theorem 4.2. Let R satisfy a permutation identity. If either of the following hold, then $\bar{R} = R/\mathbf{P}_0(R)$ is a commutative ring and $R \in \mathfrak{R}_0^2$.
(1) B^k is (B^m, B^n)-d.g., where $B = \langle \Delta(R) \rangle$, or
(2) R is d.g.
In either case, if R has d.c.c. on ideals, then each proper prime ideal of R is maximal as both a left and a right R-subgroup.

Proof. Assume (1). Then Lemma 4.1 yields R is zero symmetric and $\langle \Delta(R) \rangle \subseteq \mathbf{P}_0(R)$. So $\bar{R} = R/\mathbf{P}_0(R)$ is a distributive semiprime near-ring. Consequently $\bar{R}$ must be a ring. The permutation identity on R is inherited by the ring $\bar{R}$, and any semiprime permutation identity ring is commutative, (see [5, p.128]). So $\bar{R} \in \mathfrak{R}^2$ and hence $R \in \mathfrak{R}_0^2$.

Assume (2). Then $B = \langle R' \rangle$ satisfies the hypotheses of Lemma 4.1 and hence $\langle R' \rangle \subseteq \mathbf{P}_0(R)$. So $\bar{R}$ is a d.g. near-ring with abelian additive group — a ring. Then proceed as above.

If for either (1) or (2) we have the d.c.c. on ideals in R, then the commutative semiprime ring $\bar{R}$ is a finite direct sum of fields. So for any proper prime ideal P of R, then R/P is a field and hence P is maximal in a left or right R-subgroup of R.

This above theorem and its proof are, except for some minor modifications, that given by [7, Theorem 2.3], and [7, Corollary 2.4]. It was also observed that if R is any permutation identity near-ring given by a permutaion σ of length n with $\sigma(n) = k < n$, then R is $\left(R^{k-1}, R^{n-k}\right)$-distributive. Thus for such near-rings 4.2(1) holds (See [7, Proposition 2.5]). Several types of examples of such permutation identity near-rings are given in [4, Example 4.9 and p.108]. A general method for constructing them is given by Birkenmeier and Heatherly [3], as a special case of a universal algebraic formulation.

For permutation identity near-rings in general, prime ideals need not be completely prime. Examples of this can be found in [7, Examples 2.1, 2.2]. Of course the localized distributivity alone will not guarantee that all primes are completely prime, since it is not true for the class of rings. So Theorem 4.2 shows an interesting interplay of the two concepts. It is worth emphasizing that in that theorem the distributivity can be confined to a "small" subset.

A final illustration of the results obtained concerning prime ideals and prime radicals under localized distributivity conditions is the following result due to Enoch Lee [38], the proof of which will appear elsewhere in due course.

Theorem 4.3 ([38, Proposition 3.1.20]). Let S be a two-sided R-subgroup of a zero symmetric near-ring R. If S^k is (S^m, S^n)-distributive, for some $k, n \geq 1$, $m \geq 0$, then $\mathbf{P}_\nu(S) = S \cap \mathbf{P}_\nu(R)$, where $\nu = 1, 2$.

The equality $\mathbf{P}_1(S) = S \cap \mathbf{P}_1(R)$ does not hold in general. Groenewald [26, p.2672] gave an example of a near-ring on the group C_8 which has an ideal for which the equality fails to hold; however, this example is not zero symmetric. It remains an open question whether or not there is a zero symmetric example. It was established (independently) in [26, Theorem 4.7] and [10, Theorem 5.2] that $\mathbf{P}_1(I) \subseteq I \cap \mathbf{P}_1(R)$ for any ideal I. The next example shows that the equality $\mathbf{P}_1(S) = S \cap \mathbf{P}_1(R)$ may fail to hold even when S is a subnear-ring of a finite d.g. near-ring with unity of the most natural kind.

Example 4.4. Let S be any finite zero symmetric, non-nilpotent near-ring for which $\mathbf{P}_1(S) = S$. (For example, let S be a Malone trivial near-ring on a finite simple group $(G, +)$ with at least one left unity and at least one left annihilator element other than zero). Then S can be embedded as a subnear-ring in some $E(\mathcal{S}_n)$, where $\mathcal{S}_n$ is a symmetric group of degree $n \geq 5$. (Use Malone's [41] technique to embed S in some $E(\mathcal{A}_n)$, where $\mathcal{A}_n$ is the alternating group on n letters, and then observe that $E(\mathcal{A}_n)$ embeds in $E(\mathcal{S}_n)$. Without loss of generality one can take $n \geq 5$). As noted in [10, Example 3.5], $E(\mathcal{S}_n)$ is in $\mathfrak{R}_0^1$ and $\mathbf{P}_1(E(\mathcal{S}_n)) = B$, where B is a non-zero nilpotent ideal of $E(\mathcal{S}_n)$. The assumption that $\mathbf{P}_1(S) = S \cap \mathbf{P}_1(E(\mathcal{S}_n))$ implies $S = S \cap B$, or $S \subseteq B$, contrary to S not nilpotent. (Aside: If one does not require that R be d.g., then there is an easier way to obtain the desired counter-example. Use the embedding technique from Malone and Heatherly [42] to embed S in some $M_0(G)$ and observe that $\mathbf{P}_1(M_0(G)) = 0$. For this to yield a contradiction one only needs $\mathbf{P}_1(S) \neq 0$).

5 Primitivity and the J_ν - radicals

The concepts of ν-primitive ideals and the associated J_ν radicals are well-known for $\nu = 0, 1, 2$ and it is safe to assume the reader is familiar with them (See [47, Chapters 2–6] or [51, Chapters 4, 5]). However, for $\nu = 3$ there is no monograph level treatment, so the requisite definitions are given here. (The concepts of 3-

primitive and the J_3 radical were introduced by Holcombe and Walker [31] in 1978. Also see Holcombe [30], Kaarli [33] and De Stefano [17]).

For this section R will always be zero symmetric. Let G be a monogenic right R-subgroup such that G has no proper, non-zero right R-subgroups. Then G is said to be of *type* 3 if whenever $g_1, g_2 \in G$ such that $g_1 r = g_2 r$ for each $r \in R$, then $g_1 = g_2$. An ideal I of R is a 3-*primitive ideal* of R if $I = \{r \in R : Gr = 0\}$ for some right R-subgroup G of type 3. Then $J_3(R)$ is the intersection of all the 3-primitive ideals of R.

It is well known that a ν-primitive near-ring is $\nu - 1$ primitive, for $\nu = 1, 2, 3$, and that $\mathbf{P}_0(R) \subseteq J_0(R) \subseteq J_1(R) \subseteq J_2(R) \subseteq J_3(R)$. De Stefano [17] showed that for distributive near-rings all the primitivities, $\nu = 0, 1, 2, 3$, coincide. Laxton [36] showed every 1-primitive ideal of a finite d.g. near-ring with right unity is 2-primitive. Kaarli [33] pointed out that $J_1(R) = J_3(R)$ if R is d.g. or if R has unity. The principal results of this section show that several types of primitivity are equivalent under certain localized distributivity conditions. These results, due to Enoch Lee [38], extend those by De Stefano, Laxton, and Kaarli just mentioned. The proofs of these results will be given in a subsequent paper.

Theorem 5.1 ([38, Section 4.2]). Let R^k be (R^m, R)-distributive for some $k \geq 1$, $m \geq 0$. Then each 0-primitive ideal of R is 3-primitive and $J_0(R) = J_3(R)$.

De Stefano's result for distributive near-rings is an immediate corollary. Groenewald [26, Proposition 3.1] has shown that every 3-primitive ideal of a zero symmetric near-ring is a type one prime ideal. So $\mathbf{P}_1(R) \subseteq J_3(R)$ for any zero symmetric R. Thus if R satisfies the hypothesis of Theorem 5.1 we have $\mathbf{P}_0(R) \subseteq J_0(R) = \mathbf{P}_1(R) = J_3(R)$. Of course $\mathbf{P}_0(R)$ may be properly contained in $J_0(R)$, even for R a commutative ring. (See [18, Example 10, p.103]).

Theorem 5.2 ([38, Section 4.2]). If R^k is (R^m, R)-d.g., for some $k \geq 1$, $m \geq 0$, then each 1-primitive ideal is 3-primitive and $J_1(R) = J_3(R)$.

As immediate corollaries we have the results for d.g. near-rings by Laxton and Kaarli which are referenced above. For more on the relationships between various types of primeness and primitivities see Groenewald [26], Holcombe [29] and Lee [38]. One avenue for improving the results of this section is to restrict the localized distributivity to smaller sets. This is currently under investigation by the author and Enoch Lee.

6 Refencences

[1] V. Andrunakievic, "Radicals in associative rings *I*", *Mat. Sb.* **44** (1958), 199–212.

[2] J. Beidleman, " Strictly prime distributively generated near-rings", *Math. Zeit.* **100** (1967), 97–105.

[3] G. Birkenmeier and H. Heatherly, "Operation inducing systems", *Algebra Universalis* **24** (1987), 137–148.

[4] G. Birkenmeier and H. Heatherly, " Medial near-rings", *Monatsh. Math.* **107** (1989), 89–110.

[5] G. Birkenmeier and H. Heatherly, "Permutation identity rings and the medial radical", in *Proc. Conf. on Non-Commutative Ring Theory (Athens, Ohio), Lecture Notes in Mathematics No. 1448* (Springer-Verlag, Berlin, 1990).

[6] G. Birkenmeier and H. Heatherly, "Left self distributive near-rings", *J. Austral. Math. Soc. (Ser. A)* **49** (1990), 273–296.

[7] G. Birkenmeier and H. Heatherly, "Permutation identity near-rings and localized distributivity conditions", *Monatsh. Math.* **111** (1991), 265–285.

[8] G. Birkenmeier and H. Heatherly, "Minimal ideals in near-rings", *Comm. Algebra* **20** (1992), 457–468.

[9] G. Birkenmeier and H. Heatherly, "Minimal ideals in near-rings and localized distributivity conditions", *J. Austral. Math Soc. (Ser. A)* **54** (1993), 156–168.

[10] G. Birkenmeier, H. Heatherly and E. Lee, "Prime ideals in near-rings", *Results Math.* **24** (1993), 27–48.

[11] G. Birkenmeier, H. Heatherly and E. Lee, "Completely prime ideals and associated radicals", in *Ring Theory, Proceeding of the Ohio St. -Denison Conference*, ed. S. K. Jain and S. T. Rizvi (World Scientific, Singapore, 1993).

[12] G. Birkenmeier, H. Heatherly and E. Lee, "Prime ideals and prime radicals in near-rings", *Monatsh. Math.* (to appear 1994).

[13] D. Blackett, *Simple and Semi-simple Near-Rings, Doc. Diss.* (Princeton Univ., 1950).

[14] F. Borceux and G. Van den Bossche, *Algebra in a Localized Topos with Applications to Ring Theory* (Springer-Verlag, Berlin, 1983).

[15] J. R. Clay, "The near-rings on groups of low order", *Math. Zeit.* **104** (1968), 364–371.

[16] J. R. Clay, *Nearrings. Geneses and Applications* (Oxford Univ. Press, Oxford, 1992).

[17] S. De Stefano, "Remarks on quasi-regularity in a distributive near-ring", in *Near-Rings and Near-Fields*, ed. G. Ferrero and C. Ferrero-Cotti (San Benedetto del Tronto, 1981), 143–146.

[18] N. Divinsky, *Rings and Radicals* (Univ. Toronto Press, Ontario, 1965).

[19] L. Esch, *Commutator and Distributor Theory in Near-Rings, Doc. Diss.* (Boston Univ., 1974).

[20] G. Ferrero, "Sulla struttura aritmetica dei quasi-anelli finiti", *Atti Accad. Sci. Torino Cl. Sci. Fis. Mat. Nat.* **97** (1962-63), 1114–1130.

[21] A. Fröhlich, "Distributively generated near-rings I. Ideal theory", *Proc. London Math. Soc.* **8** (1958), 76–94.

[22] A. Fröhlich, "The near-ring generated by the inner automorphisms of a finite simple group", *J. London Math. Soc.* **33** (1958), 95–107.

[23] N. Groenewald, "Note on the completely prime radical in near-rings", in *Near-Rings and Near-Fields*, ed. G. Betsch (Amsterdam, 1987), pp.97–100.

[24] N. Groenewald, "The completely prime radical in near-rings", *Acta Math. Hung.* **51** (1988), 301–305.

[25] N. Groenewald, "Strongly prime near-rings", *Proc. Edinburgh Math. Soc.* **31** (1988), 337–343.

[26] N. Groenewald, "Different prime ideals in near-rings", *Comm. Algebra* **19** (1991), 2667–2675.

[27] H. Heatherly and H. Olivier, "Near integral domains", *Monatsh. Mat.* **78** (1974), 215–222.

[28] H. Heatherly and S. Ligh, "Pseudo-distributive near-rings", *Bull. Austral. Math. Soc.* **12** (1975), 449–456.

[29] W. M. Holcombe, *Primitive Near-Rings, Doc. Diss.* (University of Leeds, 1970).

[30] W. M. Holcombe, "A hereditary radical for near-rings", *Studia Math. Hungar.* **17** (1982), 453–456.

[31] W. M. Holcombe and R. Walker, "Radicals in categories", *Proc. Edinburgh Math. Soc.* **24** (1978), 111–128.

[32] K. Kaarli, "Minimal ideals in near-rings", *Tartu Riikl Ül. Toimetised* **336** (1975), 105–142 (in Russian).

[33] K. Kaarli, "Survey on the radical theory of near-rings", *Contributions to General Algebra* **4**, *Proc. Krems Conf.* (Teubner, Stuttgart, 1987).

[34] K. Kaarli, "On Jacobson type radicals of near-rings", *Acta Math. Hung.* **50** (1987), 71–78.

[35] K. Kaarli, "On minimal ideals of distributively generated near-rings", *Contributions to General Algebra* **7** (1991), 201–204.

[36] R. R. Laxton, "Primitive distributively generated near-rings", *Mathematika* **8** (1961), 143–158.

[37] R. R. Laxton, "Prime ideals and the ideal radical of a distributively generated nearring", *Math. Zeitschr.* **83** (1964), 8–17.

[38] E. Lee, *Prime Ideals and Prime Radials in Near-Rings, Doc. Diss.* (Univ. Southwestern LA., Lafayette, LA., 1993).

[39] S. Ligh, "Near-rings with descending chain conditions", *Compositio Mathematica* **21** (1969), 162–166.

[40] S. Ligh, "A note on matrix near-rings", *J. London Math. Soc. (2)* **11** (1975), 383–384.

[41] J. Malone, "A near-ring analogue of a ring embedding theorem", *J. Algebra* **16** (1970), 237–238.

[42] J. Malone and H. Heatherly, "Some near-ring embeddings", *Quart. J. Math. (Oxford) (2)* **20** (1969), 81–85.

[43] G. Mason, "On pseudo-distributive near-rings", *Proc. Edinburgh Math. Soc.* **28** (1985), 133–142.

[44] J. C. McConnell and J. C. Robson, *Noncommutative Noetherian Rings* (John Wiley & Sons, New York, 1987).

[45] N. McCoy, "Prime ideals in general rings", *Amer. J. Math.* **71** (1949), 823–833.

[46] N. McCoy, *Theory of Rings* (Chelsea, Bronx, New York, 1973).

[47] J. D. P. Meldrum, *Near-Rings and Their Links to Groups* (Pitman, Marsh Field, MA, 1985).

[48] J. D. P. Meldrum, "Les généralizations de la distributivité dans les presque - anneaux", *Rend. Sem. Mat. è Fis. Milano, LIX* (1989) (1992), 9–24.

[49] J. D. P. Meldrum and A. Van Der Walt, "Matrix near-rings", *Arch. Math.* **47** (1986), 312–319.

[50] A. Oswald, "Near-rings in which every *N*-subgroup is principal", *Proc. London Math. Soc. (3)* **28** (1974), 67–88.

[51] G. Pilz, *Near-Rings*, rev. ed. (North-Holland, Amsterdam, 1983).

[52] D. Ramakotaiah and G. Koteswara Rao, "IFP near-rings", *J. Austral. Math. Soc. (Ser. A)* **27** (1979), 365–370.

[53] I. Roberts, *Generalized Distributivity in Near-Rings, M. Phil. Thesis* (Univ. Edinburgh, 1982).

[54] R. Scapellato, "Sui quasi-anelli verificanti semigruppali C-mobili", *Boll. Un. Mat. Ital. B(6)* **4** (1985), 789–799.

[55] R. Scapellato, "Semigroup identities in near-rings", *Riv. Mat. Univ. Parma* **14** (1988), 315–319.

[56] S. D. Scott, "Minimal ideals in near-rings with minimal conditions", *J. London Math. Soc. (2)* **8** (1974), 8–12.

[57] O. Taussky, "Rings with commutative addition", *Bull. Calcutta Math. Soc.* **28** (1936), 245–246.

[58] A. Van Der Walt, "Prime ideals and nil radicals in near-rings", *Arch. Math. (Basel)* **15** (1964), 408–414.

[59] H. Weinert, "Ringe mit nichtkommutativer Addition I", *Jahresber. Dt. Math. Ver.* **77** (1975), 10–27.

[60] H. Weinert, "Ringe mit nichtkommutativer Addition II", *Acta Math. Sci. Hung.* **26** (1975), 295–310.

[61] H. Wielandt, "Über Bereiche aus Gruppenabbildungen", *Deutsche Mathematik* **3** (1938), 9–10.

ENDOMORPHISM NEAR-RINGS
THROUGH THE AGES

J. J. Malone

Department of Mathematical Sciences, Worcester Polytechnic Institute

Worcester, MA 01609-2280, USA

ABSTRACT

For some 35 years problems concerning morphism near-rings have been studied. In this paper, three areas of such investigation are surveyed, and significant unsolved or partially solved problems from each area are discussed. The first area is centered about the question as to whether an arbitrary distributively generated near-ring can be embedded in some $E(G)$. The motivating problem of the second area is the determination of the groups G such that $E(G)$ is a ring. The third area is typified by the problem of finding the groups G such that $I(G) = E(G)$. The first two areas seem to lie fallow at the moment; the third area is one of current investigation.

1 Introduction

One can argue that the era of sustained interest in near-rings began in the 1950s with the appearance of Donald Blackett's dissertation and subsequent papers and with the appearance of several papers by A. Fröhlich. These prepared the way for the blossoming that took place in the 1960s, the period that Gerhard Betsch referred to in his talk as "the time of dissertations." It is this era from the 60s to now that constitutes the "ages" mentioned in the title of this talk. I will review one problem from the late 60s and early 70s, and another that was prominent through the 70s and into the mid 80s. Finally, I will discuss a problem, with roots in the late 60s, which is currently an active area of investigation. I quickly admit that these three areas aren't the only sources of interesting problems for endomorphism near-rings, but they are problems that have held my interest for a number of years.

The two older problems are not brought up just for the sake of reviewing the old. Rather, they have some aspects that are still open and. in discussing them, I hope to entice some of you to take a look - or another look - at these problems. The presentation of the current problem will include a summary of recent work by several authors.

In what follows, all near-rings will be left near-rings.

2 Embeddings

The first problem that I wish to call to your attention is an embedding problem. We know that the motivation for distributively generated near-rings (dgnr) comes out of consideration of the system generated by the endomorphisms of a nonabelian

Y. Fong et al. (eds.), Near-Rings and Near-Fields, 31–43.

group. Once the concept of a dgnr is formulated, then naturally we are led to the question of whether an arbitrary dgnr can be embedded in the prototype dgnr, the endomorphism near-ring. We are also prompted to ask about this embedding because of a result we see in ring theory: Any ring can be embedded in some ring of endomorphisms.

As is well-known, the ring embedding is accomplished by first embedding the given ring in a ring with identity and then embedding this second ring into a ring of endomorphisms. So, do the techniques used in this ring embedding process work for near-rings? The answer is no and yes. In the first embedding, the distributive law of the given dgnr does not carry over to the "near-ring" with identity that is constructed. However, the second step of the process is valid for dgnr - a dgnr with left identity can be embedded in an endomorphism near-ring.

What then has been done on the problem of embedding an arbitrary dgnr in some $E(G)$? (Here, $E(G)$ is of course the endomorphism near-ring of the group G. Analogous meanings are attached to $A(G)$ and $I(G)$.) Before answering this, let's recall two results. The first is a famous result by Fröhlich.

Theorem 1 *([13], 1958)   If G is a finite simple nonabelian group, then $I(G) = M_0(G)$. (Hence, $I(G) = A(G) = E(G) = M_0(G)$.)*

The second result is a kind of near-ring "Cayley" theorem given by Malone and Heatherly.

Theorem 2 *([31], 1969)   Let $(R, +, \cdot)$ be a near-ring. If $(G, +)$ is any group containing $(R, +)$ as a proper subgroup, then $(R, +, \cdot)$ can be embedded in $M(G)$.*

The proof consists in showing that the set

$$A = \{f_a \text{ in } M(G) : (x)f_a = xa \text{ if } x \text{ in } R, \ (x)f_a = a \text{ if } x \text{ in } G/R\}$$

is a near-ring isomorphic to $(R, +, \cdot)$. Also note that, in particular, a 0-symmetric near-ring gets embedded in $M_0(G)$.

Now we are in a position to look at a theorem of mine that settles part of the problem of embedding a dgnr.

Theorem 3 *([26], 1970)   Let $(G, +)$ be a finite group. There exists a finite simple group $(S, +)$ such that any 0-symmetric near-ring whose group part is $(G, +)$ can be embedded in $I(S)$.*

PROOF   Let G have order n. Embed $(G, +)$ in $(S_n, +)$. Then embed $(S_n, +)$ in $(A_{n+2}, +)$. This is done as follows. Let f be in S_n. If f is an even permutation, then the image of f in A_{n+2} is f itself (considered as fixing the integers $n+1$ and $n+2$). If f is odd, its image is f times the transposition $(n+1, n+2)$. Since it can be assumed that $n \geq 3$, $(A_{n+2}, +)$ is a simple group. Then, since $(G, +)$ is embedded in $(A_{n+2}, +)$, the Malone and Heatherly result gives $(G, +, \cdot)$ embedded in $M_0(A_{n+2})$ which, by Fröhlich, has the property that $M_0(A_{n+2}) = I(A_{n+2}) = E(A_{n+2})$. Of course, $S = A_{n+2}$. □

Since a dgnr is 0-symmetric, this theorem shows that any finite dgnr can be embedded in an $E(G)$. But, obviously, the technique of the theorem can't be extended to provide a proof for the infinite case. Moreover, there are some aspects of the proof that might raise questions. For instance, the group in which G is embedded is relatively large; if $\mid G \mid = 12$, then $\mid A_{n+2} \mid \approx 43{,}000{,}000{,}000$.

Also, the ring technique of putting a ring into a ring with identity embeds the original ring as an ideal. But my technique does not give an embedding as an ideal. However, as it turns out, this may be too much to expect. Gerhard Betsch ([2], 1967) gave examples of finite 0-symmetric near-rings (not dgnr though) which cannot be embedded as right ideals in any near-ring with identity. While this result isn't definitive, it certainly tends to shed some doubt about embedding as an ideal for dgnr. By the way, in Betsch's examples each nonzero element is a left identity.

Again, when I embed the dgnr (R, S) into an $E(G)$, it is not guaranteed that the images of S elements are right distributive. Is it possible to find another embeddinbg that will remedy this? The answer is no. John Meldrum ([39], 1973) has shown that there are dgnr (without left identity, of course) that cannot be embedded in a $M_0(G)$ in such a way that the images of the S elements are right distributive - this even for some finite dgnr. (For a discussion of groups that do have "right distributive" embeddings, see various papers by Mahmood and Meldrum such as [25] and [43].)

It follows that there are dgnr that cannot be embedded in dgnr with left identity in a way that preserves right distributivity. For, if this could be done, then the dgnr with left identity itself embeds in some $E(G)$ in a way that preserves right distributivity. That is, right distributive elements get mapped to endomorphisms of G. Of course, $E(G)$ is in $M_0(G)$, and it is true that the right distributive elements of $M_0(G)$ are precisely the endomorphisms of G. This would contradict Meldrum's result.

But, at any rate, the dgnr-into-an-$E(G)$ problem is settled in the finite case. However, the general case is still there waiting for a resolution. Remember that the problem could be settled affirmatively either by showing an embedding of a dgnr directly into some $E(G)$ or by embedding the dgnr into a dgnr with left identity.

Before closing out this first topic, I would like to point out that there is another problem lurking here; one suggested by the finite case embedding. Can any 0-symmetric near-ring be embedded in some $E(G)$? Or perhaps: Can any 0-symmetric near-ring be embedded in a dgnr? In this regard we might note that Meldrum ([38], 1971) showed that the variety of near-rings generated by the dgnr is the class of all 0-symmetric near-rings. Does that have some bearing on this problem?

So, we have looked at some embedding problems that are very basic to dgnr. They are easily stated and easily remembered. Maybe they will come to mind the next time you are looking for a topic to investigate.

3   E-Groups

It seems quite natural that attention would be focused on $E(G)$ and its properties. The other two problems I will discuss concern $E(G)$ (and indeed $A(G)$ and

$I(G)$). Both of these problems are at the seam between near-ring theory and group theory. In fact, the next problem is one that can be stated either as a near-ring problem or as a group problem.

In the late 1960s I began to wonder about this: Although we know that when G is nonabelian the endomorphisms of G do not, in general, commute under addition, was it ever possible for $E(G)$ to be a ring?

In light of A. John Chandy's 1965 dissertation [6] (done under Donald Blackett's direction), this seemed to be a good question. Chandy took up anew a line of development that went back to Burnside [3] in 1902 and Levi [19] in 1942. The title Chandy used for his dissertaion and his 1971 paper [7] tells the story: *Rings generated by the inner automorphisms of nonabelian groups.* In other words, Chandy considered nonabelian groups for which $I(G)$ is a ring. After this, one would naturally ask about the possibility of $A(G)$ and $E(G)$ being rings. But, before we get into the details of this investigation, we should note the following two results.

Result 4 *A dgnr with commutative addition is a ring.*

Result 5 *A distributive near-ring with (right or left) identity is a ring.*

These results are mentioned in the books of Pilz [48] and Meldrum [44]. From these results it follows that $I(G)$ and $A(G)$ and $E(G)$ are distributive if and only if addition is commutative. Hence, one needs to establish only one of the two "missing" properties in order to show that a morphism near-ring is a ring. However, in practice, almost all of the work on the $E(G)$-as-a-ring problem has taken the direction of trying to establish the commutativity of addition. So, if f and h are any two endomorphisms of G, it suffices to show that $f + h = h + f$. Working in this direction has seemed much easier than trying to establish the right distributive law.

Definition 6 *A group G for which $E(G)$ is a ring is called an* E–group. *(Similar definitions hold for* A–group *and* I–group.*)*

When I got interested in the problem of $E(G)$ as a ring, I wrote to Hans Zassenhaus to see what he thought. He tended to doubt that $E(G)$ could be a ring. B. H. Neumann suggested a result (see Theorem 1 of [28]) that might be helpful in working on the problem. But, real progress came after I contacted the group theorist M. Suzuki at the University of Illinois. He said that he really didn't know the answer, but he would think about it. Well, he must have discussed the problem with other people at Illinois, for not too much later (in 1971) Ralph Faudree [11] published the first examples of E–groups. Faudree's groups are p–groups on four generators and have order p^8 and exponent p^2. Also, he shows that they have these properties:

    i. $G^p = U(G) = G' = Z(G)$, where $U(G) = \{g \text{ in } G : \ |g| \ \text{divides } p\}$,

    ii. each strict endomorphism (an endomorphism that is not an automorphism) has its image in $Z(G)$,

    iii. each automorphism of G is central.

Since G' is contained in Z, it follows that these groups are nilpotent of class two. Relative to (iii), it is of interest to note that the inner automorphisms of a group that is nilpotent of class two are always central automorphisms. Faudree's construction of an E-group works for all odd primes, but not for $p = 2$ (even though this is claimed in the paper). So, this leaves us a side question as to whether there can be a pE-group with $p = 2$. (Of course, a pE-group is a p-group that is also an E-group.) We also wonder whether conditions (i), (ii) and (iii) are necessary for an E-group. Before going on, let us look at a theorem contained in [28], a 1977 paper by Malone.

Theorem 7 *In a group $(G, +)$ each element g commutes (additively) with its endomorphic images if and only if any two endomorphic images of each g commute.*

PROOF Assume that each element commutes with its endomorphic images and let f and h be endomorphisms of G and I be the identity map on G. Then

$$
\begin{aligned}
f + h &= f + h + fh + I - I - fh \\
&= (I + f) + (I + f)h - I - fh \\
&= (I + f)(I + h) - I - fh \\
&= (I + f)(h + I) - I - fh \\
&= (I + f)h + (I + f)I - I - fh \\
&= h + fh + I + f - I - fh \\
&= h + f(h + I) - fh \\
&= h + f(I + h) - fh \\
&= h + f + fh - fh \\
&= h + f.
\end{aligned}
$$

The converse is immediate since I is an endomorphism. □

Incidentally, this last theorem is essentially the result reported by S. Rao at the 1987 near-ring conference.

The following restatement of the theorem emphasizes that our near-ring E-group problem can be stated in group theoretic terms.

Theorem 8 *G is an E-group if and only if each endomorphism of G commutes additively with the identity map on G.*

We can gain some insight by backing up from the E-group problem and considering the I-group problem: When is $I(G)$ a ring? For this we look at Chandy's work. This first result greatly simplifies things, especially for finite groups.

Theorem 9 *Let $(G, +)$ be an I-group and let a, b, c be in G. Let x be the second commutator $[[a, b], c]$. Then x is in $Z(G)$ and $3x = 0$. Also, G' is abelian.*

This theorem tells us that G is nilpotent of class at most 3. Then, since a finite nilpotent group can be written as a direct sum of its Sylow subgroups, it follows that we need only consider p-groups when we look for finite I-groups and

A–groups and E–groups. Another application of the theorem shows that a finite pI–group is nilpotent of class at most 2 if p is not 3. Chandy also had some other interesting results.

Theorem 10 *If G is nilpotent of class 2, then $I(G)$ is a commutative ring.*

Theorem 11 *The group G is an I–group if and only if the centralizer of each element of G is a normal subgroup of G.*

The last theorem gives an interesting characterization of I–groups. Is there an extension of this to characterize A–groups? I haven't seen any such result. In fact, there isn't very much at all that is known about A–groups. Several people have noted the following characterization of I–groups. It is easy to establish.

Theorem 12 *A group G is an I–group if and only if it is a 2-Engel group, that is, if and only if $[x, g, g] = 0$ for each x and g in G.*

Well, Faudree's breakthrough inspired the hunt for more E–groups and attempts to characterize them. My 1977 paper [28] was probably the first try at supplying some theory.

Theorem 13 *([28], 1977) Let G be a finite nonabelian pE–group of exponent p^2 and without a direct factor. Then*

$$G^p \leq U(G) = G' = \Phi(G) \leq Z(G),$$

and G is nilpotent of class 2.

I had better confess that what is given above is the expurgated version of the theorem. In the paper I said that $G^p = U(G)$, but A. Caranti gently pointed out that the containment could be proper.

In [29] it was shown that the groups described by Jonah and Konvisser [17] are E–groups. Like Faudree's groups, these groups are p–groups of order p^8 and exponent p^2 such that $G' = Z(G)$, each strict endomorphism has its image in $Z(G)$, and the automorphisms are all central. Moreover, [29] gives us the first example of a pE–group with $p = 2$.

At this point, all known examples of E groups had all automorphisms central and strict endomorphisms with images in the center. Is this really necessary? A. Caranti came forward to offer some good insight.

Theorem 14 *([4], 1985) For odd primes, a finite E–group of exponent p^2 has strict endomorphic images in $Z(G)$ if and only if G cannot be expressed as a direct product of two nonabelian groups.*

In this paper, Caranti also gave two families of finite E–groups of exponent p^2 having some automorphisms that are not central. Overall, a very nice paper. Thanks to Caranti, we know that statements (ii) and (iii) which describe Faudree's groups are not necessary conditions for E–groups. Incidentally, Caranti ([5], 1987) also gave the first example of an infinite E–group.

Caranti's examples of groups that were E–groups (and A–groups) having some automorphisms that were not central gives us a spinoff group theory problem, the study of centralizing automorphisms. An automorphism α of a group G is a *centralizing automorphism* if, for each g in $G, g\alpha = g + c$, with c in $C(g)$, the centralizer of g.

We might ask if there are groups other than Caranti's that have centralizing automorphisms. One such is the group $[C_3 \times C_3]C_3$ given as group 27/4 in [53], *Group Tables* by Thomas and Wood. This example was first pointed out (in some unpublished notes) by Bruce McQuarrie. We might also ask, as a group theory question, which groups have all automorphisms centralizing. Of course, this is just a rephrasing of the near-ring question as to which groups are A–groups.

There is still much to be done with A–groups and E–groups. For instance, what about A– or E–groups of exponent greater than p^2? Also, what about describing the A–groups that are not E–groups? In addition to the papers already mentioned, there are three more that might be looked at by one interested in these problems. These three give a somewhat different perspective for the problems we have been considering. In particular, the first two give approaches that may not yet have been fully exploited.

- - - *Endomorphism rings of non-abelian groups* ([37], 1970) by McQuarrie and Malone. This paper explores the function λ, where $\lambda + \lambda = i$, the identity map on G.

- - - *On groups and endomorphism rings* ([35], 1971) by Maxson. Here G is looked at as a cyclic $E(G)$-module.

- - - *D.g. near-rings and rings* ([36], 1986) by Maxson and Meldrum. Some criteria for (R, S) to be a ring are discussed.

Before leaving the topic of I-, A- and E–groups, I want to point out a problem that has been completely overlooked: Are there any A–groups or E–groups that are nilpotent of class three? Recall that, in the finite case, a class three I–group can occur, essentially, only for a p–group with $p = 3$. So, we are asking if there is a 3–group that is an A–group or an E–group. As far as I know, nothing has been published on this topic. It should be possible to settle this - especially for someone with a group theory background. Here are some sources and/or approaches one might consider in looking at this problem.

If you believe that such groups exist, you might look at factor groups of free 2-Engel groups in an attempt to find such a group. Or, one might look at the Burnside groups $B(3, r)$. These are groups with r generators such that the cube of each generator is the identity element. There is considerable literature about these Burnside groups.

Also, it might be worthwhile to consider the following.

- - - Chapter 7 in Part II in [49], D. J. S. Robinson's book *Finiteness conditions and generalized soluble groups*. The cited chapter discusses 2-Engel groups in some detail.

- - - Jewell Schubert's dissertation [50] *On groups of order 3^n and class 3*. This certainly treats groups that are of interest for the problem being considered.

Now we move to the third and last of the problem areas.

4   I-A-E Groups

The kinds of problems that we have been discussing made it inevitable that additional attention would be focused on $E(G)$ and $A(G)$ and $I(G)$. Indeed, over the last 25 years or so there has been a continuing investigation of the morphism near-rings of various groups. I might add that this is a problem area for which I do not need to do as much urging for people to join in. For instance, at this conference the presentations of S. A. Syskin, M. J. Thomsen and G. Saad fall in the area we are about to survey.

Fröhlich's paper [13] from 1958 that showed $I(G) = A(G) = E(G) = M_0(G)$ for finite simple nonabelian groups is a very early paper describing the morphism near-rings for a family of groups.

Of course, this result could make one wonder if there are any other G for which $I(G) = A(G) = E(G)$ (such groups are called $I—E$ groups). Or for that matter, G for which $I(G) = A(G)$ (called $I—A$ groups) or $A(G) = E(G)$ (called $A—E$ groups). But, as it turned out, nothing happened for ten years. Then in 1968 and 1969 the master's theses of Lyons [20], Guthrie [15] and King [18] described the morphism near-rings of the dihedral group of order six, the dihedral group of order eight and the quaternion group of order eight. Then, the next two papers described below extended this by detailing the situation for the larger families to which the two dihedral groups belong.

In 1972, Malone and Lyons [32] treated the dihedral groups of order $2n$. They showed that, for odd n, these are $I - E$ groups, the order of $E(G)$ being $2n^3$. Of course, D_6 is in this family. Then, in [33] from 1973, they discussed dihedral groups of order $2n, n$ even, and found that $| I(G) |= n^3/4$ and $| A(G) |= n^3/2$. Also, $| E(G) |= n^7/64$ if 4 divides n and $| E(G) |= 2n^3(n/gcd(n,4))^4$ if 8 divides n. So, this paper described a family that contains D_8. A key technique in these two papers and in several other papers was the near-ring version of the Peirce decomposition which was introduced by Berman and Silverman ([1], 1959).

Theorem 15 *Let e be an idempotent in the near-ring R. Then $R = A_e + M_e$ with $A_e = \{r - er : r$ in $R\} = \{t$ in $R : et = 0\}$ and $M_e = \{er : r$ in $R\} = \{z$ in $R : ez = z\}$, and $A_e \cap M_e = \{0\}$. Thus r in R has the unique decomposition $r = (r - er) + er$.*

Clearly, by giving a decomposition of a near-ring, this theorem simplifies the problem of studying it. By use of this theorem, it was possible in [32] to get not just the order of $E(G)$ for the dihedral groups of order $2n$ with n odd, but a description of the functions in that morphism near-ring. However, the morphism near-rings for the dihedral groups with n even are much more complicated and need more than a single decomposition to resolve their properties. So, the following theorem on a repeated Peirce decomposition was presented in [33] and used in that situation.

Theorem 16 *Let e and f be idempotents in the near-ring R and let $ef = 0$. Then $R = D + M + N$, where $D = \{(r - er) - f(r - er) : r$ in $R\} = \{x$ in $R : ex = fx = 0\}$, $N = \{f(r - er) : r$ in $R\} = \{y$ in $R : ey = 0$ and $fy = y\}$, $M = \{er : r$ in $R\} = \{z$ in $R : ez = z\}$. Also, $D \cap N = D \cap M = N \cap M = (D + N) \cap M =*

$D \cap (N + M) = N \cap (D + M) = \{0\}$, *and D is a right ideal in R. Thus each r in R has the unique decomposition $r = [(r - er) - f(r - er)] + f(r - er) + er$.*

The next paper, in a sense, generalizes the work in King's master's thesis. Malone ([27], 1973) treats the generalized quaternion groups of order 2^n. These are $A - E$ groups since $| I(G) | = 2^{3n-5}$ and $| A(G) | = | E(G) | = 2^{3n-4}$. However, this $E(G)$ has no nontrivial idempotents, but $E(Q)$, Q the quaternion group, does have such idempotents.

Here are some additional papers that deal with the morphism near-rings.

Meldrum ([40], 1977) looked at the infinite dihedral groups and found them to be $I - A$. He also initiated ([42], 1979) an investigation of the general linear groups over a finite field. Here at Fredericton, Syskin, Saad and Thomsen have said that $GL(2,3)$ is $I - A$. Also, a preprint [51] distributed here by Sergei Syskin says that $GL(n,q)$ is $A - E$ if and only if $n \geq 2$ and $q = 2$, and that $GL(n,q)$ is $I - A$ if q is a power of a prime.

Fong and Meldrum ([12], 1980) show that S_n, $n \geq 5$, is $I - E$. At this conference Syskin, Saad and Thomsen have announced that S_4 is also $I - E$ and that S_4 is $A - E$.

Clay and Grainger ([9], 1989) discuss odd generalized dihedral groups. The dihedral groups are generalized by replacing the normal subgroup C_n by an abelian group. There are many results in this paper. One is that a generalized dihedral group is $I - A$ if and only if it is, in fact, dihedral.

Lyons and Mason ([21], 1991) initiated a study of dicyclic groups of order $4n$. If n is even, $| I(G) | = 2n^3$ and $| A(G) | = 4n^3$. But, for odd n, the group is $I - E$ and the morphism near-ring has order $4n^3$. Malone ([30], 1993) continued this study with the dicyclic groups for which n is even. If $n \equiv 0 (mod\ 4)$, then the group is $A - E$. But, if $n \equiv 2 (mod\ 4)$, $| E(G) | = 2n^7$.

In presentations at this conference, Syskin and Thomsen have added another family to the list of $I - E$ groups: finite quasicyclic groups with cyclic center. This result leads them to conclude [51] that all special linear groups $SL(n,q)$, except for $SL(2,3)$, are $I - E$ and that $SL(2,3)$ is $A - E$.

Syskin's preprint [51] also gives some more $I - E$ groups. He says that if p is an odd prime and G is the holomorph of a cyclic group of order p^n, then G is $I - E$ if and only if $n \leq p$.

For a bit of theory on morphism near-rings, check out Lyons and Peterson ([23], 1988), Peterson ([45], 1989) and ([46], 1989). Gary Peterson ([47], 1993) has also shown that $I(G) \neq A(G)$ if G is a nonabelian p-group whose automorphism group is not a p-group. Also, Meldrum ([41], 1978) is another paper that takes an approach other than the Peirce decomposition. It introduces the S-series of a group. Additionally, Lyons and Meldrum ([22], 1980) work further in the direction indicated in the Meldrum paper [41].

(As a momentary aside from the $I - A - E$ investigations, I want to acknowledge that there are some other other kinds of questions that could be asked about $E(G)$. What are the subgroups H of G such that $E(H)$ embeds in $E(G)$? If $E(G_1)$ is isomorphic to $E(G_2)$, what can be said about G_1 and G_2? I have the impression that someone has looked at these, but have not found a reference for such work.)

All the work on the various morphism near-rings leads us quite naturally to explicit statements of the questions implicit in these investigations. Is there some characterization of the groups that are $I - A$ but not $A - E$? Unfortunately, we don't have many examples from which to generalize. What about the groups that are $A - E$ but not $I - A$? Again, there aren't too many examples to work from. Does the lack of examples indicate that $I - A$ and $A - E$ groups are fairly rare? Since the $I - A$ question concerns only the automorphisms of a group, it would seem that this would be easier to deal with than the $A - E$ problem. Perhaps that is so; time will tell. By the way, Jim Clay, in his relatively new book [8], points out that seeking a characterization of $A - E$ groups is not a totally new problem. He reminds us that the abelian group theorist L. Fuchs ([14], 1963) had long since asked when it is that $Aut(G)$ generates the ring of endomorphisms of G. Of course, this is just a special case of our $A - E$ question. Paul Hill ([16], 1969) provided a valuable insight when he showed that $A(G) = E(G)$ for all abelian groups of odd order.

Is then the question of characterizing the $I - E$ groups the hardest of the $I - A - E$ problems because all the endomorphisms must be dealt with or is it the easiest one because so many conditions are loaded onto the group? I am not sure. But, at any rate, we have discovered six families of groups that are $I - E$: finite nonabelian simple groups, dihedral groups with n odd, S_n, dicyclic groups with n odd, quasisimple groups, $\mathrm{Hol}(p^n)$ with p an odd prime and $n \leq p$. What is needed next, of course, is some organizing idea that explains why all these groups are indeed $I - E$. Admittedly, this master stroke has not yet been formulated. However, Gordon Mason and I have - in some work [34] yet to appear - attempted to simplify the situation somewhat. We have been able to show that the groups of two of the families are included among those of a larger family. More specifically, we prove the following theorem.

Theorem 17 *ZS-metacyclic groups are $I - E$.*

A *Z-metacyclic group* is a group G such that $G = \langle a, b : a^n = b^m = e, ba = a^q b \rangle$ where $gcd(n, q-1) = 1$. Additionally, we impose the condition that $gcd(n, m) = 1$. These are the groups that Coxeter and Moser [10] refer to as ZS-metacyclic. The proof that these are $I-E$ uses a Peirce decomposition to demonstrate that $End(G)$ is contained in $I(G)$. Along the way, the A and M components of $I(G)$ that are generated through the Peirce decomposition are given a clever (at least we think so) description in terms of a matrix; a Vandermonde matrix at that. In turn, this gives rise to an algorithm for determining the order of $E(G)$. It is not hard to see that the "odd" dihedral and dicyclic groups are ZS-metacyclic.

So this last theorem, as advertised, simplifies the $I - E$ problem a bit. But there must be a lot more to be said. Which way can we turn? Well, here is what I think is an interesting observation. Look at the families of $I - E$ groups and note that they can be written as semidirect sums in which the normal subgroup is G' and the other subgroup is cyclic. (Yes, this is trivial in the case of the simple and quasisimple groups.) Is this a necessary condition for an indecomposable $I - E$ group? It would certainly be of interest to know the answer to that. On the other hand, Syskin's holomorph groups show that this condition is not sufficient.

At any rate, we can see that there are many interesting problems that arise in the study of the $I - A - E$ properties.

* * * * * * * * *

That concludes our trip through the ages with endomorphism near-rings. In each of three areas, we have uncovered a wealth of problems waiting for solution. At my age, I will not have time to resolve all these questions that have been raised. So, feel free to adopt one or more of the problems. I truly am looking forward to seeing your work. Perhaps we can have a flurry of results in time for the Hamburg conference.

* * * * * * * * *

Post-Fredericton Comment

Life marches on in regard to the $I - E$ problem. Since the Fredericton conference, two more preprints have been circulated. Lyons and Peterson [24] show that if G is the semidirect product $[H]K$ with $(\mid H \mid, \mid K \mid) = 1$, then G is an $I - E$ group if and only if ϵ is in $I(G)$, where ϵ is the projection map from G onto K. This paper prompted Gordon Mason and me to take another look at the hypothesis for Theorem 17. We found that some of the hypotheses are redundant, a fact that was also confirmed in a communication from Sergei Syskin. Rather than assuming that G is ZS-metacyclic, we need only start with G as the semidirect product of two cyclic groups of relatively prime orders. From this it readily follows that G is Z-metacyclic.

The second preprint is [52] by Sergei Syskin. He extends the work of Lyons and Peterson by characterizing some families of groups for which ϵ is always in $I(G)$.

References

[1] G. Berman and R. Silverman, *Near-rings*, Amer. Math. Monthly **66** (1959), 23-34.

[2] G. Betsch, *Bemerkungen zur Einbettung von Fastringen in Fastringe mit Eine*, private communication, 1967.

[3] W. Burnside, *On groups in which every two conjugate operations are permutable*, Proc. London Math. Soc., Series 1, **35** (1902-03), 28-37.

[4] A. Caranti, *Finite p-groups of exponent p^2 in which each element commutes with its endomorphic images*, J. Algebra **97** (1985), 1-13.

[5] A. Caranti, S. Franciosi and S. de Giovanni, *Some examples of infinite groups in which each element commutes with its endomorphic images*, Group theory (Lecture Notes in Math. 1281), Springer-Verlag, Berlin and New York, 1987.

[6] A. Chandy, *Rings generated by the inner automorphisms of non-abelian groups*, dissertation, Boston Univ., 1965.

[7] _____, *Rings generated by the inner-automorphisms of nonabelian groups*, Proc. Amer. Math. Soc. **30** (1971), 59-60.

[8] J. Clay, *Nearrings, geneses and applications.* Oxford Univ. Press, Oxford and New York, 1992.

[9]   J. Clay and G. Grainger, *Endomorphism nearrings of odd generalized dihedral groups*, J. Algebra **127** (1989), 320-339.

[10] H. Coxeter and W. Moser, *Generators and relations for discrete groups*, 4th ed., Springer-Verlag, Berlin and New York, 1980.

[11] R. Faudree, *Groups in which each element commutes with its endomorphic images*, Proc. Amer. Math. Soc. **27** (1971), 236-240.

[12] Y. Fong and J. Meldrum, *The endomorphism near-rings of the symmetric groups of degree at least five*, J. Austral. Math. Soc. Ser. A **30** (1980), 37-49.

[13] A. Fröhlich, *The near-ring generated by the inner automorphisms of a finite simple group*, J. London Math. Soc. **33** (1958), 95-107.

[14] L. Fuchs, *Some results and problems on abelian groups, Topics in abelian groups.* Scott, Foresman and Company, Glenview, 1963.

[15] E. Guthrie, *The endomorphism near ring on D_8*, Master's thesis, Texas A&M Univ., 1969.

[16] P. Hill, *Endomorphism rings generated by units*, Trans. Amer. Math. Soc. **141** (1969), 99-105.

[17] D. Jonah and M. Konvisser, *Some non-abelian p-groups with abelian automorphism groups*, Arch. Math. (Basel) **26** (1975), 131-133.

[18] M. King, *The endomorphism near ring on the quaternion group*, Master's thesis, Texas A&M Univ., 1969.

[19] F. Levi, *Groups in which the commutator operation satisfies certain algebraic conditions*, J. Indian Math. Soc. **6** (1942), 87-97.

[20] C. Lyons, *Endomorphism near ring on the non-commutative group of order 6*, Master's thesis, Texas A&M Univ., 1968.

[21] C. Lyons and G. Mason, *Endomorphism near-rings of dicyclic and generalised dihedral groups*, Proc. Roy. Irish Acad. Sect. A **91A** (1991), 99-111.

[22] C. Lyons and J. Meldrum, *Reduction theorems for endomorphism near-rings*, Monatsh. Math. **89** (1980), 301-313.

[23] C. Lyons and G. Peterson, *Local endomorphism near-rings*, Proc. Edinburgh Math. Soc. **31** (1988), 409-414.

[24] _____, *Semidirect products of I-E groups*, submitted.

[25] S. Mahmood, *Limits and colimits in categories of d. g. near-rings*, Proc. Edinburgh Math. Soc. **23** (1980), 1-8.

[26] J. Malone. *A near ring analogue of a ring embedding theorem*, J. Algebra **16** (1970), 237 238.

[27] _____ *Generalised quaternion groups and distributively generated near-rings*, Proc. Edinburgh Math. Soc. **18** (1973), 235-238.

[28] _____, *More on groups in which each element commutes with endomorphic images*, Proc. Amer. Math. Soc. **65** (1977), 209-214.

[29] _____, *A non-abelian 2-group whose endomorphisms generate a ring, and other examples of E-groups*, Proc. Edinburgh Math. Soc. **23** (1980), 57-59.

[30] _____, *More on endomorphism near-rings of dicyclic groups*, Proc. Roy. Irish Acad. Sect. A **93A** (1993), 107-110.

[31] J. Malone and H. Heatherly, *Some near-ring embeddings*, Quart. J. Math. Oxford Ser. (2) **20** (1969), 81-85.

[32] J. Malone and C. Lyons, *Finite dihedral groups and d.g. near-rings I*, Compositio Math. **24** (1972), 305-312.

[33] _____ *Finite dihedral groups and d.g. near-rings II*, Compositio Math. **26** (1973), 249-259.

[34] J. Malone and G. Mason, *ZS-metacyclic groups and their endomorphism near-rings*, Monatsh. Math., to appear.

[35] C. Maxson, *On groups and endomorphism rings*, Math. Z. **122** (1971), 294-298.

[36] C. Maxson and J. Meldrum, *D.g. near-rings and rings*, Proc. Roy. Irish Acad. Sect. A **86A** (1986), 147-160.

[37] B. McQuarrie and J. Malone, *Endomorphism rings of non-abelian groups*, Bull. Austral. Math. Soc. **3** (1970), 349-352.

[38] J. Meldrum, *Varieties and d.g. near-rings*, Proc. Edinburgh Math. Soc. **17** (1971), 271-174.

[39] _____, *The representation of d.g. near-rings*, J. Austral. Math. Soc. **16** (1973), 467-480.

[40] _____, *The endomorphism near-ring of an infinite dihedral group*, Proc. Roy. Soc. Edinburgh **76A** (1977), 311-321.

[41] _____, *On the strucure of morphism near-rings*, Proc. Roy. Soc. Edinburgh **81A** (1978), 287-298.

[42] _____, *The endomorphism near-rings of finite general linear groups*, Proc. Roy. Irish Acad. Sect. A **79A** (1979), 87-96.

[43] _____, *Presentations of faithful d.g. near-rings*, Proc Edinburgh Math. Soc. **23** (1980), 49-56.

[44] _____, *Near-rings and their links with groups* Pitman Publishing Co. (Reseach Notes in Math. No. 134), Boston and London, 1985.

[45] G. Peterson, *Automorphism groups emitting local endomorphism near-rings*, Proc. Amer. Math. Soc. **105**, (1989), 840-843.

[46] _____ *On the structure of an endomorphism near-ring*, Proc. Edinburgh Math. Soc. **32** (1989), 223-229.

[47] _____, *Endomorphism near-rings of p-groups generated by the automorphism and inner automorphism groups*, Proc. Amer. Math. Soc. to appear.

[48] G. Pilz, *Near-rings*, revised edition, North-Holland Pub. Co. (Mathematics Studies No. 23), Amsterdam and New York, 1983.

[49] D. Robinson, *Finiteness conditions and generalised soluble groups*, Part 2, Springer-Verlag, New York and Heidelberg, 1972.

[50] J. Schubert, *On groups of order 3^n and class 3*, dissertation, Univ. of Illinois, 1950.

[51] S. Syskin, *Endomorphism near-rings of finite solvable groups*, preprint.

[52] _____, *Projection endomorphisms on finite groups*, preprint.

[53] A. Thomas and G. Wood, *Group tables*, Shiva Publishing, Orpington, 1980.

ON REGULAR NEAR-RING MODULES

J. Ahsan

Department of Mathematical Sciences
King Fahd University of Petroleum and Minerals
Dhahran 31261, Saudi Arabia

1. Introduction and Preliminaries

In this paper we introduce the notion of a regular near-ring module by extending the usual elementwise definition of a regular near-ring to arbitrary near-ring modules. We characterize these modules in terms of certain restricted injectivity properties (Proposition 2.9). Using this characterization we deduce several characterizations of regular near-rings (Theorem 2.10). We also determine a characterization of strictly semisimple near-rings among near-rings with no nonzero nilpotent elements (Theorem 2.13). Throughout, R will denote a right near-ring with identity 1 such that $x0 = 0$ for all $x \in R$, and all modules (that is, near-ring modules) over R are left unital. The symbol $_R M$ will denote a left near-ring module M over R, and the term *R-subgroup of M* will mean a subgroup of $(M, +)$ which is closed under left R-multiplication. By an *R-submodule of M* we shall mean a normal subgroup A of $(M, +)$ satisfying $r(m + a) - rm \in A$ for all $r \in R$, $m \in M$, $a \in A$. Submodules of $_R R$ are *left ideals* of R. If A is a left ideal of R and $AR = \{ar \mid a \in A, r \in R\} \subseteq A$, then A is an *ideal* of R. For any subset B of an R-module M the set $\{r \in R \mid rB = (0)\}$ is called the *left annihilator* of B, denoted by $\ell(B)$. If $B = \{x\}$ we write $\ell(x)$ instead of $\ell(\{x\})$. For all subsets B of $_R M$, $\ell(B)$ is a left ideal of R. If B is an R-subgroup, $\ell(B)$ is an ideal. Near-ring homomorphisms and R-homomorphisms (that is, near-ring module homomorphisms) are defined in the usual manner. The set of all R-homomorphisms between left R-modules M and N will be denoted by $\mathrm{Hom}_R(M, N)$. An R-homomorphism $f : M \to N$ is *normal* if $f(M)$ is an R-submodule of N. An exact sequence $M \xrightarrow{f} N \longrightarrow 0$ *splits* if there exists a normal $g : N \to M$ such that $fg = 1_N$. The short exact sequence (s.e.s.) $0 \to L \to M \to N \to 0$ *splits* if the sequence $M \to N \to 0$ splits. The exact sequence $0 \to L \xrightarrow{f} M$ splits if there exists $g : M \to L$ such that $gf = 1_L$. The exact sequence $0 \to L \xrightarrow{h} M$ splits if and only if the s.e.s. $0 \to L \xrightarrow{h} M \xrightarrow{f} N \to 0$ splits [2, Lemma 2.1]. If L and N are submodules of M such that $M = L + N$ and $L \cap N = (0)$ we write $M = L \oplus N$. A left R-module M is *monogenic* if there exists $a \in M$ such that $Ra = M$; M is *right cancellative* if for each nonzero $m \in M$ and $r, r' \in R$ the identity $rm = r'm$ implies $r = r'$. Thus R is *right cancellative* if $_R R$ is right cancellative. An R-module M is called *irreducible* if it has no proper nonzero R-subgroups. M is *simple* if it contains no proper nonzero R-submodules. M is *semisimple* (also called *completely reducible*) if M is a direct sum of simple submodules. A near-ring R is

45

Y. Fong et al. (eds.), Near-Rings and Near-Fields, 45–51.

called *semisimple* if $_RR$ is semisimple. Therefore, a near-ring R is semisimple if and only if R is the direct sum of simple left ideals. A module M is said to be the *semi-direct sum* of its R-subgroups A and B, and write $M = A \dot{+} B$ if A is an R-submodule, $M = A + B$ and $A \cap B = (0)$. In this case B is called a *semi-direct summand* of M. Then every $m \in M$ can be expressed uniquely as $a + b$ for some $a \in A$, $b \in B$. Moreover the canonical projection $p : M \to B$ given by $p(a+b) = b$ is an R-homomorphism (see Mason [2, p. 46]). R is called *strictly semisimple* if R is a direct sum of irreducible left ideals. As noted in [2, Theorem 3.5], R is strictly semisimple if and only if every R-subgroup of a monogenic R-module is a semi-direct summand (equivalently, if L_1, L_2 are R-subgroups of R and $0 \to L_1 \to L_2$ is exact, then it splits). A near-ring R is *regular* if for each $a \in R$, $\exists\ x \in R$ such that $a = axa$. It can be shown that R is regular if and only if, for all $a \in R$, $\exists$ an idempotent $e = e^2 \in R$ such that $Ra = Re$ [3, p. 346]. The *socle* of R is the sum of all minimal left ideals of R and is denoted by soc(R). For undefined terms and notations used in the sequel, we refer to Pilz [3].

2.   Results

Definition 2.1. Let Q be a left R-module. An element $x \in Q$ is *R-divisible* in Q if for each (nonzero) $r \in R$, $\exists\ y \in Q$ such that $x = ry$; Q is called *R-divisible* if and only if $rQ = Q$ for all nonzero $r \in R$.

Remark. The additive group of rational numbers, considered as a module over the near-ring Z of integers, is Z-divisible. More generally, in the category of groups, any group which is *n-injective* in the sense of Mason [2, Def. 2.5(a)] is Z-divisible.

Proposition 2.2. *Let Q be an R-divisible module. Then for each monogenic R-subgroup I of R and the R-homomorphism $\phi : I \to Q$, there exists an R-homomorphism $\overline{\phi} : R \to Q$ which extends ϕ.*

Proof. Let $I = Ra$ for some $a \in R$. Suppose $\phi(a) = x$ for $x \in Q$. For each $r \in R$, we have $\phi(ra) = r\phi(a) = rx$. Since Q is R-divisible, there exists $y \in Q$ such that $x = ay$. Define $\overline{\phi} : R \to Q$ by $\overline{\phi}(r) = ry$ for $r \in R$. Clearly $\overline{\phi}$ is an R-homomorphism. Moreover, $\overline{\phi}(1) = y$. Thus $\overline{\phi}(ra) = ray = rx = \phi(ra)$. Hence $\overline{\phi}$ extends ϕ.

The above proposition motivates the following definition.

Definition 2.3. Let Q be a left module over a near-ring R. Then Q is called *P-injective* if for each monogenic R-subgroup I of R and the R-homomorphism $\phi : I \to Q$, there exists an R-homomorphism $\overline{\phi} : R \to Q$, which extends ϕ. More generally, for any fixed left R-module M, Q is called *PM-injective* if each R-homomorphism ϕ from a monogenic R-subgroup of M to Q extends to an R-homomorphism $\overline{\phi} : M \to Q$. Thus PR-injective modules are P-injective. We shall say that R is *self P-injective* if $_RR$ is P-injective.

Proposition 2.4. *Let R be a right cancellative near-ring. Then the following assertions are equivalent:*

(1) Q *is a P-injective left R-module.*

(2) Q *is R-divisible.*

Proof. (1) $\Rightarrow$ (2): Let $x \in Q$ and $(0 \neq)r \in R$. Define $\phi : Ra \to Q$ by $\phi(ra) = rx$. Clearly ϕ is a well-defined R-homomorphism, since R is right cancellative. Hence ϕ extends to an R-homomorphism $\overline{\phi} : R \to Q$, since Q is P-injective. Thus we have $x = \phi(a) = \overline{\phi}(a) = \overline{\phi}(a.1) = a\overline{\phi}(1)$. As $\overline{\phi}(1) \in Q$, x is R-divisible. Hence Q is R-divisible.

(2) $\Rightarrow$ (1): This is Proposition 2.2.

Definition 2.5. A left R-module M is called *pre-regular* if for each $a \in M$, there exists an R-homomorphism $f \in \mathrm{Hom}_R(Ra, R)$ such that $a = f(a) \cdot a$. We shall say that R is *pre-regular* if $_RR$ is pre-regular.

Proposition 2.6. *Let R be a near-ring with a unique nonzero idempotent. For a left R-module M the following conditions are equivalent:*

(1) M *is pre-regular.*

(2) M *is right cancellative.*

Proof. (1) $\Rightarrow$ (2): Suppose for $r, r' \in R$ and $(0 \neq)$ $m \in M$, we have $rm = r'm$. Since M is pre-regular, so there exists an R-homomorphism $f : Rm \to R$ such that $m = f(m) \cdot m$. This implies that $f(m) = f(f(m) \cdot m) = f(m) \cdot f(m)$. Hence $f(m)$ is a nonzero idempotent element of R, and so by the uniqueness of the nonzero idempotent element, it follows that $f(m) = 1$. Hence the equation $rm = r'm$ implies that $r = r \cdot 1 = r \cdot f(m) = f(rm) = f(r'm) = r'f(m) = r' \cdot 1 = r'$, that is, $r = r'$. Therefore, M is right cancellative.

(2) $\Rightarrow$ (1): Let $_RM$ be a right cancellative module. We show that M is pre-regular. Let $m \in M$. Define an R-homomorphism $g : Rm \to R$ by $g(rm) = r$ for $r \in R$. As $g(m) = g(1 \cdot m) = 1$, it follows that $m = 1 \cdot m = g(m) \cdot m$. Hence M is pre-regular.

Corollary. *R is a pre-regular near-ring with a unique nonzero idempotent if and only if R is right cancellative.*

Proposition 2.7. *Let R be a d.g. (that is, distributively generated) near-ring. Then the following assertions for a left R-module M are equivalent:*

(1) M *is pre-regular.*

(2) *Each monogenic R-subgroup of M is projective (in the usual categorical sense [1]).*

Proof. (1) $\Rightarrow$ (2): Let $m \in M$. We show that the monogenic R-subgroup Rm of M is projective. Define $g : R \to Rm$ by $g(r) = rm$ for $r \in R$. Since M is pre-regular, there exists $f \in \mathrm{Hom}_R(Rm, R)$ such that $m = f(m) \cdot m$. Thus for $r \in R$, we have $f(rm) \cdot m = (rf(m))m = r(f(m) \cdot m) = rm$. Hence $g(f(rm)) = f(rm) \cdot m = rm$. Thus $g \circ f = id_{Rm}$. Hence Rm is a retract of R. Moreover, since $1 \in R$, $_RR$ is free and hence projective (cf. [1]). As retracts of projective objects are projective, so Rm is projective.

(2) $\Rightarrow$ (1): Let $m \in M$. We define $g : R \to Rm$ by setting $g(r) = rm$ for $r \in R$. Clearly g is a surjective R-homomorphism. By hypothesis, Rm is projective. Hence there exists $f \in \operatorname{Hom}_R(Rm, R)$ such that the following diagram is commutative.

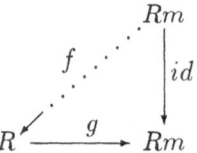

Thus $g \circ f = id$. Hence $m = (g \circ f)(m) = g(f(m)) = f(m) \cdot m$. Therefore M is pre-regular.

Definition 2.8. A left R-module M is called *regular* if for each $m \in M$, there exists an R-homomorphism $g \in \operatorname{Hom}_R(M, R)$ such that $m = g(m) \cdot m$. Thus if $_R R$ is regular, then for each $a \in R$, there exists $g \in \operatorname{Hom}_R(R, R)$ such that $a = g(a) \cdot a = g(a \cdot 1) \cdot a = ag(1) \cdot a \in aRa$. Hence R is regular in the usual sense. An obvious consequence of this definition is that a submodule of a regular module is regular. Thus every left ideal of a regular near-ring is regular as an R-module (but not necessarily as a near-ring).

Proposition 2.9. *For a left R-module M, the following assertions are equivalent:*

(1) *M is regular.*

(2) *M is pre-regular and R is PM-injective.*

Proof. (1) $\Rightarrow$ (2): Assume that M is a regular module. Then for each $m \in M$, there exists $g \in \operatorname{Hom}_R(M, R)$ such that $m = g(m) \cdot m$. In particular, there exists $f \in \operatorname{Hom}_R(Rm, R)$ such that $m = f(m) \cdot m$. Hence M is pre-regular. We now show that R is PM-injective. For $m \in M$, let there be an R-homomorphism $f : Rm \to R$. Since M is regular, there exists an R-homomorphism $g : M \to R$ such that $m = g(m) \cdot m$. Define $\overline{f} : M \to R$ by setting $\overline{f}(a) = g(a) \cdot f(m)$ for $a \in M$. It is easily checked that $\overline{f}$ is an R-homomorphism. Moreover $\overline{f}(rm) = r(\overline{f}(m)) = r(g(m) \cdot f(m)) = r(f(g(m) \cdot m)) = r(f(m)) = f(rm)$. Thus $\overline{f}$ is an extension of f. Hence R is PM-injective.

(2) $\Rightarrow$ (1): Assume that M is pre-regular and R is PM-injective. We show that M is regular. Since M is pre-regular, so for each $m \in M$, there exists $f \in \operatorname{Hom}_R(Rm, R)$ such that $m = f(m) \cdot m$. On the other hand, since R is PM-injective, so there exists an R-homomorphism $g : M \to R$ which extends f. Hence $m = f(m) \cdot m = g(m) \cdot m$, showing that M is regular.

Taking $M = {}_R R$ in the above proposition, we obtain the following characterization of regular near-rings.

Corollary. *A near-ring R is regular if and only if R is pre-regular and self P-injective.*

We now prove the following characterization theorem for regular near-rings.

Theorem 2.10. *The following conditions are equivalent:*

(1) *Every left R-module M is P-injective.*

(2) *$_R R$ is completely P-injective (that is, every homomorphic image of R, considered as an R-module, is P-injective).*

(3) *Every monogenic left R-module $M = Rm$ is P-injective.*

(4) *Every monogenic R-subgroup Ra of $_R R$ is P-injective.*

(5) *R is regular.*

(6) *R is pre-regular and self P-injective.*

Proof. Since the equivalence of the last two conditions has just been proved, we only need to verify the following implications: $(1) \Rightarrow (2) \Rightarrow (3) \Rightarrow (4) \Rightarrow (5) \Rightarrow (1)$.

$(1) \Rightarrow (2)$: Obvious.

$(2) \Rightarrow (3)$: Rm is the image of R under the epimorphism $f : R \to Rm$, defined by $f(r) = rm$, for all $r \in R$.

$(3) \Rightarrow (4)$: Put $m = a$.

$(4) \Rightarrow (5)$: Let $i : Ra \to Ra$ be the identity map. As Ra is a P-injective R-module, there exists an R-homomorphism $g : R \to Ra$ such that $a = g(a) = g(a \cdot 1) = ag(1)$, and $g(1) \in Ra$. Hence we can write $g(1) = xa$ for some $x \in R$. Hence $a = axa$, and so R is regular.

$(5) \Rightarrow (1)$: Let M be a left R-module. We show that M is P-injective. For $a \in R$, consider an R-homomorphism $f : Ra \to M$. Since R is regular, so $\exists \, e = e^2 \in R$ such that $Ra = Re$. Define $\overline{f} : R \to M$ by $\overline{f}(r) = f(re)$. Clearly $\overline{f}$ is an R-homomorphism. Moreover, $\overline{f}(re) = f(re \cdot e) = f(re)$. Hence $\overline{f}$ is an extension of f. Hence M is P-injective.

Proposition 2.11. *Let R be a near-ring with no nonzero nilpotent elements and M be a maximal left ideal of R. If R/M is a regular left R-module, then M is a monogenic left R-module generated by a central idempotent.*

Proof. Consider the element $(1 + M) \in R/M$. Since R/M is regular, so there exists $f \in \mathrm{Hom}_R(R/M, R)$ such that $(f(1 + M))(1 + M) = (1 + M)$. This means that $f(1+M) - 1 \in \ell(1+M)$. Note that $\ell(1+M) = \{r \in R \mid r(1+M) = M\} = M$. Hence $f(1 + M) - 1 = m \in M$. For any $b \in M, f(b(1 + M)) = b(f(1 + M)) = b(m + 1)$. But $f(b(1 + M)) = f(b + M) = f(M) = 0$. Hence $b(m + 1) = 0$. Now $[(m + 1)b]^2 = (m + 1)b(m + 1)b = (m + 1)(b(m + 1)) \cdot b = 0$. Since R has no nonzero nilpotent elements, it follows that $(m + 1)b = 0$. Thus $mb + b = 0$, that is, $b = -mb = (-m)b$ for all $b \in M$. Thus $-m$ is an idempotent element. Since R has no nonzero nilpotent elements, it follows from [3, Prop. 9.43, p. 304] that m is a central idempotent. Hence $b = b(-m)$ for all $b \in M$. This implies that $M = R(-m)$. Hence M is monogenic left R-module generated by a central idempotent.

Proposition 2.12. *Let R be a near-ring with no nonzero nilpotent elements. If each maximal left ideal M of R is a monogenic left R-module generated by a central idempotent, then $R = soc(R)$.*

Proof. Let S be the socle of R. Suppose $R \neq S$. Then since $1 \in R$, there exists a maximal left ideal M of R containing S. By the assumption, $M = Re$, for some central idempotent $e \in M$. Hence by [3, Prop. 3.47, p. 93], we can write $R = Re \oplus \ell(e)$. This implies that $R/Re \cong \ell(e)$ is a simple left R-module, and hence $\ell(e)$ is a minimal left ideal of R. Consequently, $\ell(e) \subseteq S \subseteq M$. Since $1 - e \in \ell(e)$ and $e \in M$, it follows that $1 = (1 - e) + e \in M$, a contradiction. Hence $R = S$ as we were to show.

Corollary. *Let R be a near-ring with no nonzero nilpotent elements. If for each maximal left ideal M of R the left R-module R/M is regular, then R is the direct sum of a finite number of simple left ideals.*

Proof. The preceding two propositions show that $R = soc(R)$. By [3, Thm. 2.48, p. 55] this means that R is a direct sum of simple left ideals with finite number of summands, since R contains an identity (see [3], Th. 2.30).

The assumptions of the above corollary are satisfied if each simple left R-module is regular (since for each maximal left ideal M of R, the left R-module R/M is simple) or if each monogenic left R-module is regular (since for each maximal left ideal M of R the left R-module R/M is monogenic, with generator $1 + M$ where 1 is the identity of R). Moreover in the latter case we have a stronger consequence.

Theorem 2.13. *Let R be a near-ring with no nonzero nilpotent elements. Then the following conditions are equivalent:*

(1) *Each monogenic left R-module is regular.*

(2) *R is strictly semisimple.*

Proof. (1) $\Rightarrow$ (2): By the previous corollary, R is semisimple. On the other hand, since $_R R$ is a monogenic R-module generated by 1, it is regular. Therefore R is a regular near-ring. As R has no nonzero nilpotent elements, it follows from ([3], Thm. 9.158 and Corollary 9.159) that R-subgroups of R coincide with left ideals of R. Hence R is strictly semisimple.

(2) $\Rightarrow$ (1): Let m be an element of a monogenic left R-module M. Define $f : R \to Rm$ by $f(r) = rm$ for $r \in R$. Clearly f is an R-homomorphism. Since the sequence $0 \to \ell(m) \xrightarrow{i} R$, where i denotes the canonical inclusion, is exact and since R is strictly semisimple, it follows from [2, Thm. 3.5(d)] that the above sequence splits. Hence the s.e.s. $0 \to \ell(m) \xrightarrow{i} R \xrightarrow{f} Rm \to 0$ splits by [2, Lemma 2.1]. Hence there exists a normal $g : Rm \to R$ such that $fg = 1_{Rm}$.

Moreover since Rm is an R-subgroup of the monogenic R-module M and R is strictly semisimple, it follows from [2, Thm. 3.5(b)] that Rm is a semi-direct summand of M. Hence we can write $M = A \dotplus Rm$ for some R-submodule A of M. Moreover, the projection map $\pi : M \to Rm$ defined by $\pi(a + b) = b$ for $a \in A$ and $b \in Rm$ is an R-homomorphism [2]. Let $\alpha = g\pi$. Then $\alpha : M \to R$ is an R-homomorphism, and $m = id_{Rm}(m) = fg(m) = f(g(m)) = g(m) \cdot m = g(\pi(0+m)) \cdot m = [g\pi(0+m)] \cdot m = [\alpha(0+m)] \cdot m = \alpha(m) \cdot m$. Hence M is regular. This completes the proof.

Acknowledgement. The author would like to take this opportunity to express his gratitude to the referee for his helpful suggestions.

REFERENCES

[1] A. Fröhlich; On groups over distributively generated near-rings I, Sum constructions and free R-groups; Quart. J. Math. Oxford Ser. II **11** (1960), 193–210.

[2] G. Mason; Injective and projective near-ring modules; Compositio Mathematica **33** (1976), 43–54.

[3] G. Pilz; NEAR-RINGS; North Holland Publishing Company (1983).

DOES R PRIME IMPLY $M_R(R^2)$ IS SIMPLE? [1]

Scott W. Bagley

Department of Mathematics, Texas A & M University
College Station, TX 77843, U. S. A.

ABSTRACT

In [1] Maxson and van der Walt prove that when $M_R(R^2)$ is simple, then R must be prime. In this paper we present a class of examples which proves the converse need not be true. That is, R prime does not imply $M_R(R^2)$ is simple.

1990 Mathematics Classification Number 16Y30.

Introduction

Let R be a ring and $M_R(R^2)$ the set of all maps on $R^2 = R \oplus R$ with the property that scalar multiples can be pulled out. That is, $M_R(R^2) = \{f \mid f : R^2 \to R^2$ and $f[\begin{smallmatrix} gr \\ hr \end{smallmatrix}] = (f[\begin{smallmatrix} g \\ h \end{smallmatrix}])r$ for all $[\begin{smallmatrix} g \\ h \end{smallmatrix}] \in R^2$ and $r \in R\}$. Under function compostion and pointwise addition, $M_R(R^2)$ forms a right near-ring. In [1] it was shown that if $M_R(R^2)$ is simple, then R must be a prime ring, and in the case that R was known to be commutative, R prime implies $M_R(R^2)$ simple. It was left open whether the converse was true. That is whether or not R prime implies $M_R(R^2)$ is simple.

We first recall that a ring R is prime if and only if for all ideals J and I of R, $IJ = \{0\}$ or $JI = \{0\}$ implies either I or J is $\{0\}$. Herein we shall exhibit a class of primitive (and hence prime) rings for which $M_R(R^2)$ is *not* simple, thereby settling the question negatively.

Throughout this work we shall denote the image space of a homomorphism c by $\Im c$. The next lemma is basic.

Lemma 1. *Let V be an infinite–dimensional vector space over a division ring D and let $c_1, c_2 \in \mathrm{Hom}_D(V, V)$ such that $\dim(\Im c_1)$ and $\dim(\Im c_2)$ are strictly less then the dimension of V. Then $\dim (V/(\ker c_1 \cap \ker c_2)) = [V : \ker c_1 \cap \ker c_2]$ is strictly less then the dimension of V. In particular, when $\dim(\Im c_1) < \infty$, and $\dim(\Im c_2) < \infty$, then $[V : \ker c_1 \cap \ker c_2] < \infty$.*

Proof: Let $I_1 = \dim \Im c_1$, and $I_2 = \dim \Im c_2$. Without loss of generality there exist linearly independent sets X_1 and X_2 in V with the following properties:(e.g. X_1 and X_2 are selected such that $c_1(X_1)$ is a basis for $\Im c_1$, and $c_2(X_1 \cup X_2)$ contains a basis for $\Im c_2$.)

(i) $|X_1| = I_1, |X_2| \leq I_2$,
(ii) $\Im c_1 = c_1(\mathrm{span}\ X_1)$,

---

[1] This work is part of the author's doctoral dissertation at Texas A & M University.

Y. Fong et al. (eds.), Near-Rings and Near-Fields, 53–56.

(iii) $\Im c_2 = c_2(\text{span } (X_1 \cup X_2))$,

(iv) $X_2 \subset \ker c_1$,

(v) $c_1(X_1)$ and $c_2(X_2)$ are linearly independent sets, and

(vi) $X_1 \cup X_2$ is a linearly independent set.

We shall show that for all $v \in V$, v can be written as a sum $v_1 + k$ where $v_1 \in \text{span } (X_1 \cup X_2)$ and $k \in \ker c_1 \cap \ker c_2$. Since $I_1 < \dim V$ and $I_2 < \dim V$ this implies

$$\dim(\text{span } (X_1 \cup X_2)) \leq I_1 + I_2 < V$$

which will yield the result.

By properties (ii) and (v), $c_1(X_1)$ is a basis for $\Im c_1$. By (iii), (v), and (vi), $c_2(X_2)$ can be extended to a basis for $\Im c_2$ in such a way that there exists a set $X_3 \subset X_1$ such that for all $v \in V$, $c_2(v) = \sum a_i c_2(y_i) + \sum b_j c_2(z_j)$ where $a_i, b_j \in D, y_i \in X_2, z_j \in X_3 \subset X_1$, $\{c_2(X_2) \cup c_2(X_3)\}$ is a basis for $\Im c_2$, and $X_1 \backslash X_3 \subset \ker c_2$.

Define $c_1' : \Im c_1 \to \text{span } (X_1)$ by $c_1'(v) = \sum a_i x_i$ if $v = \sum a_i c_1(x_i), a_i \in D, x_i \in X_1$. Similarly, define $c_2' : \Im c_2 \to \text{span } (X_2 \cup X_3)$ by

$$c_2'(v) = \sum a_i y_i + \sum b_j z_j$$

if

$$v = \sum a_i c_2(y_i) + \sum b_j c_2(z_j), \quad a_i, b_j \in D, y_i \in X_2, z_j \in X_3.$$

Let $v \in V$. Let $k = v - c_1' c_1(v) - c_2'(c_2(v - c_1' c_1(v)))$. We claim $k \in \ker c_1 \cap \ker c_2$. By construction, $c_2(c_2' c_2(x)) = c_2(x)$ for all $x \in V$ so

$$c_2(k) = c_2(v - c_1' c_1(v)) - c_2(v - c_1' c_1(v)) = 0.$$

Also

$$\begin{aligned} c_1(k) &= c_1(v) - c_1(v) - c_1 c_2'(c_2(v) - c_2 c_1' c_1(v)) \\ &= -c_1 c_2'(c_2(v) - c_2 c_1' c_1(v)). \end{aligned}$$

Claim:

$$c_2'(c_2(v) - c_2 c_1' c_1(v)) \in \text{span } X_2$$

which implies

$$- c_1 c_2'(c_2(v) - c_2 c_1' c_1(v)) = 0$$

by property (iv).

Noting that $c_1' c_1(v) = \sum b_j z_j + \sum d_k w_k, b_j, c_k \in D, z_j \in X_3, w_k \in X_1 \backslash X_3$, then

$$c_2 c_1' c_1(v) = \sum b_j c_2(z_j)$$

and

$$c_2(v) - c_2 c_1' c_1(v) = \sum a_i c_2(y_i), \quad a_i \in D, y_i \in X_2$$

which implies

$$c_2'(c_2(v) - c_2 c_1' c_1(v)) \in \text{span } X_2$$

as claimed.

Therefore $v = v_1 + k, k \in \ker c_1 \cap \ker c_2$ with

$$v_1 = c_1' c_1(v) + c_2'(c_2(v - c_1'(c_1(v)))) \in \text{span}(X_1 \cup X_2)$$

which completes the result.

We briefly note that for the case where $[V : \ker c_1] < \infty$ and $[V : \ker c_2] < \infty$, the result follows much more easily by noting that

$$[V : \ker c_1 \cap \ker c_2] = [V : \ker c_1][\ker c_1 : \ker c_1 \cap \ker c_2] \text{ and } [V : \ker c_1] < \infty.$$

If $[\ker c_1 : \ker c_1 \cap \ker c_2]$ were infinite, then by considering an inverse image of a basis for $\ker c_1/(\ker c_1 \cap \ker c_1)$ we can easily construct an infinite basis for $V/\ker c_2$, which contradicts that $[V : \ker c_2] < \infty$.

We shall use the following notations. Let V be an infinite–dimensional vector space over a division ring D and let R be the endomorphism ring $\text{Hom}_D(V, V)$ of V. We recall that R is a primitive (and hence prime) ring. We shall show that $M_R(R^2)$ is not simple even though R is prime. More formally we have

Theorem 2. *Let $R = \text{Hom}_D(V, V)$ as above. Let $I = \{f \in M_R(R^2)| \text{ if } f[\begin{smallmatrix} x \\ y \end{smallmatrix}] = [\begin{smallmatrix} c_1 \\ c_2 \end{smallmatrix}], \text{ then } \dim(\Im c_1) < \infty \text{ and } \dim(\Im c_2) < \infty \text{ for all } [\begin{smallmatrix} x \\ y \end{smallmatrix}] \in R^2\}$; then I is a proper ideal in $M_R(R^2)$.*

Proof: Since $(R^2, +)$ is abelian, it is trivial to see that $(I, +)$ is a normal subgroup and for all $f \in I, g \in M_R(R^2)$ we have $fg \in I$. Since the identity map is not in I, then $I \neq M_R(R^2)$. We note that if $S = \{c \in R| \dim(\Im c) < \infty\}$, then S is an ideal in R; hence

$$\left\{ \begin{bmatrix} a & b \\ c & d \end{bmatrix} \middle| a, b, c, d \in S \right\}$$

is in $M_R(R^2)$ with $[\begin{smallmatrix} a & b \\ c & d \end{smallmatrix}][\begin{smallmatrix} x \\ y \end{smallmatrix}] = [\begin{smallmatrix} ax+by \\ cx+dy \end{smallmatrix}]$, and so $I \neq \{0\}$.

We now turn our attention to the remaining left ideal characteristic. Let $n, g \in M_R(R^2)$, $f \in I$ and $[\begin{smallmatrix} x' \\ y' \end{smallmatrix}] \in R^2$ such that $f[\begin{smallmatrix} x' \\ y' \end{smallmatrix}] = [\begin{smallmatrix} c_1 \\ c_2 \end{smallmatrix}]$, and $g[\begin{smallmatrix} x' \\ y' \end{smallmatrix}] = [\begin{smallmatrix} x \\ y \end{smallmatrix}]$. Then

$$(n(f + g) - ng)\begin{bmatrix} x' \\ y' \end{bmatrix} = n\left(\begin{bmatrix} c_1 \\ c_2 \end{bmatrix} - \begin{bmatrix} x \\ y \end{bmatrix}\right) - n\begin{bmatrix} x \\ y \end{bmatrix} = \begin{bmatrix} \tilde{x} \\ \tilde{y} \end{bmatrix}.$$

We shall show that

$$\ker c_1 \cap \ker c_2 \subset \ker \tilde{x} \text{ and}$$

$$\ker c_1 \cap \ker c_2 \subset \ker \tilde{y}.$$

Then by Lemma 1, $\dim(\Im \tilde{x}) < \infty$ and $\dim(\Im \tilde{y}) < \infty$, yielding the required result.

Let $r \in R$ such that r is the projection map of V onto $\ker c_1 \cap \ker c_2$. Then

$$\begin{bmatrix} \tilde{x} \\ \tilde{y} \end{bmatrix} r = n\left(\begin{bmatrix} c_1 \\ c_2 \end{bmatrix} r + \begin{bmatrix} x \\ y \end{bmatrix} r \right) - n \begin{bmatrix} x \\ y \end{bmatrix} r$$

$$= n\left(0 + \begin{bmatrix} x \\ y \end{bmatrix} r \right) - n \begin{bmatrix} x \\ y \end{bmatrix} r = 0 \ .$$

So we have $\tilde{x}r = 0, \tilde{y}r = 0$. Let $a \in \ker c_1 \cap \ker c_2$. Then

$$\tilde{x}a = \tilde{x}(ra) = (\tilde{x}r)a = 0.$$

Similarly, $\tilde{y}a = 0$ and the result follows.

Corollary 3. *Let R be a nonsimple dense subring of* $\mathrm{Hom}_D(V,V)$, *for some infinite-dimensional vector space V. If*

$$R \cap \{c \in \mathrm{Hom}_D(V,V) | \dim(\Im c) < \dim V\} \neq \{0\}$$

then $M_R(R^2)$ is not simple.

Proof: Let $I = \{f \in M_R(R^2) \mid \text{for all } \begin{bmatrix} x \\ y \end{bmatrix} \in R^2, \text{ if } f\begin{bmatrix} x \\ y \end{bmatrix} = \begin{bmatrix} c_1 \\ c_2 \end{bmatrix} \text{ then } \dim(\Im c_1) < \dim V \text{ and } \dim(\Im c_2) < \dim V\}$.

Here again the only difficulty is the left ideal property. Let $B = \{e_i \mid i \in \mathcal{I}\}$ be a basis for $\ker c_1 \cap \ker c_2$. Then if

$$\begin{bmatrix} \tilde{x} \\ \tilde{y} \end{bmatrix} = (n(f+g) - ng) \begin{bmatrix} x' \\ x' \end{bmatrix}$$

as before, we need only show $\tilde{x}(e_i) = 0$, and $\tilde{y}(e_i) = 0$ for all $e_i \in B$.

Fix e_i. Since R is dense in $\mathrm{Hom}_D(V,V)$, there exists $\psi_i \in R$ such that $\psi_i(e_i) = e_i$. Then

$$\begin{bmatrix} \tilde{x} \\ \tilde{y} \end{bmatrix} e_i = \left(\begin{bmatrix} \tilde{x} \\ \tilde{y} \end{bmatrix} \psi_i \right) e_i = \left(n\left(\begin{bmatrix} c_1\psi_i \\ c_2\psi_i \end{bmatrix} + \begin{bmatrix} x\ \psi_i \\ y\ \psi_i \end{bmatrix} \right) - n\begin{bmatrix} x\psi_i \\ y\psi_i \end{bmatrix} \right) e_i$$

$$= n\left(\begin{bmatrix} 0 \\ 0 \end{bmatrix} + \begin{bmatrix} x\ a_i \\ y\ e_i \end{bmatrix} \right) - n\begin{bmatrix} x\ e_i \\ y\ e_i \end{bmatrix} = \begin{bmatrix} 0 \\ 0 \end{bmatrix}$$

since $c_1\psi_i e_i = c_2\psi_i e_i = 0$. Hence $\begin{bmatrix} \tilde{x} \\ \tilde{y} \end{bmatrix} \in I$ and the result is complete.

Thus by Theorem 2 and Corollary 3 we have exhibited a class of prime rings R for which $M_R(R^2)$ is not simple.

References

[1] C. J. Maxson and A. P. J. van der Walt, *Centralizer near-rings over free ring modules*, J. Austral. Math. Soc. (Series A) **50** (1991), 279–296

ESSENTIAL NILPOTENCY IN NEAR-RINGS

Gary F. Birkenmeier

Department of Mathematics, University of Southwestern Louisiana
Lafayette, LA 70504, U.S.A.

ABSTRACT

In this paper we begin the development of a theory of essential nilpotence for near-rings. We show that the prime radical is essentially nilpotent and there exists a unique largest essentially nilpotent ideal $EN(R)$ in any zerosymmetric near-ring R. Basic properties of $EN(R)$ are determined and examples are provided to illustrate and delimit our theory.

Throughout this paper R is a left near-ring and $P(R)$ is its prime radical. As is well known one can build $P(R)$ from nilpotent ideals of R via transfinite induction as in Kaarli and Kriis [7] (this proceedure is just the near-ring version of Baer's lower radical construction for rings as indicated in Divinsky [4]). Thus it is natural to ask the following questions: (1) How "close" is $P(R)$ to being nilpotent? (2) How "densely" are the nilpotent ideals of R packed into $P(R)$? In 1970, Fisher [6] gave an answer to these questions for rings by defining an essentially nilpotent ideal and showing that $P(R)$ is essentially nilpotent. In 1992, Eslami and Stewart [5] extended Fisher's results by showing that there is a unique largest essentially nilpotent ideal $EN(R)$ for any ring R, and they investigated its behavior. In 1993, the author in [2] developed a theory of essential supernilpotency for rings which encompassed the results of Fisher, Eslami, and Stewart. Unfortunately, much of the essential supernilpotent theory for rings cannot be extended to near-rings unless one imposes some form of Andrunakievic's Lemma [1] or [4]. However significant parts of the essentially nilpotent theory of rings can be carried over to near-rings. In this paper we initiate the theory of essential nilpotency for near-rings.

For general terminology we follow Meldrum [8] or Pilz [9] (with the appropriate modifications for left near-rings). For $X \subseteq R$ we use the following notation: $\langle X \rangle_R$ for the ideal of R generated by X, $(0 : X)$ for the right annihilator of X in R, and $I \triangleleft R$ for I is an ideal of R. Let $X, Y \triangleleft R$ such that $X \subseteq Y$. Then X is *essential in* Y (or Y an *essential extension* of X) if X has nonzero intersection with every nonzero ideal of R contained in Y. It is easy to see that "is essential in" is a transitive relation. A *complement* of X in Y is an ideal C of R contained in Y which is maximal among ideals of R contained in Y having zero intersection with X. By a routine Zorn's Lemma argument X always has a complement C in Y, furthermore $X \oplus C$ is essential in Y. Following Fisher [6], Y is *essentially nilpotent* if there exists some nilpotent ideal X of R such that X is essential in Y. A basic property of essential nilpotency is that if A is essentially nilpotent and is essential in B then B is essentially nilpotent. There is an abundance of essentially nilpotent near-rings

57

(e.g., subdirectly irreducible near-rings with nilpotent heart, prime radical near-rings (see Proposition 4), and local near-rings with J_2 nilpotent). Our first result shows that every nonsemiprime near-ring is either essentially nilpotent or has a nontrivial homomorphic image which is essentially nilpotent. Finally, $EN(R)$ denotes the sum of all essentially nilpotent ideals of R.

PROPOSITION 1. *If R is not semiprime, then there exists an ideal S of R such that R/S is essentially nilpotent and $S \neq R$.*

Proof. If R is not semiprime, there exists a nonzero nilpotent ideal I of R. Let S be a complement of I. If $S = 0$ then I is essential in R, so R is essentially nilpotent. If $S \neq 0$ then $I \simeq \dfrac{I \oplus S}{S}$. Let $0 \neq M/S \lhd R/S$. Then, by the modular law, $M \cap (S \oplus I) = S \oplus M \cap I$. Since $S \oplus I$ is essential in R, then $M \cap (S \oplus I)$ is essential in M. Since S is a complement ideal, $M \cap I \neq 0$. So $\dfrac{I \oplus S}{S} \cap M/S \neq 0$. Therefore $\dfrac{I \oplus S}{S}$ is essential in R/S, consequently R/S is essentially nilpotent.

LEMMA 2. *Let T be a sum of nilpotent ideals of R. Then there exists X an ideal of R such that $X^2 = 0$ and X is essential in T.*

Proof. Let $F = \{X_\alpha \mid \alpha \in \Lambda\}$ be the family of all nilpotent ideals of R contained in T which have nilpotent index two. Let us call a subset J of the index set Λ *direct* if $\sum X_j$ $(j \in J)$ is a direct sum. Consider a family $\{J_k \mid k \in K\}$ of direct sets which is simply ordered under inclusion and let U be its union. Then U is a direct set. Applying Zorn's Lemma we obtain a maximal direct set $M \subseteq \Lambda$. Let $X = \sum X_m (m \in M)$. Then $X^2 = 0$. Let I be a nonzero ideal of R such that $I \subseteq T$ and $X \cap I = 0$. Let $0 \neq y \in I$. Then $\langle y \rangle_R$ is a nilpotent ideal of R. Let $Y = \langle y \rangle_R \cap (0 : \langle y \rangle_R)$. Hence $Y \neq 0$, but $Y^2 = 0$. So $X \cap Y = 0$, contradicting the maximality of X. Therefore X is essential in T.

Early in our development, the example in Kaarli and Kriis [7] highlights the divergence between the theory of essential nilpotence for rings versus near-rings. For if R is a ring and T is an ideal of R which is a sum of *its* nilpotent ideals then by the Andrunakievic Lemma [1] T is a sum of nilpotent ideals of R. However this is not true for zerosymmetric near-rings. Kaarli and Kriis [7] provide an example of a zerosymmetric *prime* near-ring with an ideal I of R, where I is a sum of *its* nilpotent ideals.

PROPOSITION 3. *Let R be zerosymmetric and T a sum of essentially nilpotent ideals of R. Then there exists X an ideal of R such that $X^2 = 0$ and X is essential in T.*

Proof. Let Y be the sum of all nilpotent ideals of R contained in T. Let I be a nonzero ideal of R such that $I \subseteq T$ and $0 \neq v \in I$. Assume $Y \cap I = 0$. Then there exists a finite set $\{T_1, T_2, \cdots, T_n\}$ of essentially nilpotent ideals of R such that $T_j \subseteq T$, for all $j \in J = \{1, 2, \cdots, n\}$, and $\langle v \rangle_R \subseteq \sum T_j$. If $\langle v \rangle_R \cap T_j = 0$, for all $j \in J$, then since R is zerosymmetric $\langle v \rangle_R^2 = 0$, a contradiction. So there exists

$j \in J$ such that $\langle v \rangle_R \cap T_j \neq 0$. Since T_j is essentially nilpotent, then $\langle v \rangle_R$ contains a nonzero nilpotent ideal of R, again a contradiction. Thus Y is essential in T. The result is now a consequence of Lemma 2 and the fact that "is essential in" is a transitive relation.

In Birkenmeier, Heartherly, and Lee [3] an element $x \in R$ is said to be *strongly nilpotent* provided every sequence $x_0, x_1, \cdots$, such that $x_0 = x$ and $x_{n+1} \in \langle x_n \rangle_R^2$, is ultimately zero. In Proposition 3.2 [3] they proved that $P(R)$ is equal to the set of all strongly nilpotent elements of R. Using this characterization of the prime radical, we are able to extend to near-rings Fisher's Theorem 2.3 [6] that $P(R)$ is an essentially nilpotent ideal of R, where R is a ring.

PROPOSITION 4.   *The prime radical of R is essentially nilpotent.*

Proof.   Suppose $0 \neq y \in P(R)$ and T is the sum of all nilpotent ideals of R. Note that $T \subseteq P(R)$ by Theorem 2.105 [9]. We claim that $\langle y \rangle_R \cap T \neq 0$. If $\langle y \rangle_R^2 = 0$, we are finished. So assume $\langle y \rangle_R^2 \neq 0$, then there exists $0 \neq y_1 \in \langle y \rangle_R^2$. If $\langle y_1 \rangle_R^2 = 0$, we are finished. Otherwise continue the above procedure. Since y is strongly nilpotent we ultimately get a nonzero $y_{n+1} \in \langle y_n \rangle_R^2$ such that $\langle y_{n+1} \rangle_R^2 = 0$. So the claim is proved. Thus T is essential in $P(R)$. By Lemma 2, T is essentially nilpotent and the result follows from basic properties of essential nilpotency.

The following result shows that $EN(R)$ (i.e., the sum of all essentially nilpotent ideals of R) is the unique largest essentially nilpotent ideal in R, where R is zerosymmetric.

THEOREM 5.   *Let R be zerosymmetric. Then:*

(i) $EN(R)$ *is essentially nilpotent.*

(ii) $P(R)$ *is essential in $EN(R)$.*

(iii) $EN(R)$ *is a semiprime ideal of R.*

(iv) $EN(R) = V$, *where* $V = \{x \in R \mid$ *if* $I \triangleleft R$ *such that* $\langle x \rangle_R \cap I \neq 0$, *then* I *contains a nonzero nilpotent ideal of R*$\}$.

Proof.   Parts (i) and (ii) are consequences of Propositions 3 and 4 and the transitive property of essential ideals.

(iii) Let K be an ideal of R such that $K^2 \subseteq EN(R)$. By Proposition 3, there exists X an ideal of R such that $X^2 = 0$ and X is essential in $EN(R)$. Let $0 \neq y \in K$. We claim $\langle y \rangle_R \cap X \neq 0$. If $\langle y \rangle_R^2 = 0$, we are finished. Assume $\langle y \rangle_R^2 \neq 0$. Then there exist $y_1, y_2 \in \langle y \rangle_R$ such that $y_1 y_2 \neq 0$. Hence $0 \neq \langle y_1 y_2 \rangle_R \subseteq EN(R)$. Thus $\langle y_1 y_2 \rangle_R \cap X \neq 0$. So the claim is proved. Thus, every element y of K has $\langle y \rangle_R$ containing a nonzero nilpotent ideal. Therefore the sum of the nilpotent ideals of R contained in K is essential in K. By Lemma 2, the sum of nilpotent ideals is essentially nilpotent and so $K \subseteq EN(R)$.

(iv) By part (i), $EN(R) \subseteq V$. Let $0 \neq v \in V$. Let $0 \neq B \triangleleft R$ such that $B \subseteq \langle v \rangle_R$. Then B contains a nonzero nilpotent ideal of R. Hence the sum of all nilpotent ideals of R which are contained in $\langle v \rangle_R$ is essential in $\langle v \rangle_R$. By Lemma 2, as before, $\langle v \rangle_R$ is essentially nilpotent. So $\langle v \rangle_R \subseteq EN(R)$ and $V \subseteq EN(R)$ from which the result follows.

COROLLARY 6.   *Let R be zerosymmetric. The following statements are equivalent:*

(i) *R is essentially nilpotent.*

(ii) *$EN(R) = R$.*

(iii) *There exists an ideal X of R such that $X^2 = 0$ and X is essential in R.*

THEOREM 7.   *Let R be zerosymmetric and S a complement of $EN(R)$. Then:*

(i) *S is maximal among ideals of R which contain no nonzero nilpotent ideals of R.*

(ii) *$EN(R)$ has no essential extension in R.*

(iii) *$EN(R)$ is the right annihilator of S.*

(iv) *Let $(L, +)$ be a nonzero subgroup of $(R, +)$ such that $RL \subseteq L$. Then $L \cap (S \oplus EN(R)) \neq 0$. In particular, $S \oplus EN(R)$ is essential in R.*

(v) *R/S is essentially nilpotent.*

Proof.   Parts (i) and (ii) are consequences of Theorem 5 (i), definitions and basic properties of essential nilpotency.

(iii) Observe $EN(R) \subseteq (0:S)$, and $(0:S) \cap S = 0$. Let $0 \neq K \triangleleft R$ such that $K \subseteq (0:S)$ and $K \cap EN(R) = 0$. By the maximality of S, $(K \oplus S) \cap EN(R) \neq 0$. Hence there exists $0 \neq k \in K$ and $0 \neq s \in S$ such that $0 \neq k + s \in EN(R)$. So $0 \neq s \in (0:S) \cap S$, a contradiction. Thus $EN(R)$ is essential in $(0:S)$. Hence by part (ii) $(0:S) = EN(R)$.

(iv) Assume $L \cap (S \oplus EN(R)) = 0$. Then $L \oplus EN(R) \subseteq (0:S) = EN(R)$. Hence $L = 0$, a contradiction. Therefore $L \cap (S \oplus EN(R)) \neq 0$.

(v) The proof of this part is similar to the proof of Proposition 1 (use $EN(R)$ in place of I).

THEOREM 8.   *Let R and T be the direct product and direct sum of the zerosymmetric near-rings R_α ($\alpha \in \Lambda$), respectively. Then $EN(R) = \prod EN(R_\alpha)$ and $EN(T) = \oplus EN(R_\alpha)$.*

Proof.   We will only consider the direct product case since the direct sum case has a similar proof. Let $x \in EN(R)$, $p_\alpha(x) = x_\alpha$ (the cannonical projection), and $K \triangleleft R_\alpha$. Assume $H \neq 0$, where $H = \langle x_\alpha \rangle_{R_\alpha} \cap K$. Let $\overline{H}$ be the image of the cannonical injection of H into R. First assume $\overline{H} \cap \langle x \rangle_R \neq 0$. Let $Y = \overline{H} \cap \langle x \rangle_R$. Then Y is a nonzero ideal contained in $EN(R)$ so Y contains a nontrivial nilpotent ideal and $Y \cap P(R) \neq 0$. As is well known $P(R) \subseteq \prod P(R_\alpha)$ (see p.115 of Meldrum [8]). Hence $Y \cap \prod P(R_\alpha) \neq 0$. Therefore $0 \neq K \cap P(R_\alpha)$. Hence K contains a nonzero nilpotent ideal of R_α. Now assume $\overline{H} \cap \langle x \rangle_R = 0$. Then $\overline{H} \langle x \rangle_R = 0$. So $\overline{H}^2 = 0$. Hence $H^2 = 0$. Again K contains a nonzero nilpotent ideal of R_α. By Theorem 5(iv), $EN(R) \subseteq \prod EN(R_\alpha)$.

To show $\prod EN(R_\alpha) \subseteq EN(R)$ assume $I \lhd R$ such that $I \subseteq \prod EN(R_\alpha)$ and $EN(R) \cap I = 0$. Let $\overline{EN(R_\alpha)}$ be the cannonical injection of $EN(R_\alpha)$ into R. $\overline{EN(R_\alpha)}$ is an essentially nilpotent ideal of R and so $\overline{EN(R_\alpha)} \subseteq EN(R)$ and $I \cap \overline{EN(R_\alpha)} = 0$, for all $\alpha \in \Lambda$. Thus I is in a complement of every $\overline{EN(R_\alpha)}$ and by Theorem 7 (iii), $\overline{IEN(R_\alpha)} = 0$. But $I \subseteq \prod EN(R_\alpha)$ so $I^2 = 0$. Thus $I \subseteq EN(R)$ so $I = EN(R) \cap I = 0$. Therefore $EN(R)$ is essential in $\prod EN(R_\alpha)$ and by Theorem 7 (ii), $EN(R) = \prod EN(R_\alpha)$.

ACKNOWLEDGEMENT

The author wishes to thank the referee for his suggestions which improved the exposition of this paper.

Reference

[1] V. Andrunakievic, *Radicals in associative rings I*, Mat. Sb. **44** (1958), 179–212.

[2] G. F. Birkenmeier, *Rings which are essentially supernilpotent*, Comm. Algebra **22** (1994), 1063–1082.

[3] G. F. Birkenmeier, H. E. Heatherly, and E. K. Lee, *Prime ideals and prime radicals in near-rings*, Monat. für Math., to appear.

[4] N. J. Divinsky, *Rings and Radicals*, (University of Toronto Press, Toronto, 1965).

[5] E. Eslami and P. Stewart, *Two-sided essential nilpotence*, Internat. J. Math. & Math. Sci. **15** (1992), 351–354.

[6] J. W. Fisher, *On the nilpotency of nil subrings*, Can. J. Math. **22** (1970), 1211–1216.

[7] K. Kaarli and T. Kriis, *Prime radical of near-rings*, TRÜ Toimetised **764** (1987), 23–29.

[8] J. D. P. Meldrum, *Near-Rings and Their Links to Groups*, Pitman, Marshfield, MA, 1985.

[9] G. Pilz, *Near-Rings*, second rev. ed., North-Holland, Amsterdam, 1983.

COMPLETELY PRIME IDEALS AND RADICALS
IN NEAR-RINGS

G. Birkenmeier, H. Heatherly and E. Lee

Department of Mathematics, University of Southwestern Louisiana
Lafayette, Louisiana 70504, U.S.A.

ABSTRACT

An ideal I of a near-ring R is *2-primal* if the prime radical of R/I equals the set of nilpotent elements of R/I. We show that if $IR \subseteq I$, then I is a 2-primal ideal of R if and only if each minimal prime ideal containing I is a completely prime ideal. A complete classification of the subdirectly irreducible zero symmetric near-rings with a 2-primal heart is provided. Zero symmetric near-rings with each prime ideal completely prime are classified in terms of the 2-primal condition. Various chain conditions are invoked on 2-primal near-rings to obtain decompositions and additive group information.

0 Introduction

This paper continues our investigation of the interrelations between various types of prime ideals and prime radicals in near-rings [1], [2], [4], [5]. The basic results for prime ideals (also called type 0 prime ideals by Ramakotaiah and Koteswara Rao [15]) can be found in Meldrum [13] and Pilz [14]. We use the notation $P_0(R)$ and $P_2(R)$ for the intersection of all the prime ideals, respectively completely prime ideals, of the near-ring R. The study of completely prime ideals in near-rings goes back at least to 1979, [15], where such an ideal was called a "prime ideal of type 2". (An ideal I of R is a completely prime ideal if $xy \in I$ implies $x \in I$ or $y \in I$). Groenewald [7], [8] carried out the first major investigation of the radical P_2. (He used different notation, however). Since any completely prime ideal is prime, we have $P_0(R) \subseteq P_2(R)$. We [5] began a study of when $P_0(R)$ and $P_2(R)$ coincide and of 2-primal ideals and 2-primal near-rings. We continue this investigation and look at some further consequences thereof. We say an ideal I is *2-primal* if $P_0(R/I) = N(R/I)$, where $N(X)$ is the set of all nilpotent elements of X. If the zero ideal is a 2-primal ideal of R, then we say R is a *2-primal near-ring*. (Here R will always be a near-ring and "near-ring" will mean left near-ring. Except where otherwise specified we follow the terminology and conventions found in Meldrum [13]). We showed in [5, Theorem 2.3] that if I is an ideal of R and $IR \subseteq I$, then I is completely prime if and only if I is both prime and 2-primal as an ideal of R.

In section 1 we compare the condition that an ideal is 2-primal to the condition that it is semi-symmetric [16] and the condition that it is completely reflexive [12]. We show that an ideal I which is also a right R-subgroup is a 2-primal ideal if and only if every minimal prime ideal which contains I is completely prime. (Given any ideal I there is always a minimal prime ideal of R which contains I, [14, 2.75]).

Y. Fong et al. (eds.), Near-Rings and Near-Fields, 63–73.
© 1995 Kluwer Academic Publishers.

Let $\mathcal{C} = \{X_1, \cdots, X_n\}$ be any set of 2-primal ideals of R. We show $\cap\mathcal{C}$ is a 2-primal ideal of R and if $X_j R \subseteq X_j$, for each j, then $< X_1 \cdots X_n >_R$ is a 2-primal ideal of R. (Here $< A >_R$ is the ideal of R generated by the subset A; if no ambiguity will arise we use $< A >$ for $< A >_R$). We give an example, provided by E. P. Armendariz, which shows that the intersection of an infinite set of 2-primal ideals need not be 2-primal, answering a question raised by the authors [6].

For some near-rings every prime ideal is completely prime. The class of all such near-rings we denote by $\mathfrak{R}^2$. The study of this class began by the authors [5], [6] for near-rings and rings, respectively, where - *inter alia* - numerous natural examples and construction methods are given. Let $\mathcal{C} = \{X_1, \cdots, X_n\}$ be a set of ideals of R such that $X_j R \subseteq X_j$, and $R/X_j \in \mathfrak{R}^2$, for each j; then $R/\cap\mathcal{C}$ and $R/ < \Pi\mathcal{C} >$ are in $\mathfrak{R}^2$.

In section 2 we investigate the structure of $R/P_2(R)$ and $R/P_0(R)$ under various finiteness conditions. Playing key roles in this are conditions which guarantee there are "enough" distributive elements to make certain homomorphic images near-fields or conditions that guarantee these images are non-trivial finite integral near-rings. We use these structural results to obtain information about the additive group of R. Several classes of natural examples are given to illustrate the results of this section.

In any near-ring R, we use $\mathfrak{p}(I)$ for the intersection of all prime ideals containing I; similarly we use $\mathfrak{p}_2(I)$ for the intersection of all completely prime ideals containing I. Observe that $\mathfrak{p}(I) \subseteq \mathfrak{p}_2(I)$ and $\mathfrak{p}(I)$ is a semiprime ideal, while $\mathfrak{p}_2(I)$ is a completely semiprime ideal.

1 2-primal ideals

The notion of a completely prime ideal is stronger than that of a prime ideal. So naturally one seeks conditions on an ideal which when coupled with prime would be equivalent to completely prime. Several such conditions are already in the literature on rings and near-rings. Mason [12] called an ideal I of a ring "completely reflexive" if $ab \in I$ implies $ba \in I$. He observed that an ideal is completely prime if and only if it is prime and completely reflexive. Reddy and Murty [16] introduced "semi-symmetric" ideals for zero symmetric near-rings: an ideal I is *semi-symmetric* if $x^n \in I$ implies $< x >^n \subseteq I$. They showed that for zero symmetric near-rings an ideal is completely prime if and only if it is both prime and semi-symmetric. We showed in [5, Theorem 2.3] that if I is an ideal of R such that $IR \subseteq I$ (for example, if R is zero symmetric), then I is completely prime if and only if I is prime and is a 2-primal ideal of R.

In this section we will show that the 2-primal condition is weaker than either the completely reflexive condition or the semi-symmetric condition, and we continue the investigation of the 2-primal condition. Extending the definition given by Mason [12] we say an ideal I of a near-ring R is *completely reflexive* if $ab \in I$ implies $ba \in I$. Thus any ideal of R which contains $< R, R > = < \{ab - ba : a, b \in R\} >$ is completely reflexive. Also, if I is a completely semiprime ideal and $IR \subseteq I$, then I is completely reflexive. Recall that R is an *IFP near-ring* if $a, b \in R$ such that $ab = 0$, then $aRb = 0$, [14, p.288].

Proposition 1.1. *If I is a completely reflexive ideal of R, then:*

(1) *I is a right R-subgroup;*

(2) *if $x_1 \cdots x_n \in I$, then $< x_1 > \cdots < x_n > \subseteq I$;*

(3) *I is a 2-primal ideal of R;*

(4) *if I is a prime (semiprime) ideal, then I is completely prime (completely semiprime) ideal; and*

(5) *R/I is an IFP near-ring.*

Proof. Part (1) is immediate. For part (2), take $x_1 \cdots x_n \in I$; so $R \cdot (x_1 \cdots x_n) \subseteq I$ and hence $(x_2 \cdots x_n)(Rx_1) \subseteq I$. A calculation shows $(x_2 \cdots x_n) < x_1 > \subseteq I$. Repeat the process to get $< x_1 > \cdots < x_n > \subseteq I$.

Observe that an ideal T of R is 2-primal if and only if $x^n \in T$ implies $x \in \mathfrak{p}(T)$. This with part (2) yield part (3). Part (4) and (5) are immediate consequences of (2).

Note that an ideal I is completely prime if and only if I is a type one prime ideal and R/I is an IFP near-ring. (Following Ramakotaiah and Rao [15], I is a type one prime ideal if $xRy \subseteq I$ implies $x \in I$ or $y \in I$. See Birkenmeier, Heatherly, and Lee [4] for further results on type one prime).

¿From Proposition 1.1(2) we see that every completely reflexive ideal is semi-symmetric. Moreover, it is immediate from the definitions that a semi-symmetric ideal is a 2-primal ideal. It is shown in [6, Example 1.4] that the 2-primal condition is strictly weaker than the semi-symmetric condition on ideals. Our next result generalizes a theorem of Reddy and Murty, [16, Theorem 2.2]. In fact the "semi-symmetric" condition can be replaced by the "2-primal" condition wherever it is invoked in Reddy and Murty [16].

Theorem 1.2. *Let I be an ideal of R such that $IR \subseteq I$. Then I is a 2-primal ideal of R if and only if each minimal prime ideal containing I is a completely prime ideal of R.*

Proof. Take I to be 2-primal. A proof that each minimal prime ideal containing I is completely prime can be had by using the scheme in the proof of Theorem 2.2 of [16], with the changes of using Lemma 2.2(v) of [5] in place of the Reddy and Murty Theorem 1.3 and using Proposition 1.1 above in place of their Lemma 2.1, observing that if I is completely semiprime then I is completely reflexive.

For the converse, take $x^n \in I$ and let P be a minimal prime ideal containing I. Since $\mathfrak{p}(I)$ is the intersection of all such P, [18], we have $x \in \mathfrak{p}(I)$. Thus every nilpotent element in R/I is contained in $P_0(R/I)$.

The condition "$IR \subseteq I$" cannot be deleted from Theorem 1.2, as shown in [5, Example 2.8]. This condition is strictly weaker than the zero symmetric condition. (For example, the set A_c of constant elements in an abstract affine near-ring A is an ideal with $A_cA \subseteq A_c$, but A is not zero symmetric).

Corollary 1.3. *If R is zero symmetric, then the following are equivalent:*

(1) *R is 2-primal;*

(2) *every minimal prime ideal of R is a completely prime ideal;*

(3) *$P_0(R) = P_2(R)$.*

The proof follows immediately from Theorem 1.2 and basic properties of P_0 and P_2.

Corollary 1.4. *Let I be an ideal of R such that $IR \subseteq I$. Then I is completely semiprime if and only if $\mathfrak{p}_2(I) = I$.*

Proof. The implication that $\mathfrak{p}_2(I) = I$ implies I is completely semiprime, is immediate. So take I to be completely semiprime. Then I is 2-primal. Invoke Theorem 1.2 to obtain $\mathfrak{p}(I)$ is equal to the intersection of a set of completely prime ideals of R, and hence $\mathfrak{p}_2(I) \subseteq \mathfrak{p}(I)$. But for any ideal $\mathfrak{p}(I) \subseteq \mathfrak{p}_2(I)$, giving the desired equality.

This corollary is an extension of a result due to Groenewald [8, Theorem 3.1] , where it is proved for zero symmetric near-rings, using results about m-systems.

Proposition 1.5. *Let $\mathcal{C}$ be a finite, nonempty set of 2-primal ideals of R. Then:*

(1) *$\cap \mathcal{C}$ is a 2-primal ideal of R; and*
(2) *if $\mathcal{C} = \{X_1, \cdots, X_n\}$ and $X_j R \subseteq X_j$ for each j, then $< X_1 \cdots X_n >_R$ is a 2-primal ideal of R.*

Proof.

(1) First take $\mathcal{C} = \{A, B\}$. We show that $\bar{R} = R/A \cap B$ is a 2-primal near-ring and hence $A \cap B$ is a 2-primal ideal of R. Identify $\bar{A}$ and $\bar{B}$ as the images in $\bar{R}$ of A and B, respectively, and observe that $\bar{A} \cap \bar{B} = 0$. So $\bar{R}$ is isomorphic to a subnear-ring of the direct sum $\bar{R}/\bar{A} \oplus \bar{R}/\bar{B}$. Since $\bar{R}/\bar{A} \approx R/A$ and $\bar{R}/\bar{B} \approx R/B$, we have that $\bar{R}/\bar{A}$ and $\bar{R}/\bar{B}$ are 2-primal near-rings. The direct sum of 2-primal near-rings is 2-primal, [5, Proposition 4.1]; so $\bar{R}/\bar{A} \oplus \bar{R}/\bar{B}$ is 2-primal. Any subnear-ring of a 2-primal near-ring is a 2-primal near-ring, [5, Proposition 3.4]; so $\bar{R}$ is 2-primal, which yields $A \cap B$ is a 2-primal ideal of R. A straightforward induction argument then yields the desired result for any finite $\mathcal{C}$.
(2) We show that if $x^n \in < X_1 \cdots X_n >$, then x is in each prime ideal which contains $< X_1 \cdots X_n >$, from which it follows that $< X_1 \cdots X_n >$ is a 2-primal ideal, (see [5, Lemma 2.2]). Let P be a prime ideal of R which contains $< X_1 \cdots X_n >$. So $X_1 \cdots X_n \subseteq P$ and hence some X_j is contained in P. Since X_j is 2-primal and $X_j \subseteq P$, we have $x \in P$.

Note that using the argument in the proof of (2) above, for any ideal I of R and any natural number n we have: I is 2-primal if and only if $< I^n >_R$ is 2-primal. The argument for this does not require $IR \subseteq I$.

If the members of $\mathcal{C}$ are also right R-subgroups, then a shorter, more self contained, argument can be given for Proposition 1.5(1).

An immediate consequence of Proposition 1.5 is that a finite subdirect product of 2-primal near-rings is itself 2-primal. The authors [6] raised the question of whether a subdirect product of 2-primal rings is 2-primal. This is equivalent to asking if an arbitrary intersection of 2-primal ideals in a ring is again a 2-primal ideal. We [5] observed that while the direct sum of an arbitrary number of 2-primal near-rings is 2-primal, the situation for direct products was unsolved. An example showing the answer to these questions is "no" was given to us by E. P. Armendariz.

Example 1.6. Let R be the matrix ring $\begin{bmatrix} \mathbb{Z} & 2\mathbb{Z} \\ 2\mathbb{Z} & \mathbb{Z} \end{bmatrix}$ and let $I = \begin{bmatrix} 2\mathbb{Z} & 2\mathbb{Z} \\ 2\mathbb{Z} & 2\mathbb{Z} \end{bmatrix}$. Then R is prime, but not 2-primal; so the zero ideal of R is a prime ideal which is not a 2-primal ideal. Since R/I is isomorphic to $\mathbb{Z}_2 \oplus \mathbb{Z}_2$, we have that I is a 2-primal ideal. So $< I^n >_R$ is a 2-primal ideal of R for each n. However, $\cap < I^n >_R$, where n runs over the natural numbers, is the zero ideal.

Proposition 1.7. *Let* $\mathcal{C} = \{X_1, \cdots, X_n\}$ *be a set of ideals of R which are also right R-subgroups. If R/X is in $\mathfrak{R}^2$ for each $X \in \mathcal{C}$, then $R/ < \Pi\mathcal{C} >$ and $R/\cap\mathcal{C}$ are in $\mathfrak{R}^2$. (Here $\Pi\mathcal{C} = X_1 \cdots X_n$).*

Proof. Any prime ideal of $R/ < \Pi\mathcal{C} >$ has the form $P/ < \Pi\mathcal{C} >$, where P is a prime ideal of R and $< \Pi\mathcal{C} > \subseteq P$. So some $X \in \mathcal{C}$ is contained in P. Note that the function given by $r + < \Pi\mathcal{C} > \rightarrow r + X$ is a homomorphism from $R/ < \Pi\mathcal{C} >$ onto R/X. Since P/X is a prime ideal of R/X, we have P/X is completely prime. Then the inverse image of P/X in $R/ < \Pi\mathcal{C} >$ is $P/ < \Pi\mathcal{C} >$, and this ideal is completely prime in $R/ < \Pi\mathcal{C} >$. So $R/ < \Pi\mathcal{C} >$ is in $\mathfrak{R}^2$.

Since $< \Pi\mathcal{C} > \subseteq \cap\mathcal{C}$, we have that $R/\cap\mathcal{C}$ is a homomorphic image of $R/ < \Pi\mathcal{C} >$. Since $\mathfrak{R}^2$ is closed under homomorphic images, [5, Proposition 4.2], we have that $R/\cap\mathcal{C}$ is in $\mathfrak{R}^2$.

An immediate consequence of Proposition 1.7 is that the class of all zero symmetric near-rings which are in $\mathfrak{R}^2$ is closed under finite subdirect products.

In the sequel we will use N for $N(R)$ when no ambiguity will arise.

Proposition 1.8. *If R is zero symmetric and is not 2-primal, then for each nonzero 2-primal ideal X of R none of $< N >X$, XN, nor $N \cap X$ can be zero.*

Proof. Since R is not 2-primal there exists a minimal prime ideal K of R such that $N \not\subseteq K$. Assume $< N > X = 0$. Since $< N > \not\subseteq K$ and K is prime, we have $X \subseteq K$. Thus K is a minimal prime ideal containing the 2-primal ideal X and hence, by Theorem 1.2, K is completely prime. This yields the contradictory $N \subseteq K$.

Assume $XN = 0$. Then $X < N > = 0$. Proceed as in the previous case to arrive at a contradiction. Since $XN \neq 0$, there exist $x \in X$, $y \in N$ such that $xy \neq 0$. Let m be a positive integer such that $xy^{m+1} = 0$ and $xy^m \neq 0$. Since $(y^m x y^m)(y^m x y^m) = 0$, we have $y^m x y^m \in N \cap X$. Now assume $N \cap X = 0$. Then $y^m x y^m = 0$ and hence $(xy^m)(xy^m) = 0$. So $xy^m \in N \cap X$, yielding the contradictory $xy^m = 0$.

Lemma 1.9. *Let I be a minimal ideal in a zero symmetric near-ring R. Then $I \subseteq P_0(R)$ if and only if $I^2 = 0$.*

Proof. Take $I \subseteq P_0(R)$. If $I^2 \neq 0$, then $(0 : I) = \{r \in R : Ir = 0\}$ is a prime ideal by Proposition 2 of [3]; hence $I \subseteq (0 : I)$, a contradiction. So $I^2 = 0$. The converse is immediate.

It is useful at this point to introduce some terminology. An ideal I of a near-ring R is a *reduced ideal* if $N(I) = 0$. An ideal X of R is *essential* in R if X has nonzero intersection with each nonzero ideal of R.

Lemma 1.10. *If I is a reduced ideal of a zero symmetric near-ring R, then:*

(1) $I < N(R) > = 0$;

(2) *if I is essential in R, then R is reduced.*

Proof.

(1) Take $y \in N(R)$ and $x \in I$, with $y^n = 0$, $n > 1$. Using an argument much like that in the proof of Proposition 1.8, we get xy^{n-1} is a nilpotent element in I, and hence is zero. Repeat this argument to eventually get $xy = 0$. So $I \cdot N(R) = 0$; consequently $I < N(R) > = 0$.

(2) Take I to be essential. If $N(R) \neq 0$, then there is some nonzero element x in $I \cap < N(R) >$. So $x^2 \in I < N(R) >$ and thus $x^2 = 0$, a contradiction to I is reduced. So R is reduced.

Proposition 1.11. *If R is a zero symmetric, 2-primal near-ring and I is a minimal ideal of R with $I^2 \neq 0$, then $R/(0 : I)$ is a subdirectly irreducible integral near-ring whose heart is isomorphic to I.*

Proof. Use Proposition 2 of [3] to get $(0 : I)$ is a prime ideal of R and $\bar{R} = R/(0 : I)$ is a subdirectly irreducible near-ring with heart $\bar{I}$ isomorphic to I. Lemma 1.9 yields $I \cap P_0(R) = 0$ and hence I is a reduced ideal. Thus $\bar{I}$ is a reduced ideal of $\bar{R}$. Since the heart is essential, use Lemma 1.10 to obtain that $\bar{R}$ is reduced, and hence $\bar{R}$ is 2-primal. Proposition 2.7 of [5] gives that a 2-primal near-ring which is prime must be integral.

Proposition 1.12. *Let R be a subdirectly irreducible, zero symmetric near-ring. If the heart, H, is a 2-primal ideal, then every minimal prime ideal containing H is completely prime and exactly one of the following holds:*

(1) R *is a 2-primal near-ring with $H^2 = 0$;*

(2) R *is an integral near-ring;*

(3) R *is a prime near-ring, $H^2 \neq 0$, and $H \cap N(R) \neq 0$.*

Proof. From Theorem 1.2 every minimal prime containing H is completely prime. Now consider the case where R is not a prime near-ring. Then from Lemma 1.9, $H^2 = 0$, and $P_0(R) \neq 0$. Thus $\mathfrak{p}(H) = P_0(R)$. Recall $P_0(R)$ is the intersection of all minimal prime ideals of R, [14, 2.75]. Then use Theorem 1.2 to get that each of these minimal ideals is completely prime; so $P_0(R) = P_2(R)$ and R is 2-primal.

Next consider R to be prime. So $H^2 \neq 0$. If $N(R) = 0$, then R is reduced. All subdirectly irreducible reduced (zero symmetric) near-rings are integral, [14, 9.38]. If $N(R) \neq 0$, note that $H \cap N(R) \neq 0$; for otherwise H is a reduced ideal and by Lemma 1.10 we would have $H < N(R) > = 0$, contrary to (0) is a prime ideal.

In [5, Theorem 4.4] we showed that a zero symmetric near-ring is in $\mathfrak{R}^2$ if and only if every ideal of the near-ring is 2-primal. Our final result of this section improves that theorem and indicates how the subclass of zero symmetric near-rings in $\mathfrak{R}^2$ "fits" into the class of 2-primal near-rings.

Theorem 1.13. *For a zero symmetric near-ring R the following are equivalent:*

(1) $R \in \mathfrak{R}^2$;

(2) *R is 2-primal and every essential ideal of R is a 2-primal ideal;*

(3) *R is 2-primal and every essential prime ideal of R is completely prime.*

Proof. Assume (1); then (2) follows from [5, Theorem 4.4]. In [5, Theorem 2.3] we showed that every 2-primal ideal which is prime must be completely prime; so (3) follows immediately from (2). Finally, assume (3). Let P be a prime ideal of R. It suffices to consider the case where P is not essential. So there exists an ideal X of R such that $P \cap X = 0$. If K is a prime ideal of R such that $K \subseteq P$, then since $P \cdot X = (0) \subseteq K$, we have either $P \subseteq K$ (and hence $P = K$), or $X \subseteq K$. The latter contradicts $K \cap X = 0$; so P is a minimal prime ideal of R. Since R is 2-primal, P must be completely prime by Corollary 1.3. So $R \in \mathfrak{R}^2$.

Observe that part of the argument given above can be used to show that any prime ideal in a zero symmetric near-ring is either essential or a minimal prime.

2 Applications and Examples

Much can be said about reduced integral near-rings which satisfy certain chain conditions and which have "enough" distributive elements. (See [14, pp. 301-312]). Even more can be said in the finite case. (See [9], [10]). Since $R/P_2(R)$ is a subdirect product of integral near-rings, this gives us an opportunity to apply our previous results. Also in this section we give some examples to help illustrate the results obtained.

Deviating slightly from the definition given in 9.46 of [14], herein we allow the ring $\mathbb{Z}_2$ to be a non-trivial integral near-ring.

We use $\mathcal{D}(R)$ for the set of all distributive elements of R and $\Delta(R) = \{(x + y)r - yr - xr : x, y, r \in R\}$. So $R/< \Delta(R) >$ is a distributive near-ring. We use R' to denote the commutator subgroup of the additive group R^+.

Proposition 2.1. *Let P be a completely prime ideal of R such that there exists $d \in \mathcal{D}(R)$ with $d \notin P$. Then:*

(1) *$\bar{R} = R/P$ is a non-trivial integral near-ring with characteristic either zero or a prime p;*

(2) *for each prime q, $q \neq p$, every Sylow q-subgroup of R^+ is contained in P;*

(3) *if $\bar{R}$ has d.c.c. on right $\bar{R}$-subgroups, then $\bar{R}$ is a near-field, P is maximal as a one-sided R-subgroup and as a one-sided ideal of R, and $R' \subseteq P$.*

Proof. The integral near-ring $\bar{R}$ contains the nonzero distributive element $d + P$; so $\bar{R}$ is non-trivial and 9.57 of [14] yields $\bar{R}$ has characteristic zero or some prime p. From this (2) is immediate. When $\bar{R}$ has d.c.c. on right R-subgroups, use [14, 9.48] to obtain $\bar{R}$ is a near-field. The rest of (3) is immediate. (Recall: near-fields have commutative addition).

If in the above proposition we assume R is d.g., then for part (3) we obtain $\bar{R}$ is a skewfield, using 9.48 of [14]. If R is finite and d.g. in the proposition, then

$\bar{R}$ is a field and $R' \cup \Delta(R) \cup < R, R > \subseteq P$. (Note: In Proposition 2.1 $\bar{R}$ is zero symmetric, although R may not be; however, we do have $0 \cdot R \subseteq P$).

The hypothesis $d \in \mathcal{D}(R)$, $d \notin P$, is satisfied if R has a unity (or a right unity). In such a case each completely prime ideal has such a distributive element associated with it. The next proposition carries this further.

Observe that every near-ring in $\mathfrak{R}^2$ is 2-primal, however the converse is not true (see [4, Example 2.9 and Theorem 4.4]). In the first part of our next result we provide a condition which guarantees that a 2-primal near-ring is in $\mathfrak{R}^2$.

Proposition 2.2. *Let $\hat{R} = R/P_2(R)$ have d.c.c. on right $\hat{R}$-subgroups.*

(1) *If R is zero symmetric, then R is 2-primal if and only if $R \in \mathfrak{R}^2$.*

(2) *If for each proper minimal prime ideal $\hat{P}$ of $\hat{R}$ there exists $d \in \mathcal{D}(\hat{R})$ with $d \notin \hat{P}$ (here d can depend on $\hat{P}$) and $P_2(R) \neq R$, then $\hat{R}$ is a finite direct sum of near-fields and $R' \cup J_2(R) \subseteq P_2(R)$.*

Proof. First assume $P_2(R) \neq R$. Since $P_2(\hat{R}) = 0$, there exists a family $\mathcal{F}$ of completely prime ideals of $\hat{R}$, which are also minimal prime ideals of $\hat{R}$, such that $\cap \mathcal{F} = 0$. Thus $\hat{R}$ is a subdirect product of integral near-rings of the form $\hat{R}/\hat{P}$, where $\hat{P} \in \mathcal{F}$. Using either the hypothesis of (2) and Proposition 2.1 or using the hypothesis of (1) directly, we have that each $\hat{R}/\hat{P}$ is zero symmetric and hence so is $\hat{R}$. Thus $\hat{R}$ is a reduced, zero symmetric near-ring with d.c.c. on right $\hat{R}$-subgroups and consequently is a finite direct sum $\oplus \hat{R}/X$ of integral 2-primitive near-rings with the analogous chain condition [14, 9.39, 9.41]. Each such summand is simple, [13, Theorem 7.9]. Since simple integral near-rings are in $\mathfrak{R}^2$ and since the direct sum of zero symmetric near-rings from $\mathfrak{R}^2$ is back in $\mathfrak{R}^2$, [5, Proposition 4.2], we have $\hat{R} \in \mathfrak{R}^2$.

Take R to be 2-primal. Then from Corollary 1.3 we have $P_0(R) = P_2(R)$. So $R/P_0(R)$ is in $\mathfrak{R}^2$. For any near-ring R, if $R/P_0(R)$ is in $\mathfrak{R}^2$, then R is in $\mathfrak{R}^2$ by [5, Corollary 4.3].

A moments reflection reveals that if T is any zero symmetric near-ring which is a direct sum of integral near-rings, then the direct sum with exactly one summand removed is a minimal prime ideal of T. Now assume the hypothesis of (2). Then each summand $\hat{R}/X$ is a simple integral near-ring with X a minimal prime ideal of $\hat{R}$. So by Proposition 2.1, each $\hat{R}/X$ is a near-field. Thus $(\hat{R}, +)$ is abelian and $R' \subseteq P_2(R)$. Since $J_2(\hat{R}) = 0$, we have $J_2(R) \subseteq P_2(R)$.

Finally, if $P_2(R) = R$ and R is 2-primal and zero symmetric, then $P_0(R) = P_2(R) = R$, by Corollary 1.3. Hence $R \in \mathfrak{R}^2$.

Note that in the above proposition, R need not be zero symmetric, but $\hat{R}$ is forced to be zero symmetric in the argument.

Theorem 2.3. *Let R be zero symmetric with $P_0(R) \neq R$. Assume $\bar{R} = R/P_0(R)$ has d.c.c. on right $\bar{R}$-subgroups and for each proper minimal prime ideal $\bar{P}$ of $\bar{R}$ there exists $d \in \mathcal{D}(R)$ with $d \notin \bar{P}$. If R is 2-primal, then:*

(1) *$\bar{R}$ is a finite direct sum of near-fields;*

(2) *$R' \subseteq P_0(R)$ and $P_0(R) = P_2(R) = J_2(R)$;*

(3) *every proper prime ideal of R is maximal as a right (left) R-subgroup;*

(4) *either $R \approx \bar{R} \oplus P_0(R)$, or each maximal ideal M of R satisfies one of* (a) $P_2(R) \subseteq M$ *or* (b) $R^2 \subseteq M$ *and M is maximal as a normal subgroup of R^+; and*

(5) *each minimal ideal of R is either square zero or is a near-field.*

Proof. Since R is 2-primal, Theorem 1.2 yields $P_0(R) = P_2(R)$. So $R/P_0(R) = R/P_2(R)$. Use Proposition 2.2 to obtain (1) and (2). For (3), let P be a proper prime ideal of R. Then $\bar{R}$ maps homomorphically onto R/P, forcing R/P to be a near-field, (use (1)). Thus P is maximal as a right (left) R-subgroup. For (4), let M be a maximal ideal of R. If $M \cap P_2(R) = 0$, then $R = M + P_2(R)$ and $\bar{R} = R/P_2(R) \approx M$. In this case M is a finite direct sum of near-fields. Take $M \cap P_2(R) \neq 0$. If $P_2(R) \not\subseteq M$, then $P_2(R/M) \neq 0$. Use that $\mathfrak{R}^2$ is closed under taking homomorphic images, [5, Proposition 4.2], and $R \in \mathfrak{R}^2$ to obtain $P_2(R/M) = P_0(R/M)$. Since R/M is a simple near-ring this forces $(R/M)^2 = 0$ and hence $R^2 \subseteq M$.

If I is a minimal ideal of R and $I^2 \neq 0$, then Lemma 1.9 yields $I \not\subseteq P_0(R)$. Thus $I \cap P_0(R) = 0$ and $\bar{I}$, the homomorphic image of I in $\bar{R}$, is a minimal ideal of $\bar{R}$ and is isomorphic to I. Consequently I is a near-field.

Observe that if in the above theorem R has d.c.c on right R-subgroups, then $J_2(R)$ is nilpotent. Also note that the condition of the distributive elements could have been placed in R instead of $\bar{R}$.

Corollary 2.4. *Let $\bar{R} = R/P_0(R)$ have d.c.c. on right $\bar{R}$-subgroups and $P_0(R) = P_2(R)$. If for each minimal prime ideal $\bar{P}$ of $\bar{R}$ there is some $d \in \mathcal{D}(\bar{R})$ such that $d \notin \bar{P}$, then R is zero symmetric and the results of Theorem 2.3 hold.*

Proof. It suffices to show R is zero symmetric. If $P_0(R) = R$, then R is nil and hence R is zero symmetric. Take $P_0(R) \neq R$. Then $R/P_0(R) = R/P_2(R)$ is isomorphic to a subdirect product of integral near-rings, each with a nonzero distributive element. So each of these component near-rings is zero symmetric (because they are non-trivial). Thus $R/P_0(R)$ is zero symmetric. Then $0 \cdot R \subseteq P_0(R)$. Since $P_0(R)$ is nil, we have $0 \cdot R = 0$.

Corollary 2.5. *Let R be a d.g. near-ring with d.c.c. on right R-subgroups. If R is 2-primal, then either R is nilpotent, or R is a finite direct sum of skewfields, every minimal ideal of R is either square zero or is a skewfield, and $R' \cup \Delta(R) \subseteq P_0(R)$.*

Proof. This is immediate from Theorem 2.3 and that the d.g. near-ring $R/P_0(R)$ is a finite direct sum of skewfields, [14, 9.48].

Corollary 2.6. *Let R have a right identity and have d.c.c. on right R-subgroups. If either of the following hold, then R satisfies the hypothesis of Theorem 2.3:*

(1) *R is zero symmetric and 2-primal; or*

(2) *$P_0(R) = P_2(R)$.*

It is of interest to see these results realized in concrete, natural examples.

Example 2.7. Let $(G, +)$ be either a finite dihedral group or the generalized quaternion group of order 2^n and let R be any one of the endomorphism near-rings $I(G)$, $A(G)$, or $E(G)$. (We will exclude $E(G)$ below whenever G is dihedral with order divisible by four). In each case $R/P_0(R)$ is a commutative ring − and hence 2-primal; more precisely it is a finite direct sum of $\mathbb{Z}_p$ for various distinct primes p. (See [13, pp. 199-202] for details on these near-rings).

Example 2.8. Let $(G, +)$ be a finite, non-abelian p-group and let $R = I(G)$. Then $P_0(R) = J_2(R) = P_2(R)$ and $R/P_0(R)$ is a finite field (see [11, Theorem 2.2] and [5, Example 5.6]). This carries over to more general setting of any local, tame endomorphism near-ring which has d.c.c. on right ideals. (See Birkenmeier, Heatherly, and Lee [5] for further details).

Without the 2-primal condition the conclusion of Theorem 2.3 need not hold, even in the finite, d.g. case with unity. For example $R = E(G)$, where G is the symmetric group of degree four, has $P_0(R) = J_2(R)$, but $R/P_0(R)$ is not 2-primal and R has a maximal ideal M such that R/M is isomorphic to the full 2-by-2 matrix ring over the field $\mathbb{Z}_2$. (See [13, pp. 207-208]).

At a key stage in the main results above we used the decomposition of a reduced, zero symmetric near-ring into a subdirect product of integral near-rings. It is well-known that this cannot be carried out in general if the zero symmetric hypothesis is dropped, even when the near-ring is finite. A good example of this is the near-ring of all affine transformations on the four element matrix ring $\{ \begin{bmatrix} 1 & 0 \\ 0 & 0 \end{bmatrix}, \begin{bmatrix} 0 & 0 \\ 0 & 1 \end{bmatrix},$ $\begin{bmatrix} 1 & 0 \\ 0 & 1 \end{bmatrix}, \begin{bmatrix} 0 & 0 \\ 0 & 0 \end{bmatrix} \}$, over $\mathbb{Z}_2$. This sixteen element near-ring is reduced, but has no integral near-ring as a proper homomorphic image.

Proposition 2.9. *Let P be a completely prime ideal of R such that R/P is finite and either (i) there exists $d \in \mathcal{D}(R)$ with $d \notin P$ or (ii) there exists $a, b \in R$ with $a \notin P$ and $ab - b \notin P$.*

    (1) *If P^+ is solvable, then R^+ is solvable.*
    (2) *If there exists a nilpotent normal subgroup S of R^+ such that $P \subseteq S'$, then R^+ is nilpotent.*

Proof. The key point is R/P is a finite, non-trivial integral near-ring and hence by Ligh's Theorem, [14, 9.51], $(R/P, +)$ is nilpotent. Then (1) follows from elementary group theory and (2) follows from Hall's Criterion, [17, p.129].

Observe that in Proposition 2.9 we have R/P has no proper nonzero right R/P-subgroups and there exists a fixed integer n, $n > 1$, such that $(r + P)^n = r + P$ for each $r \in R$, [9, Theorem 5.1]. Consequently, P is maximal as a right ideal (right R-subgroup) of R and $r^n - r \in P$ for each $r \in R$. Note that if R/P has order $p + 1$, where p is an odd prime, then $R/P \approx GF(2^m)$, where $p + 1 = 2^m$, [9, Corollary 4.2]. Using the fact that $(R/P, +)$ has a fixed-point-free automorphism of prime order, [14, 9.51], other results can be obtained about the order of R/P, and hence about the order of R, when P is finite and the order of P is given. (See [10, Section 3]).

Proposition 2.10. *Let R be 2-primal and let $R/\boldsymbol{P}_0(R)$ be finite and nonzero. Assume either : (i) for each proper minimal prime ideal of P of R there exists $d \in \mathcal{D}(R)$ such that $d \notin P$, or (ii) for each proper minimal prime ideal P of R there exists $a, b \in R$ such that $a \notin P$ and $ab - b \notin P$.*

(1) If $(\boldsymbol{P}_0(R), +)$ is solvable, then R^+ is solvable.

(2) If there exists a nilpotent normal subgroup S of R^+ such that $\boldsymbol{P}_0(R) \subseteq S'$, then R^+ is nilpotent.

Proof. Note that $R/\boldsymbol{P}_0(R)$ is finite and reduced. So $R/\boldsymbol{P}_0(R)$ is a subdirect product of a finite number of near-rings of the form R/P, where P is a completely prime ideal of R. If (i) holds, then each of these R/P is a near-field and hence $(R/\boldsymbol{P}_0(R), +)$ is abelian. If (ii) holds, then each of the R/P is a non-trivial, finite integral near-ring, and hence $(R/P, +)$, and thus $(R/\boldsymbol{P}_0(R), +)$, are nilpotent. Then (1) and (2) follow as in Proposition 2.9.

Proceeding as in the observation following Proposition 2.9, in Proposition 2.10 with condition (ii) holding we have there exists a fixed $k > 1$ such that $r^k - r \in \boldsymbol{P}_0(R)$ for each $r \in R$.

References

1. G. Birkenmeier and H. Heatherly, *Medial near-rings*, Monatsh. Math. **107** (1989), 89–110.
2. _____, *Permutation identity near-rings and "localized" distributive conditions*, Monatsh. Math. **111** (1991), 265–285.
3. _____, *Minimal ideals in near-rings*, Comm. Algebra **20** (1992), 457–468.
4. G. Birkenmeier, H. Heatherly, and E. Lee, *Prime ideals in near-rings*, Results in Math. **24** (1993), 27–48.
5. _____, *Prime ideals and prime radicals in near-rings*, Monatsh. Math. (1994) (to appear).
6. _____, *Completely prime ideals and associated radicals*, Ring Theory, Proc. Ohio St.-Denison Conf. (S.K. Jain and S.T. Rizvi, eds.), World Scientific, Singapore, 1993, pp. 102–129.
7. N. Groenewald, *Note on the completely prime radical in near-rings*, Near-Rings and Near-Fields (G. Betsch, ed.), North-Holland, Amsterdam, 1987, pp. 97–100.
8. _____, *The completely prime radical in near-rings*, Acta Math. Hung. **51** (1988), 301–305.
9. H. Heatherly and H. Olivier, *Near integral domains*, Monatsh. Math. **78** (1974), 215–222.
10. _____, *Near integral domains II*, Monatsh. Math. **80** (1975), 85–92.
11. C. Lyons and G. Peterson, *Local endomorphism near-rings*, Proc. Edinburgh Math. Soc. **31** (1988), 409–414.
12. G. Mason, *Reflexive ideals*, Comm. Algebra **9** (1981), 1709–1724.
13. J.D.P. Meldrum, *Near-Rings and Their Links to Groups*, Pitman, Marshfield, MA, 1985.
14. G. Pilz, *Near-Rings*, 2nd ed., North-Holland, Amsterdam, 1983.
15. D. Ramakotaiah and G. Koteswara Rao, *IFP near-rings*, J. Austral. Math. Soc. (Series A) **27** (1979), 365–370.
16. Y.V. Reddy and C.V.L.N. Murty, *Semi-symmetric ideals in near-rings*, Indian J. Pure Appl. Math. **16** (1985), 17–21.
17. D.J.S. Robinson, *A Course in the Theory of Groups*, Springer-Verlag, New York, 1982.
18. A.P.J. van der Walt, *Prime ideals and nil radicals in near-rings*, Arch. Math. **15** (1964), 408–414.

CONNECTING SEMINEARRINGS TO PROBABILITY GENERATING FUNCTIONS

Donald W. Blackett

Math. Department, Boston University, Charles River Campus,
Boston, Mass. 02215, USA

ABSTRACT

Are seminearrings important to other areas of mathematics? The purpose of this paper is to show an example of an application in the theory of probability.

Over one hundred years ago Francis Galton posed the problem of determining the probability that a family surname becomes extinct (See [1], solution in [3], further references in [2]). The theory starts from two facts:

- If $G_1(x)$ is the probability generating function for the number of sons of one man and $G_2(x)$ is the probability generating function for the number of sons of the second man, then $G_1(x)G_2(x)$ is the probability generating function for the number of sons of the two men.

- If the probability generating function for the sons of a man is $G(x)$ and the distributions of the number of sons of each of the man's sons are independent and each has the probability generating function $F(x)$, then the probability generating function for the number of grandsons of the first man is $G(F(x))$.

Definition 1 *The probability generating function (abbreviated p.g.f.) for a non-negative-integer-valued random variable A which takes the value n with probability a_n is*

$$A(x) = a_0 + a_1 x + a_2 x^2 + \ldots + a_n x^n + \ldots = \sum_{n=0}^{\infty} a_n x^n \tag{1}$$

$$\text{where } a_i \geq 0 \text{ for } i = 0, 1, \ldots, n, \ldots \text{ and } \sum_{n=0}^{\infty} a_n = 1 \tag{2}$$

Let $B(x) = b_0 + b_1 x + \ldots + b_n x^n + \ldots$ be the p.g.f. for a second random variable B. The product

$$A(x)B(x) = a_0 b_0 + (a_0 b_1 + a_1 b_0)x + \ldots + (\sum_{k=0}^{n} a_k b_{n-k})x^n + \ldots \tag{3}$$

is the p.g.f. for the sum $A + B$ of the two random variables. If $A(x)$ is the p.g.f. for the number of "sons" in the first stage of an experiment and $B(x)$ is the p.g.f. for the number of "grandsons" from each of the "sons", then

$$B(A(x)) = b_0 + b_1 A(x) + b_2 (A(x))^2 + \ldots + b_n (A(x))^n + \ldots \tag{4}$$

is the p.g.f. for the number of "sons of sons" in the second stage.

Y. Fong et al. (eds.), Near-Rings and Near-Fields, 75–81.
© *1995 Kluwer Academic Publishers.*

Definition 2 *The* pseudosum $A \oplus B$ *of two p.g.f.s is the p.g.f. defined by* $(A \oplus B)(x) = A(x) \cdot B(x)$. *The* pseudoproduct $B \otimes A$ *of two p.g.f.s is the p.g.f. defined by* $(B \otimes A)(x) = B(A(x))$.

Definition 3 *A "ring" $(R, +, \cdot)$ with addition $+$ and a multiplication $\cdot$ such that the elements form a group under addition, the multiplication is associative and only one distributive law is required, is called a (right)* nearring. *If the axioms are further relaxed so that the additive inverses and the additive identity are not required, the system R is a right* seminearring.

A natural example of a seminearring is the collection of all maps from the naturals into itself, under addition and composition of functions.

Theorem 1 *Under the operations "pseudosum" and "pseudoproduct" the set Ω of probability generating functions forms a right seminearring $(\Omega, \oplus, \otimes)$ with a commutative addition and an additive identity.*

Proof: The axioms to be verified are

$$A \oplus (B \oplus C) = (A \oplus B) \oplus C \tag{5}$$
$$A \oplus B = B \oplus A \tag{6}$$
$$I \oplus A = A \text{ where } I \text{ is the function } I(x) = 1 \tag{7}$$
$$A \otimes (B \otimes C) = (A \otimes B) \otimes C \tag{8}$$
$$(A \oplus B) \otimes C = (A \otimes C) \oplus (B \otimes C) \tag{9}$$

The proofs of these are very straightforward and we will simply illustrate the proofs by verifying the right distributive law

$$(A \oplus B) \otimes C = (A \otimes C) \oplus (B \otimes C) \tag{10}$$

and showing why the left distributive law

$$A \otimes (B \oplus C) = (A \otimes B) \oplus (A \otimes C) \tag{11}$$

is not a valid computation rule. First, the valid distribution law:

$$((A \oplus B) \otimes C)(x) = (A \oplus B)(C(x)) \tag{12}$$
$$= A(C(x)) \cdot B(C(x)) \tag{13}$$
$$= (A \otimes C)(x) \cdot (B \otimes C)(x) \tag{14}$$
$$= ((A \otimes C) \oplus (B \otimes C))(x) \tag{15}$$

Second, the invalid distibution law:

$$(A \otimes (B \oplus C))(x) = A((B \oplus C)(x)) \tag{16}$$
$$= A(B(x) \cdot C(x)) \tag{17}$$

and

$$((A \otimes B) \oplus (A \otimes C))(x) \quad = \quad (A \otimes B)(x) \cdot (A \otimes C)(x) \qquad (18)$$
$$= \quad A(B(x)) \cdot A(C(x)) \qquad (19)$$

If

$$A(x) = 1/2 + 1/2x, \quad B(x) = x, \text{ and } \quad C(x) = x, \qquad (20)$$

then

$$A(B(x) \cdot C(x)) \quad = \quad 1/2 + 1/2x^2 \text{ and} \qquad (21)$$
$$A(B(x)) \cdot A(C(x)) \quad = \quad (1/2 + 1/2x)(1/2 + 1/2x) \qquad (22)$$
$$= \quad 1/4 + 1/2x + 1/4x^2 \qquad (23)$$

Thus the left distributive law is not valid in general. $\qquad\qquad$ $\square$

How can the algebra of the right seminearring of p.g.f.s help us understand the probability theory of non-negative integer-valued random variables? Theorem 6 will illustrate the usefullness of the algebra by showing that any such random variable can be approximated with a prescribed accuracy from a subseminearring generated by combining the results of a finite number of "tosses" of a "fair" coin. In particular a computer program to generate random numbers in accordance with an arbitrary p.g.f. can be manufactured using only a random chip which selects each of the two outcomes with probability $1/2$.

The probability generating function for the number of "heads" in the toss of a "fair" coin is

$$T(x) = 1/2 + 1/2x \qquad (24)$$

A natural question is what is the subseminearring on p.g.f.s generated by T. First consider the set of p.g.f.s for all success-failure events. By a success-failure event we mean a random variable with probability generating function

$$S_p(x) = (1 - p) + px, 0 \le p \le 1 \qquad (25)$$

where p is the probability of success.

Theorem 2 *The seminearring of p.g.f.s generated by the set of all p.g.f.s for success-failure events is the set of all p.g.f.s which are polynomials. Every p.g.f. is the limit of a sequence of p.g.f.s which are generated from success-failure events.*

Proof: The constant p.g.f. $S_0(x)$ and the linear p.g.f.s $S_p(x)$ for $0 \le p \le 1$ are already success-failure events. Assume as an induction hypothesis that every p.g.f. which is a polynomial of degree $\le n$ can be constructed from success-failure events. Let

$$A_{n+1}(x) = a_0 + a_1 x + \ldots + a_{n+1} x^{n+1} \qquad (26)$$

be of degree $n + 1$. Then

$$A_{n+1}(x) = a_0 + (a_1 + \ldots + a_{n+1}) \left[\left(\frac{a_1}{a_1 + \ldots + a_{n+1}} + \ldots \right. \right. \tag{27}$$

$$\left. \ldots + \frac{a_2}{a_1 + \ldots + a_{n+1}} x + \frac{a_{n+1}}{a_1 + \ldots + a_{n+1}} x^n \right) x \right] \tag{28}$$

$$= a_0 + (a_1 + \ldots + a_{n+1}) \left[B_n(x) \cdot x \right] \tag{29}$$

$$= S_{(a_1 + \ldots + a_{n+1})} \left[B_n(x) \cdot S_1(x) \right] \tag{30}$$

$$= \left[S_{(a_1 + \ldots + a_{n+1})} \otimes [B_n \oplus S_1] \right](x) \tag{31}$$

where

$$b_j = \frac{a_{j+1}}{a_1 + \ldots + a_{n+1}} \quad \text{for } j = 0, \ldots, n \text{ and } B_n(x) = b_0 + b_1 x + \ldots + b_n x^n \tag{32}$$

This proves by induction that every polynomial p.g.f. can be constructed from success-failure events and the seminearring of polynomial p.g.f.s is generated by the success-failure events.

Let $A(x) = \sum_{j=0}^{\infty} a_j x^j$ be any p.g.f.. Clearly

$$A(x) = \lim_{n \to \infty} \sum_{j=0}^{n} a_{j,n} x^j \tag{33}$$

where

$$a_{j,n} = \begin{cases} \frac{a_j}{\sum_{k=0}^{n} a_k} & \text{if } \sum_{k=0}^{n} a_k \neq 0 \\ 0 & \text{otherwise} \end{cases} \tag{34}$$

Thus every p.g.f. is the limit of p.g.f.s in the seminearring generated by success-failure events.   □

Do we need all success-failure events to generate a seminearring that approximates an arbitrary p.g.f.? First, we will show as a side issue that if the only approximation property we require is that the mean of the distribution can be arbitrarily closely approximated, the seminearring generated by $1/2 + 1/2x$ is adequate to approximate every possible value of the mean. Second, we will consider a p.g.f. seminearring N which contains $1/2 + 1/2x$ and has the property that if the p.g.f. $(1 - p) + px$ of any success-failure event is in the seminearring, then $p + (1 - p)x$, the p.g.f. of the negation of the first success-failure event, is also in the seminearring. We will prove that every p.g.f. can be approximated arbitrarily closely in the seminearring N.

If

$$A(x) = a_0 + a_1 x + \ldots + a_n x^n + \ldots, \tag{35}$$

then

$$A'(x) = a_1 + 2a_2 x + 3a_3 x^2 + \ldots + n a_n x^{n-1} + \ldots \tag{36}$$

and

$$E(A) = A'(1) = a_1 + 2a_2 + 3a_3 + \ldots + na_n + \ldots = \sum_{n=0}^{\infty} na_n \qquad (37)$$

is the mean of the distribution with $A(x)$ as the p.g.f..

Lemma 1 *The mapping $\tau : A \to A'(1)$ is a seminearring homomorphism from the subseminearring of the p.g.f.s with finite mean to the semiring of non-negative real numbers.*

Proof:

$$
\begin{aligned}
\tau(A \oplus B) &= D_x((A \oplus B)(x)]_{x=1} & (38)\\
&= D_x((A(x)B(x))]_{x=1} & (39)\\
&= (A'(x)B(x) + A(x)B'(x))]_{x=1} & (40)\\
&= A'(1)B(1) + A(1)B'(1) & (41)\\
&= A'(1) + B'(1) & (42)\\
&= \tau(A) + \tau(B) & (43)\\
\tau(A \otimes B) &= D_x(A(B(x)))]_{x=1} & (44)\\
&= (A'(B(x))B'(x))]_{x=1} & (45)\\
&= A'(1)B'(1) & (46)\\
&= \tau(A)\tau(B) & (47)
\end{aligned}
$$

$\square$

Lemma 2 *If $T(x) = 1/2 + 1/2x$, then*

$$T(T(\ldots(T(x))\ldots)) = T^{(k)}(x) = (1 - 2^{-k}) + 2^{-k}x, k = 1, 2, \ldots, n, \ldots \qquad (48)$$

Proof: If

$$T^{(k)}(x) = (1 - 2^{-k}) + 2^{-k}x, \qquad (49)$$

then

$$
\begin{aligned}
T^{(k+1)}(x) &= T(T^{(k)}(x)) = \frac{1}{2} + \frac{1}{2}((1 - 2^{-k}) + 2^{-k}x) & (50)\\
&= (\frac{1}{2} + \frac{1}{2} - 2^{-(k+1)}) + 2^{-(k+1)}x & (51)\\
&= (1 - 2^{-(k+1)}) + 2^{-(k+1)}x & (52)
\end{aligned}
$$

and with

$$T^{(1)}(x) = T(x) = (1 - \frac{1}{2}) + \frac{1}{2}x, \qquad (53)$$

we get the lemma.

$\square$

Theorem 3 *Given any non-negative real number r and a specified accuracy requirement, there is a distribution in the seminearring generated by $T(x) = 1/2 + 1/2x$ such that the mean of the distribution approximates r within the specified accuracy.*

Proof: Let

$$k = k_1 2^{-1} + k_2 2^{-2} + \ldots + k_n 2^{-n} \tag{54}$$

be a binary expansion of r such that the k_is are nonnegative integers and $r \le k < r + 2^{-n}$. Note that the k_i are not restricted to 0 and 1 so that a term 2^{-j} can be replaced by $2 \cdot 2^{-(j+1)}$ and a term $2\lfloor r \rfloor 1/2$ takes care of $\lfloor r \rfloor$ the greatest integer less than or equal to r. By Lemma 3 the distribution

$$(T^{(1)}(x))^{k_1} (T^{(2)}(x))^{k_2} \ldots (T^{(n)}(x))^{k_n} = \tag{55}$$

$$\left[\overbrace{(T \oplus \ldots \oplus T)}^{k_1 \text{ terms}} \oplus \overbrace{((T \otimes T) \oplus \ldots \oplus (T \otimes T))}^{k_2 \text{ terms}} \oplus \ldots \right. \tag{56}$$

$$\left. \ldots \oplus \overbrace{((\underbrace{T \otimes \ldots \otimes T}_{n \text{ factors}}) \oplus \ldots \oplus (\underbrace{T \otimes \ldots \otimes T}_{n \text{ factors}}))}^{k_n \text{ terms}} \right] (x) \tag{57}$$

will have mean k which approximates r to an accuracy of 2^{-n}. (The multiplicity of choices for the binary expansion of r allows for an effort to achieve a simultaneous matching of the variance of the approximating distribution and the given distribution).   □

Definition 4 *The negation of the success-failure event with p.g.f. $S_p(x) = (1 - p) + px$ is the success-failure event with p.g.f. $S_{(1-p)}(x) = p + (1 - p)x$.*

Theorem 4 *Each probability generating function can be approximated arbitrarily closely in the smallest seminearring which contains the p.g.f. $T(x) = 1/2 + 1/2x$ and is closed under negation of success-failure events.*

Proof: Let $S_p(x) = (1 - p) + px$ be the p.g.f. for a success-failure event. Now

$$T^{(n)}(x) = (1 - 2^{-n}) + 2^{-n}x = S_{2^{-n}}(x) \tag{58}$$

and the negation of $T^{(n)}(x)$ is $S_{(1-2^{-n})}(x)$. The mth iterate of $S_{(1-2^{-n})}$ is

$$S_{(1-2^{-n})}^{(m)} = \left[1 - \left(\frac{2^n - 1}{2^n} \right)^m \right] + \left(\frac{2^n - 1}{2^n} \right)^m x. \tag{59}$$

Since the successive powers of $\frac{2^n-1}{2^n}$ decrease by steps of less than 2^{-n} from 1 when $m = 0$ towards the limit 0, the value p is approximated by $(\frac{2^n-1}{2^n})^m$ with error less than 2^{-n} for some positive integer m. By Theorem 2 every p.g.f. is the limit of some sequence of p.g.f.s generated from success-failure events. But each of the success-failure p.g.f.s used in the sequence of p.g.f.s is the limit of a sequence of p.g.f.s generated from $S_{1/2}(x)$. This proves Theorem 6. □

Theorem 6 shows that every probability distribution over the nonnegative integers can be approximated as accurately as desired by the results of an appropriate pattern of "tosses" of a "fair" coin. An alternative interpretation is that every probability distribution over the nonnegative integers may be approximated arbitrarily closely by the probability distribution of the output of an electrical network such that at each branch point there are two choices of the open circuit with the open branch of the continuing circuit being selected with probability 1/2.

References

[1] F. Galton. Problem 4001. *Educational Times*, page 17, April 1873.

[2] T. Harris. *The Theory of Branching Processes*. Dover, 1990.

[3] Watson and F. Galton. On the probability of extinction of families. *Anthropological Institute of Great Britain*, 4:138–144, 1874.

NILPOTENCY AND SOLVABILITY IN CATEGORIES

S. G. Botha

Department of Mathematics, University of South Africa

POB 392, Pretoria 0001, South Africa

ABSTRACT

Nilpotent and solvable ideals are discussed in categories. It is shown that these two concepts differ in the category of near-rings although they coincide in the category of rings. Certain generalisations of theorems in rings, groups, near-rings, Lie algebras, etc. are also proven.

AMS Subject classification: Primary 18D35; Secondary 16A76, 16A90, 17B30, 20F16, 20F18, 20J15, 20K99.

1. INTRODUCTION

Botha and Buys [2] defined and investigated nilpotent and solvable ideals and objects. In section 3 of this paper the discussion is continued. In the category of rings the concepts of nilpotency and solvability coincide. In this paper, however, it is shown that these concepts are not the same in the category of near-rings. An example is given of an ideal of a near-ring which is solvable, but not nilpotent. Now all theorems on solvability which are proved in categories in this paper will give something new in the category of near-rings. Botha and Buys [2] also proved some theorems on solvability which also give something new in near-rings, e.g.:

THEOREM 1.1.
Let $\alpha : I \to A$ be an ideal. If A is solvable of class n, then A/I is solvable of class m with $m \leq n$.

THEOREM 1.2.
Let $\alpha : I \to A$ and $\beta : J \to A$ be ideals.

(a) If β and I are both solvable, then $I + J$ is solvable.

(b) If A/I and A/J are both solvable, then $A/(I \cap J)$ is solvable.

In this paper the general theorems are given in a category $\mathcal{C}$ satisfying certain conditions. When $\mathcal{C} = \Gamma$-*rings*, for example, Theorem 3.8 gives a well-known result, but when $\mathcal{C} = groups$ Theorem 3.8 apparently gives a new result.

2. PRELIMINARIES

Let $\mathcal{C}$ be a category. Suppose the following axioms hold for $\mathcal{C}$:

Y. Fong et al. (eds.), Near-Rings and Near-Fields, 83–88.

(1) C has a zero object.

(2) Every morphism α in C admits an image $\alpha = \nu\mu$, where ν is a normal epi-morphism and μ a monomorphism.

(3) C has a direct and a free product for an arbitrary family of objects.

(4) The subobjects and factor objects of any object form a set.

(5) If $\alpha : A \to B$ is a monomorphism, $\beta : B \to C$ a normal epimorphism, and $\alpha\beta$ admits an image $\alpha\beta = \nu'\mu'$, then

    (a) α normal implies μ' normal,

    (b) if μ denotes the kernel of β, then $\mu \leq \alpha$ and μ' normal imply that α is normal.

Categories satisfying 1 to 5 are, for example, the categories of groups, rings, Γ-rings (as defined by Coppage and Luh [4]), near-rings and Lie algebras (over a commutative ring with identity).

DEFINITION 2.1 [6]
A normal subobject of the object A is called an *ideal* of the object A.

Notation.
The zero morphism will be denoted by ω, and the cokernel of $I \to A$ by $A \to A/I$.

Remark.
For the direct product $A \xleftarrow{\pi_1} A \times B \xrightarrow{\pi_2} B$, the projections π_1 and π_2 are normal epimorphisms and there exist uniquely determined monomorphisms $\sigma_1 : A \to A \times B$ and $\sigma_2 : B \to A \times B$ such that $\sigma_i \pi_i = 1$ and $\sigma_i \pi_j = \omega$, for $i \neq j$, $i = 1, 2$; $j = 1, 2$.

Huq [5] defined the commutator ideal (α, β) of two morphisms α and β as follows:

DEFINITION 2.2.
(a) Two morphisms $\gamma : C \to Y$ and $\delta : D \to Y$ *commute* if there exists a morphism $\gamma \circ \delta : C \times D \to Y$ such that the diagram

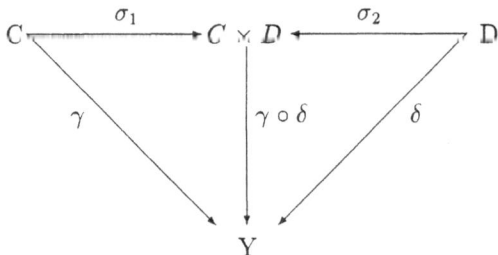

commutes, where σ_1 and σ_2 denote the monomorphisms of the direct product.

(b) Let $\alpha : A \to X$ and $\beta : B \to X$ be two morphisms of $\mathcal{C}$. Let ϵ be the maximal normal epimorphism of X for which $\alpha\epsilon$ and $\beta\epsilon$ commute, then $(\alpha, \beta) = \ker \epsilon$ is called the *commutator ideal* of α and β.

The following concepts will also be used:

DEFINITION 2.3.

(a) An ideal $\alpha : I \to A$ is a *prime ideal* of A if $(\beta_1, \beta_2) \leq \alpha$ implies $\beta_1 \leq \alpha$ or $\beta_2 \leq \alpha$, where β_1 and β_2 are ideals of A.

(b) An object A is a *prime object* if $(\beta_1, \beta_2) = \omega$ implies $\beta_1 = \omega$ or $\beta_2 = \omega$, where β_1 and β_2 are ideals of A.

(c) An ideal $\alpha : I \to A$ is a *semiprime ideal* of A if $(\beta, \beta) \leq \alpha$ implies $\beta \leq \alpha$, where β is an ideal of A.

Remark.
Botha and Buys [1] proved that ideal $\alpha : I \to A$ is a prime ideal if and only if A/I is a prime object.

DEFINITION 2.4 [3].
Define the *annihilator*, *Ann* β, β an ideal of A, as follows:
    Ann $\beta = \Sigma\{\alpha_i : \alpha_i$ an ideal of A such that $(\alpha_i, \beta) = \omega\}$.

Remark.
Let $\alpha : I \to A$ be an ideal. Then it is true that $(Ann \; \alpha, \alpha) = \omega$ in the categories of rings, Γ-rings, groups and Lie algebras. In the category of near-rings, however, it is not necessarily true that $(Ann \; \alpha, \alpha) = \omega$, as has been shown by Botha and Buys [3].

Notation.
Let $\alpha : I \to A$ be an ideal. Then α^n [$\alpha^{(n)}$ resp.], $n \in \mathbb{N}$, denotes the following:

$$\alpha^1 = \alpha \qquad [\alpha^{(1)} = \alpha]$$
$$\alpha^n = (\alpha^{n-1}, \alpha) \quad [\alpha^{(n)} = (\alpha^{(n-1)}, \alpha^{(n-1)})] \quad \text{for } n > 1.$$

Notation.
Let $\theta : A \to B$ be a normal epimorphism, and let $\alpha : I \to A$ be an ideal of A. Then $(Im \; \alpha)_\theta$ denotes the image of α under θ.

3. NILPOTENCY AND SOLVABILITY

DEFINITION 3.1. [2].

(1) An ideal $\alpha : I \to A$ is *nilpotent* if there exists an $n \in \mathbb{N}$ such that $\alpha^n = \omega$.

(2) An object A is *nilpotent* if $1_A : A \to A$ is nilpotent.

DEFINITION 3.2. [2].

(1) An ideal $\alpha : I \to A$ is *solvable* if there exists an $n\epsilon\mathbb{N}$ such that $\alpha^{(n)} = \omega$.

(2) An object A is *solvable* if 1_A is solvable.

EXAMPLES:
In what follows let $\alpha : I \to A$ be an ideal of A.

(1) Let C be the category of *groups*. Then the commutator ideal of $\alpha : I \to A$ is given by the commutator subgroup $[I, I]$ of A. Thus the usual definitions for solvability and nilpotency for normal subgroups are equivalent to the definitions given in this paper.

(2) In the categories of *rings* and Γ-*rings* the concepts of nilpotency and solvability coincide. Let C be the category of rings. Then the usual definition for a nilpotent ideal of a ring is equivalent to definition 3.1. Also, let C be the category of Γ-rings. Then the definition of a strongly nilpotent ideal of a Γ-ring [4] is equivalent to definition 3.1.

(3) Let C be the category of *near-rings*:

  (a) Botha and Buys [2] proved that 1_I^2 is not necessarily equal to α^2. Hence the definition of a nilpotent ideal of a near-ring is not equivalent to the categorical definition.

  (b) All nilpotent ideals are solvable, but a solvable ideal is not necessarily nilpotent: Consider the near-ring A formed by the group D_8 with addition and multiplication defined as in Pilz [7], p.346 no. 80

+	0	1	2	3	4	5	6	7
0	0	1	2	3	4	5	6	7
1	1	2	3	0	5	6	7	4
2	2	3	0	1	6	7	4	5
3	3	0	1	2	7	4	5	6
4	4	7	6	5	0	3	2	1
5	5	4	7	6	1	0	3	2
6	6	5	4	7	2	1	0	3
7	7	6	5	4	3	2	1	0

·	0	1	2	3	4	5	6	7
0	0	0	0	0	0	0	0	0
1	4	4	4	6	4	6	4	4
2	0	0	0	0	0	0	0	0
3	4	4	4	6	4	6	4	4
4	4	4	4	4	4	4	4	4
5	0	0	0	2	0	2	0	0
6	4	4	4	4	4	4	4	4
7	0	0	0	2	0	2	0	0

Now $J = \{0, 2\}$ and $I = \{0, 2, 5, 7\}$ are ideals of A and J is an ideal of I. Let $\alpha : I \to A$ and $\beta : J \to A$. Now $\alpha^2 = < [I, I] >_A = \{0, 2\} = \beta$, and $\alpha^3 = (\alpha^2, \alpha) = (\beta, \alpha) = \{0, 2\} = \beta$ again and hence $\alpha^n \neq 0$, $n\epsilon\mathbb{N}$, i.e. α is not nilpotent. However, $\alpha^{(3)} = (\alpha^{(2)}, \alpha^{(2)}) = (\beta, \beta) = 0$ i.e. α is solvable.

(4) Let C be the category of *Lie algebras*. Then the usual definitions for nilpotent and solvable ideals are equivalent to the definitions given in this paper.

Now we can prove:

THEOREM 3.3.

Let $\alpha : I \to A$ be an ideal. Then A/I is nilpotent (solvable) of class n if and only if $(1_A)^n \leq \alpha$ $((1_A)^{(n)} \leq \alpha)$ for an $n \epsilon \mathbb{N}$.

THEOREM 3.4.

Let A be nilpotent and $\alpha : I \to A$ an ideal of A such that $1_A^2 + \alpha = 1_A$. Then $\alpha = 1_A$.

Proof.

Suppose $\alpha \neq 1_A$. Now, since A is nilpotent, there is an $n \epsilon \mathbb{N}$ such that $1_A^n = \omega \leq \alpha$ and hence A/I is nilpotent. Thus $(1_{A/I})^2 \neq 1_{A/I}$. Let θ be the cokernel of α. Then $1_{A/I} = (Im \ 1_A)_\theta = (Im \ (1_A^2 + \alpha))_\theta = (1_{A/I})^2$, which is a contradiction, and thus $\alpha = 1_A$.

THEOREM 3.5.

Let $\alpha : M \to A$ be a monomorphism. If A is solvable (nilpotent) of class n, $n \epsilon \mathbb{N}$, then M is solvable (nilpotent) of class $\leq n$.

Proof.

It follows by induction that $1_M^{(n)} \alpha \leq 1_A^{(n)}$, $n \epsilon \mathbb{N}$, and similarly, $1_M^n \alpha \leq 1_A^n$, $n \epsilon \mathbb{N}$.

Botha and Buys [2] have proved

THEOREM 3.6.

Let $\alpha : I \to A$ be an ideal. If α is solvable of class n and A/I solvable of class m, then it follows that A is solvable of class s with $n \leq s \leq n + m - 1$.

A direct consequence of Theorem 3.6 is

THEOREM 3.7.

Let (K, π_A, π_B) be the product of objects A and B. Let σ_B be the kernel of π_A. If σ_B and A are both solvable, then K is solvable.

Remark.

In the category of *groups* this theorem has been proved as follows: If the groups A and B are both solvable, then $A \times B$ is solvable. In this case, however, we had to use σ_B and not B solvable, otherwise we could not prove the theorem.

THEOREM 3.8.

For any object A it follows that $\Sigma\{\alpha_i : \alpha_i$ a solvable ideal of $A\} \leq \cap\{\mu_i : \mu_i$ a semi-prime ideal of $A\} \leq \cap\{\alpha_i : \alpha_i$ a prime ideal of $A\}$.

Proof.

Let α be a solvable ideal of A and μ any semi-prime ideal of A. Then $\alpha^{(n)} = \omega \leq \mu$, $n \epsilon \mathbb{N}$, implies $\alpha \leq \mu$.

Before we can prove the next theorem, we need a theorem by Botha and Buys [3]:

THEOREM 3.9.

A is a prime object if and only if for any ideal $\alpha : I \to A$, $\alpha \neq \omega$, it follows that $Ann \ \alpha = \omega$.

THEOREM 3.10.

Let $\theta : A \to B$ be a normal epimorphism, $\theta \neq \omega$, and let A be solvable. Then B contains an ideal $\alpha \neq \omega$ such that $Ann \; \alpha \neq \omega$.

Proof.

Use the contrapositive to prove that if A is solvable, then B cannot be a prime object: Let B be a prime object and let μ be the kernel of θ. Now μ is a prime ideal. Also, since $\theta \neq \omega$ it follows that $\mu \neq 1_A$. Thus it follows by Theorem 3.6 that 1_A is not solvable.

REFERENCES

[1] Botha, S. G. and Buys, A., Prime ideals in Categories, *Comm. Algebra.* **13** (1985), 1171–1186.

[2] Botha, S. G. and Buys, A., Nilpotency, Solvability and Radicals in Categories, *Quaestiones Math.* **14 (2)** (1991), 129-136.

[3] Botha, S. G. and Buys, A., The Annihilator in Categories, Chin. J. Math, **19 (4)**, (1991) 295–308.

[4] Coppage, W. and Luh, J., Radicals of gamma rings, *J. Math. Soc. Japan* **23** (1971), 40–52.

[5] Huq, S. A., Commutator, Nilpotency and Solvability in Categories, *Quart. J. Math. Oxford Ser.***2** (1968), 363–389.

[6] Kurosh, A.G., Livshits, A.K. and Shul'geifer, E.G., Foundations of the Theory of Categories, *Russian Math. Surveys,* **15(6)** (1960), 1-46.

[7] Pilz, Günther, *Near-Rings* (1977), North-Holland Publ. Company, Amsterdam-New York-Oxford.

CENTRALIZER NEAR-RINGS DETERMINED BY End G

G. Alan Cannon

Department of Mathematics, Texas A&M University
College Station, Texas 77843-3368, U.S.A.

ABSTRACT

Let G be a group. The structure of the centralizer near-ring $M_E(G) = \{f\colon G \to G \mid f\sigma = \sigma f \text{ for every } \sigma \in \text{End } G\}$ is investigated for the cases in which G is a finitely generated abelian, characteristically simple, symmetric or generalized quaternion group.

1. Introduction

Let $(G,+)$ be a group, not necessarily abelian, and let S be a subsemigroup of End G. The set $M_S(G) = \{f\colon G \to G \mid f\sigma = \sigma f \text{ for every } \sigma \in S\}$ forms a near-ring under pointwise addition and function composition and is called the *centralizer near-ring determined by the pair* (S, G). Since every near-ring with identity is isomorphic to an $M_S(G)$ for some pair (S, G) [2], these near-rings are quite general and are difficult to study without some restrictions on S or G.

If we let S consist of only the identity function on G, then $M_S(G) = M(G)$, the set of all functions from G to G. Similarly, if we let S consist of the identity and zero functions on G, then $M_S(G) = M_0(G)$, the set of all zero-preserving functions from G to G. The structure of these near-rings is well-known (see [2] or [9]). In this paper we are interested in the structure of the other extreme situation, in other words, when $S = \text{End } G = E$. To this end we will consider when $M_E(G)$ is simple, local, or a ring.

We recall that a near-ring N is local if the set of non-units in N is an additive subgroup. If N is finite, this condition is equivalent to saying that every element of N is either invertible or nilpotent [8]. This provides further motivation for studying $M_E(G)$. If G is finite and $M_E(G)$ is not local for some group G, then $M_S(G)$ cannot be local for any subsemigroup S of End G since $M_E(G)$ is a subnear-ring of $M_S(G)$. Studying these "smallest" near-rings, therefore, provides some information about the general case.

For a given group G, however, the problem of finding all of the endomorphisms of G is usually extremely difficult. For this reason we further restrict our study to some classes of groups for which the endomorphisms are known: finitely generated abelian, characteristically simple, symmetric, and generalized quaternion groups.

Throughout the paper we will assume that G is finite unless otherwise stated. We also denote the center of a group G by $Z(G)$, the centralizer of an element $g \in G$ by $C(g)$ and the identity function on G by id. Finally, if H is a subgroup of G, the subscript E in $M_E(H)$ will always refer to End H.

Y. Fong et al. (eds.), Near-Rings and Near-Fields, 89–111.

2. General results

In this section we develop some general results that will be used in subsequent sections with our specific groups. We begin by examining direct products of groups.

Lemma 2.1: Let $G = H_1 \times H_2 \times \cdots \times H_n$ and $f \in M_E(G)$. Then $f = (f_1, f_2, \ldots, f_n)$ with $f_i \in M_E(H_i)$ for each $i = 1, 2, \ldots, n$, and acts componentwise on G.

Proof: Let $f \in M_E(G)$ and $(h_1, \ldots, h_n) \in G$ with $f(h_1, \ldots, h_n) = (k_1, \ldots, k_n)$. Let $\pi_i \colon G \to G$ be defined by $\pi_i(g_1, \ldots, g_n) = (0, \ldots, 0, g_i, 0, \ldots, 0)$ for $i = 1, 2, \ldots, n$. Since $\pi_i \in \operatorname{End} G$ for each i, $f\pi_i(h_1, \ldots, h_n) = \pi_i f(h_1, \ldots, h_n)$. Simplifying gives $f(0, \ldots, 0, h_i, 0, \ldots, 0) = \pi_i(k_1, \ldots, k_n)$ and finally $f(0, \ldots, 0, h_i, 0, \ldots, 0) = (0, \ldots, 0, k_i, 0, \ldots, 0)$. Thus we have $f(h_1, \ldots, h_n) = \sum_{i=1}^{n} f(0, \ldots, 0, h_i, 0, \ldots, 0)$. Since $(h_1, \ldots, h_n) \in G$ is arbitrary, we can define $f_i \colon H_i \to H_i$ by $f_i(h_i) = k_i$ as given by the function f above. Then acting componentwise on G, the function $(f_1, \ldots, f_n)$ is identically f since

$$f(h_1, \ldots, h_n) = (k_1, \ldots, k_n) = (f_1(h_1), \ldots, f_n(h_n)) = (f_1, \ldots, f_n)(h_1, \ldots, h_n).$$

If we let $\sigma_i \in \operatorname{End} H_i$, $i = 1, 2, \ldots, n$, then $(\sigma_1, \ldots, \sigma_n) \in \operatorname{End} G$, acting componentwise on G. So for $(h_1, \ldots, h_n) \in G$,

$$(\sigma_1, \ldots, \sigma_n)(f_1, \ldots, f_n)(h_1, \ldots, h_n) = (f_1, \ldots, f_n)(\sigma_1, \ldots, \sigma_n)(h_1, \ldots, h_n)$$

implies that

$$(\sigma_1 f_1(h_1), \ldots, \sigma_n f_n(h_n)) = (f_1 \sigma_1(h_1), \ldots, f_n \sigma_n(h_n)).$$

In other words, $f_i \sigma_i = \sigma_i f_i$ for each i. Since $\sigma_i \in \operatorname{End} H_i$ is arbitrary, $f_i \in M_E(H_i)$ for each i. This final observation proves the result. ∎

Lemma 2.2: Let $G = A_1 \times A_2 \times \cdots \times A_n$ with $A_i \hookrightarrow A_n$ (i.e., A_i is embedded in A_n) for every $i = 1, 2, \ldots, n$. Then $M_E(G) \hookrightarrow M_E(A_n)$.

Proof: Let $\varphi_i \colon A_i \to A_n$ be a monomorphism for each $i = 1, 2, \ldots, n$. Consider the map $\psi_i \in \operatorname{End} G$ given by $\psi_i(a_1, \ldots, a_n) = (0, 0, \ldots, 0, \varphi_i(a_i))$. Let $f \in M_E(G)$. By Lemma 2.1 we have $f = (f_1, \ldots, f_n)$ with $f_i \in M_E(A_i)$ for every $i = 1, 2, \ldots, n$. Therefore $f\psi_i(0, 0, \ldots, 0, a_i, 0, \ldots, 0) = \psi_i f(0, 0, \ldots, 0, a_i, 0, \ldots, 0)$ implies that $f(0, \ldots, 0, \varphi_i(a_i)) = \psi_i(0, 0, \ldots, f_i(a_i), 0, \ldots, 0)$. But this gives

$$(0, \ldots, 0, f_n \varphi_i(a_i)) = (0, 0, \ldots, 0, \varphi_i f_i(a_i)), \text{ or } f_n \varphi_i(a_i) = \varphi_i f_i(a_i).$$

Regarding φ_i^{-1} as a function from $\varphi_i(A_i)$ to A_i, we have $f_i = \varphi_i^{-1} f_n \varphi_i$ for every $i = 1, 2, \ldots, n$. Hence $f = (\varphi_1^{-1} f_n \varphi_1, \ldots, \varphi_n^{-1} f_n \varphi_n)$. It follows directly that the map $f = (\varphi_1^{-1} f_n \varphi_1, \ldots, \varphi_n^{-1} f_n \varphi_n) \mapsto f_n$ is a near-ring monomorphism from $M_E(G)$ to $M_E(A_n)$. ∎

Since $M_E(G)$ is a subnear-ring of $M_S(G)$ for every subsemigroup S of $\operatorname{End} G$, it is useful to investigate $M_S(G)$ for some fixed semigroup S. In particular, we choose $S = \text{Inn } G \cup \{0\} = I$. The corresponding near-ring $M_I(G)$ will be particularly useful in studying centralizer near-rings on the symmetric groups.

Lemma 2.3: Let $f\colon G \to G$ be a function. Then $f \in M_I(G)$ iff for every $x, g \in G$, $f(-x + g + x) = -x + f(g) + x$.

Proof: For each $x \in G$ define $\varphi_x\colon G \to G$ by $\varphi_x(g) = -x + g + x$. Then $\varphi_x \in \text{Inn } G$. So $f \in M_I(G)$ iff for every $x, g \in G$, $f\varphi_x(g) = \varphi_x f(g)$. This last condition is just $f(-x + g + x) = -x + f(g) + x$. ∎

Lemma 2.4: Let $f \in M_I(G)$. Then $f(Z(G)) \subseteq Z(G)$.

Proof: Let $g \in Z(G)$ and $f \in M_I(G)$. By Lemma 2.3, $f(-x+g+x) = -x+f(g)+x$ for every $x \in G$. Since $g \in Z(G)$ we have $f(g) = -x + f(g) + x$. Therefore $x + f(g) = f(g) + x$ for every $x \in G$, i.e., $f(g) \in Z(G)$, and we have the result. ∎

Let A be a group of automorphisms of the group G and let $g \in G$. We define the *stabilizer of g in A* as $\text{Stab}(g) = \{\sigma \in A \mid \sigma(g) = g\}$. Stabilizers are instrumental in studying centralizer near-rings determined by automorphisms using the well-known result Betsch's Lemma [9]. We restrict our attention to I.

Lemma 2.5: The following are equivalent for $g_1, g_2 \in G$:

  (i) There exists an $f \in M_I(G)$ such that $f(g_1) = g_2$;
  (ii) $\text{Stab}(g_1) \subseteq \text{Stab}(g_2)$;
 (iii) $C(g_1) \subseteq C(g_2)$;
 (iv) $g_2 \in Z(C(g_1))$.

Proof: The equivalence of (i) and (ii) is given by Betsch's Lemma. Condition (ii) holds if and only if $\{\varphi_x \in I \mid \varphi_x(g_1) = g_1\} \subseteq \{\varphi_x \in I \mid \varphi_x(g_2) = g_2\}$, i.e., $\{x \in G \mid -x + g_1 + x = g_1\} \subseteq \{x \in G \mid -x + g_2 + x = g_2\}$. This latter condition is the same as $C(g_1) \subseteq C(g_2)$, and (ii) is equivalent to (iii).

    Now assume $C(g_1) \subseteq C(g_2)$ and let $x \in C(g_1)$. Then $x \in C(g_2)$ and $g_2 + x = x + g_2$. Since x is arbitrary, $g_2 \in Z(C(g_1))$. If we assume $g_2 \in Z(C(g_1))$, then for $x \in C(g_1)$, $x + g_2 = g_2 + x$ and $x \in C(g_2)$. Thus $C(g_1) \subseteq C(g_2)$, (iii) is equivalent to (iv), and the lemma is proved. ∎

Lemma 2.6: $M_I(G)$ is an abelian near-ring.

Proof: Let $f_1, f_2 \in M_I(G)$ and let $g \in G$. By the previous lemma, $f_1(g)$ and $f_2(g)$ are in $Z(C(g))$. Since $Z(C(g))$ is an abelian group we have $(f_1 + f_2)(g) = f_1(g) + f_2(g) = f_2(g) + f_1(g) = (f_2 + f_1)(g)$. Hence $f_1 + f_2 = f_2 + f_1$ and $M_I(G)$ is abelian. ∎

Corollary 2.7: $M_E(G)$ is an abelian near-ring.

Proof: Since $M_E(G)$ is a subnear-ring of $M_I(G)$, the result is immediate. ∎

    To facilitate the computation of $M_E(G)$ for G a symmetric or generalized quaternion group we consider the following general situation. Let G be a group with a normal subgroup N of index two. Let Y be the set of elements in G of order two. For each $y \in Y$ define $\emptyset_y \colon G \to G$ by $\emptyset_y(g) = \begin{cases} 0 & \text{if } g \in N \\ y & \text{if } g \in G \backslash N \end{cases}$. It is easy to check that $\emptyset_y \in \text{End } G$. Furthermore we get:

Lemma 2.8: Let G, N, and Y be as above and let $f \colon G \to G$ be a function with $f(0) = 0$. Then $f \emptyset_y = \emptyset_y f$ for every $y \in Y$ iff

(i) $f(N) \subseteq N$; and
(ii) Either (a) $f(G \backslash N) \subseteq N$ and $f|_Y = 0$ or (b) $f(G \backslash N) \subseteq G \backslash N$ and $f|_Y = id$.

Proof: We prove only the direct implication. The converse follows from straightforward calculations.

(i) Let $n \in N$. Then $f \emptyset_y(n) = 0$ for every $y \in Y$. Assume $f(n) \notin N$. Then $\emptyset_y f(n) = y$ implies $y = 0$, a contradiction. So $f(n) \in N$ and we have the result.

(ii) (a) If $n \notin N$ and $f(n) \in N$, then $\emptyset_y f(n) = \emptyset_y(0) = 0$ and $f \emptyset_y(n) = f(y)$. So $f(y) = 0$ for every $y \in Y$.
    (b) If $n \notin N$ and $f(n) \notin N$, then $\emptyset_y f(n) = y$ and $f \emptyset_y(n) = f(y)$. So $f(y) = y$ for every $y \in Y$.

Since only one of (a) and (b) can occur, the proof is complete.   ■

To conclude this section we prove that only certain groups can give rise to local near-rings.

Lemma 2.9: Let G be a finite group. If $M_E(G)$ is local, then G is a p-group for some prime, p.

Proof: Let $\exp G = m$. Then for $H = \langle id \rangle$, a subnear-ring of $M_E(G)$, $|H| = m$. Furthermore, H is ring isomorphic to Z_m. If $M_E(G)$ is local, then H is local and m must be a power of a prime. Since $\exp G = m$, G must be a p-group.   ■

3. G a finitely generated abelian group

We can completely determine the structure of $M_E(G)$ when G is a finitely generated abelian group. These near-rings will, in fact, provide our first examples of rings.

Lemma 3.1: Let G be a finitely generated abelian group, say

$$G = Z_{m_1} \times Z_{m_2} \times \cdots \times Z_{m_k} \times Z \times Z \times \cdots \times Z \quad (n \text{ copies of } Z).$$

If $f \in M_E(G)$, then

$$f = \begin{bmatrix} a_1 & & & & & & & \\ & a_2 & & & & \bigcirc & & \\ & & \ddots & & & & & \\ & & & a_k & & & & \\ & & & & b_1 & & & \\ & \bigcirc & & & & b_2 & & \\ & & & & & & \ddots & \\ & & & & & & & b_n \end{bmatrix}$$

where $a_i \in Z_{m_i}$ for $i = 1, 2, \ldots, k$, $b_j \in Z$ for $j = 1, 2, \ldots, n$ and acts by matrix multiplication on a column vector of G.

Proof: Let $f \in M_E(G)$ with $f\begin{pmatrix} 1 \\ 1 \\ \vdots \\ 1 \end{pmatrix} = \begin{pmatrix} a_1 \\ a_2 \\ \vdots \\ a_k \\ b_1 \\ \vdots \\ b_n \end{pmatrix}$. Let $\sigma \in \operatorname{End} Z_{m_1} \times \operatorname{End} Z_{m_2} \times$

$\cdots \times \operatorname{End} Z_{m_k} \times \operatorname{End} Z \times \cdots \times \operatorname{End} Z \subseteq \operatorname{End} G$ be given by

$$\sigma = \begin{bmatrix} c_1 & & & & & & & \\ & c_2 & & & & \bigcirc & & \\ & & \ddots & & & & & \\ & & & c_k & & & & \\ & & & & d_1 & & & \\ & \bigcirc & & & & \ddots & \\ & & & & & & d_n \end{bmatrix}$$

where $c_i \in Z_{m_i}$ for $i = 1, 2, \ldots, k$, $d_j \in Z$ for $j = 1, 2, \ldots, n$ and acting by matrix multiplication on a column vector of G. Then $f\sigma\begin{pmatrix} 1 \\ 1 \\ \vdots \\ 1 \end{pmatrix} = \sigma f\begin{pmatrix} 1 \\ 1 \\ \vdots \\ 1 \end{pmatrix}$ implies that

$$f\begin{pmatrix} c_1 \\ c_2 \\ \vdots \\ c_k \\ d_1 \\ \vdots \\ d_n \end{pmatrix} = \begin{pmatrix} c_1 a_1 \\ c_2 a_2 \\ \vdots \\ c_k a_k \\ d_1 b_1 \\ \vdots \\ d_n b_n \end{pmatrix} = \begin{bmatrix} a_1 & & & & & & \\ & a_2 & & & \bigcirc & & \\ & & \ddots & & & & \\ & & & a_k & & & \\ & & & & b_1 & & \\ & \bigcirc & & & & \ddots & \\ & & & & & & b_n \end{bmatrix}\begin{pmatrix} c_1 \\ c_2 \\ \vdots \\ c_k \\ d_1 \\ \vdots \\ d_n \end{pmatrix}.$$

Since we can choose σ arbitrarily, we get that f is completely determined by the matrix

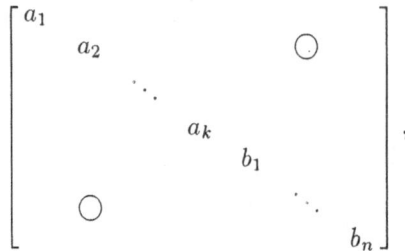

$$\left[\begin{array}{ccccccc} a_1 & & & & & & \\ & a_2 & & & & \bigcirc & \\ & & \ddots & & & & \\ & & & a_k & & & \\ & & & & b_1 & & \\ & \bigcirc & & & & \ddots & \\ & & & & & & b_n \end{array}\right],$$

hence the result. ■

Theorem 3.2: If G is a finite abelian group, say $G = Z_{m_1} \times \cdots \times Z_{m_k}$ where $m_i | m_{i+1}$ for $i = 1, 2, \ldots, k-1$, then $M_E(G) \cong Z_{m_k}$.

Proof: Let $f \in M_E(G)$. By Lemma 3.1, $f = \left[\begin{array}{cccc} a_1 & & & \bigcirc \\ & a_2 & & \\ & & \ddots & \\ \bigcirc & & & a_k \end{array}\right]$ where $a_i \in$

Z_{m_i} for $i = 1, 2, \ldots, k$. Define $\sigma_i \colon Z_{m_k} \to Z_{m_i}$ by $\sigma_i(g) = g(\bmod\ m_i)$ for $i = 1, 2, \ldots, k$. Then $\sigma_i \in \mathrm{Hom}(Z_{m_k}, Z_{m_i})$ since $m_i | m_k$. It is a simple matter to see

that $\sigma = \left[\begin{array}{cccc} \sigma_1 & & & \\ & \sigma_2 & & \\ \bigcirc & & \ddots & \\ & & & \sigma_k \end{array}\right] \in \mathrm{End}\ G$. Let $g = \begin{pmatrix} 0 \\ 0 \\ \vdots \\ 0 \\ 1 \end{pmatrix} \in G$. Then

$$\sigma f(g) = \left[\begin{array}{cccc} \sigma_1 & & & \\ & \sigma_2 & & \\ \bigcirc & & \ddots & \\ & & & \sigma_k \end{array}\right] \left[\begin{array}{cccc} a_1 & & & \bigcirc \\ & a_2 & & \\ & & \ddots & \\ \bigcirc & & & a_k \end{array}\right] \begin{bmatrix} 0 \\ 0 \\ \vdots \\ 0 \\ 1 \end{bmatrix} = \begin{bmatrix} \sigma_1(a_k) \\ \sigma_2(a_k) \\ \vdots \\ \sigma_k(a_k) \end{bmatrix} = \begin{bmatrix} a_k(\bmod\ m_1) \\ a_k(\bmod\ m_2) \\ \vdots \\ a_k(\bmod\ m_k) \end{bmatrix}$$

and

$$f\sigma(g) = \left[\begin{array}{cccc} a_1 & & & \bigcirc \\ & a_2 & & \\ & & \ddots & \\ \bigcirc & & & a_k \end{array}\right] \left[\begin{array}{cccc} \sigma_1 & & & \\ & \sigma_2 & & \\ \bigcirc & & \ddots & \\ & & & \sigma_k \end{array}\right] \begin{bmatrix} 0 \\ 0 \\ \vdots \\ 0 \\ 1 \end{bmatrix} = \begin{bmatrix} a_1 \\ a_2 \\ \vdots \\ a_k \end{bmatrix}.$$

Therefore $a_k(\mathrm{mod}\ m_i) = a_i$ for $i = 1, 2, \ldots, k$. So

$$
a_k
\begin{bmatrix}
1 & & & \bigcirc \\
& 1 & & \\
& & \ddots & \\
\bigcirc & & & 1
\end{bmatrix}
=
\begin{bmatrix}
a_k(\mathrm{mod}\,m_1) & & & \bigcirc \\
& a_k(\mathrm{mod}\,m_2) & & \\
\bigcirc & & \ddots & \\
& & & a_k(\mathrm{mod}\,m_k)
\end{bmatrix}
$$

$$
=
\begin{bmatrix}
a_1 & & & \bigcirc \\
& a_2 & & \\
& & \ddots & \\
\bigcirc & & & a_k
\end{bmatrix}
= f.
$$

Since f is arbitrary, $M_E(G) \subseteq \left\langle \begin{bmatrix} 1 & & & \bigcirc \\ & 1 & & \\ & & \ddots & \\ \bigcirc & & & 1 \end{bmatrix} \right\rangle$. The reverse inclusion is

clear, so that $M_E(G) = \left\langle \begin{bmatrix} 1 & & & \bigcirc \\ & 1 & & \\ & & \ddots & \\ \bigcirc & & & 1 \end{bmatrix} \right\rangle \cong Z_{m_k}$.   ∎

Theorem 3.3: If G is an infinite, finitely generated abelian group, say $G = Z_{m_1} \times Z_{m_2} \times \cdots \times Z_{m_k} \times Z \times Z \times \cdots \times Z$ (n copies of Z), then $M_E(G) \cong Z$.

Proof: Let $f \in M_E(G)$. From Lemma 3.1, $f = \begin{bmatrix} a_1 & & & & & & \\ & a_2 & & & & & \bigcirc \\ & & \ddots & & & & \\ & & & a_k & & & \\ & & & & b_1 & & \\ & & & & & \ddots & \\ \bigcirc & & & & & & b_n \end{bmatrix}$

where $a_i \in Z_{m_i}$ for $i = 1, 2, \ldots, k$ and $b_j \in Z$ for $j = 1, 2, \ldots, n$. Consider

$$
\sigma =
\begin{bmatrix}
& & \pi_1 \\
& & \pi_2 \\
& & \vdots \\
& \bigcirc & \pi_k \\
& & 0 \\
& & \vdots \\
& & 0
\end{bmatrix}
\quad \text{where } \pi_i \colon Z \to Z_{m_i} \text{ is given by } \pi_i(g) = g(\mathrm{mod}\,m_i).\ \text{It is}
$$

straightforward to check that $\sigma \in$ End G. Let $\begin{pmatrix} 0 \\ 0 \\ \vdots \\ 0 \\ 1 \end{pmatrix} \in G$. Since $f \in M_E(G)$ we have

$$
\sigma f \begin{pmatrix} 0 \\ 0 \\ \vdots \\ 0 \\ 1 \end{pmatrix} = \sigma \begin{pmatrix} 0 \\ 0 \\ \vdots \\ 0 \\ b_n \end{pmatrix} = \begin{pmatrix} \pi_1(b_n) \\ \pi_2(b_n) \\ \vdots \\ \pi_k(b_n) \\ 0 \\ \vdots \\ 0 \end{pmatrix} = \begin{pmatrix} b_n(\bmod m_1) \\ b_n(\bmod m_2) \\ \vdots \\ b_n(\bmod m_k) \\ 0 \\ \vdots \\ 0 \end{pmatrix}
$$

and

$$
f\sigma \begin{pmatrix} 0 \\ 0 \\ \vdots \\ 0 \\ 1 \end{pmatrix} = f \begin{pmatrix} \pi_1(1) \\ \pi_2(1) \\ \vdots \\ \pi_k(1) \\ 0 \\ \vdots \\ 0 \end{pmatrix} = f \begin{pmatrix} 1 \\ 1 \\ \vdots \\ 1 \\ 0 \\ \vdots \\ 0 \end{pmatrix} = \begin{pmatrix} a_1 \\ a_2 \\ \vdots \\ a_k \\ 0 \\ \vdots \\ 0 \end{pmatrix} .
$$

So $a_i = b_n(\bmod\ m_i)$ for $i = 1, 2, \ldots, k$, and we have a relationship between the a_i's and b_n. Now we relate the b_j's to b_n.

Consider $e_{i,j} \in \operatorname{End} G$ given by a $(k + n) \times (k + n)$ matrix with $1 \in Z$ in the $k + i$, $k + j$ position and zeros elsewhere. Let $g_j \in G$ be the $(k + n)$-tuple with $1 \in Z$ in the $k + j$ position and zeros elsewhere. Since $f \in M_E(G)$ we have $f e_{i,j}(g_j) = f(g_i) = b_i g_i$ and $e_{i,j} f(g_j) = e_{i,j}(b_j g_j) = b_j g_i$, hence $b_i = b_j$. Since we can choose i and j arbitrarily, $b_1 = b_2 = \cdots = b_n$. So

$$
b_n \begin{bmatrix} 1 & & & \bigcirc \\ & 1 & & \\ & & \ddots & \\ \bigcirc & & & 1 \end{bmatrix} = \begin{bmatrix} b_n(\bmod m_1) & & & & & \bigcirc \\ & b_n(\bmod m_2) & & & & \\ & & \ddots & & & \\ & & & b_n(\bmod m_k) & & \\ \bigcirc & & & & b_n & \\ & & & & & \ddots \\ & & & & & & b_n \end{bmatrix}
$$

$$
= \begin{bmatrix} a_1 & & & & \bigcirc \\ & a_2 & & & \\ & & \ddots & & \\ & & & a_k & \\ & & & & b_1 \\ \bigcirc & & & & & \ddots \\ & & & & & & b_n \end{bmatrix} = f.
$$

Thus, as in the finite case, $f \in \left\langle \begin{bmatrix} 1 & & & \bigcirc \\ & 1 & & \\ & & \ddots & \\ \bigcirc & & & 1 \end{bmatrix} \right\rangle$, generated additively. There-

fore $M_E(G) \subseteq \left\langle \begin{bmatrix} 1 & & & \bigcirc \\ & 1 & & \\ & & \ddots & \\ \bigcirc & & & 1 \end{bmatrix} \right\rangle$, hence equality, and thus the result.   ∎

From the previous two theorems and Lemma 2.9 we immediately get the following result.

Corollary 3.4: Let G be a finitely generated abelian group. Then

(i) $M_E(G)$ is a ring;
(ii) $M_E(G)$ is a local ring iff G is a p-group, p a prime;
(iii) $M_E(G)$ is semisimple iff G is infinite or $G = Z_{m_1} \times Z_{m_2} \times \cdots \times Z_{m_k}$ with $m_i | m_{i+1}$ for $i = 1, 2, \ldots, k-1$ and m_k is square free;
(iv) The following are equivalent:

    a. $M_E(G)$ is a simple ring;
    b. $M_E(G)$ is field;
    c. G is an elementary abelian p-group, p a prime;
    d. $M_E(G) \cong Z_p$.

4. G a finite, characteristically simple group

We now consider the case when G is a finite, characteristically simple group, i.e., a finite group with no nontrivial, proper characteristic subgroups. For our purposes a more useful characterization is given by the following:

Theorem 4.1, [4]: A characteristically simple group is the direct product of isomorphic simple groups.

To avoid cumbersome notation with isomorphic simple groups, without a loss of generality we will assume that $G = H \times \cdots \times H$ where H is a finite simple group. Moreover, the structure of $M_E(G)$ is completely determined by that of $M_E(H)$. To verify this we need to know the structure of End G. Since the abelian case is handled in the above section, we take G to be nonabelian.

Lemma 4.2, [3]: Let $G = H \times \cdots \times H$ (n copies) with H a finite, simple nonabelian group. Then End $G = \{(\alpha_{ij}): \alpha_{ij} \in \text{End } H, 1 \leq i, j \leq n$, with at most one nonzero entry in each row$\}$.

Theorem 4.3: Let $G = H \times H \times \cdots \times H$ (n copies) with H a finite, nonabelian simple group. Then $M_E(G) \cong M_E(H)$.

Proof: Let $\alpha \in \text{End } G$. Then $\alpha = (\alpha_{ij})$ with $\alpha_{ij} \in \text{End } H$, $1 \leq i, j \leq n$, and with at most one nonzero entry in each row. Let α_{i,k_i} denote the possibly nonzero element in row i. Let $f \in M_E(H)$ and consider the function $\bar{f} \colon G \to G$ given by $\bar{f} = (f, f, \ldots, f)$ (n copies) and acting componentwise on G. Therefore for $(h_1, \ldots, h_n) \in G$, we have

$$
\alpha \bar{f} \begin{pmatrix} h_1 \\ \vdots \\ h_n \end{pmatrix} = \alpha \begin{pmatrix} f(h_1) \\ \vdots \\ f(h_n) \end{pmatrix} = \begin{pmatrix} \alpha_{1,k_1} f(h_{k_1}) \\ \vdots \\ \alpha_{n,k_n} f(h_{k_n}) \end{pmatrix}
$$

$$
= \begin{pmatrix} f\alpha_{1,k_1}(h_{k_1}) \\ \vdots \\ f\alpha_{n,k_n}(h_{k_n}) \end{pmatrix} = \bar{f} \begin{pmatrix} \alpha_{1,k_1}(h_{k_1}) \\ \vdots \\ \alpha_{n,k_n}(h_{k_n}) \end{pmatrix} = \bar{f}\alpha \begin{pmatrix} h_1 \\ \vdots \\ h_n \end{pmatrix}.
$$

Thus $\bar{f} \in M_E(G)$. By Lemma 2.2, there exists a near-ring monomorphism ψ from $M_E(G)$ into $M_E(H)$. It is straightforward to see that $\psi(\bar{f}) = f$ and, thus, ψ is onto. Therefore ψ is an isomorphism and the result is established. ∎

Theorem 4.3 implies that we can restrict our study of $M_E(G)$ to the case where G is a finite, simple group. In this instance we have $E = \mathrm{Aut}\, G \cup \{0\}$ and can use some well-known results on centralizer near-rings determined by a group of automorphisms.

Theorem 4.4: For a finite, characteristically simple group G, the following are equivalent:

(i) $M_E(G)$ is simple;
(ii) $M_E(G)$ is local;
(iii) $M_E(G) \cong Z_p$, p a prime;
(iv) $G \cong Z_p \times Z_p \times \cdots \times Z_p$, p a prime.

Proof: It is obvious that (iii) implies (i) and (ii). Theorem 3.2 also gives that (iv) implies (iii).

Assume that $M_E(G)$ is local. Then for $G = H \times H \times \cdots \times H$, $M_E(G) \cong M_E(H)$ by the previous theorem, and $M_E(H)$ is local. But this implies that H is a simple p-group by Lemma 2.9, i.e., $H \cong Z_p$. So $G \cong Z_p \times \cdots \times Z_p$ and we have (ii) implies (iv).

Now assume that $M_E(G)$ is simple. As above we get $M_E(H)$ is simple. If H is abelian, then $H \cong Z_p$ and $G \cong Z_p \times \cdots \times Z_p$, hence the result. Assume H is nonabelian. Corollary 1 of [7] gives that $\mathrm{Aut}\, H$ must act fixed point free on H. But for $h \in H$, the inner automorphism $\varphi_h(x) = -h + x + h$ has $\varphi_h(h) = h$. Also, φ_h is not the identity function, for if so we have that H is abelian. So $\varphi_h \in \mathrm{Aut}\, H$ is not fixed point free, a contradiction. Therefore H must be abelian and (i) implies (iv).

Because we have shown (iii) $\Rightarrow$ (ii) $\Rightarrow$ (iv) $\Rightarrow$ (iii) $\Rightarrow$ (i) $\Rightarrow$ (iv) $\Rightarrow$ (iii), we have the result. ∎

The previous theorem addresses the simplicity and localness of $M_E(G)$. Our final result in this section characterizes when $M_E(G)$ is a ring.

Theorem 4.5: Let $G = H \times H \times \cdots \times H$ be a finite, characteristically simple group. Then $M_E(G)$ is a ring iff $H \cong A_5$ or $H \cong Z_p$, p a prime.

Proof: Since $M_E(G) \cong M_E(H)$, we only need to show the result holds for $M_E(H)$.
($\Leftarrow$): If $H \cong Z_p$, then $M_E(H) \cong Z_p$. If $H \cong A_5$, then $M_E(H) \cong Z_2 \times Z_3 \times Z_5$ by [6].
($\Rightarrow$): Assume $M_E(H)$ is a ring. Then by [6], $M_E(H)$ is a direct sum of fields and either

(i) H is a p-group of exponent p, p a prime,
(ii) H is a Frobenius group with kernel of order p^α and complement of order q where p and q are distinct primes, or
(iii) $H \cong A_5$.

We consider each case.

(i) If H is a p-group, then H has a nontrivial center. Since H is simple, $H \cong Z_p$.
(ii) Since the kernel of a Frobenius group is a proper, normal subgroup, then $\alpha = 0$, and we have case (a) again.
(iii) Immediate. ∎

5. $G = S_n$, the Symmetric Group with $n \neq 6$

In this section we let $G = S_n$, the symmetric group on n objects. Our main goal is to characterize those functions in $M_E(G)$ by determining how to construct them. To do this we first need End G.

In Suzuki [11], one finds that for $n \neq 4$, any nontrivial normal subgroup of S_n contains A_n. So for $n \neq 4$, A_n is the only proper, nontrivial normal subgroup of S_n. Therefore if $\psi \in$ End S_n with Ker $\psi = A_n$, then Im $\psi = \{0, y\}$ for some $y \in S_n$ with $2y = 0$. Also for $n \geq 3$, $n \neq 6$, each automorphism of S_n is inner. This information leads us to:

Lemma 5.1: For $n \geq 3$ and $n \neq 4, 6$, the endomorphisms of S_n are:

(i) 0, the zero map;
(ii) φ_x, for all $x \in S_n$ the inner automorphism determined by x, i.e., $\varphi_x(g) = -x + g + x$ for all $g \in S_n$; and
(iii) $\emptyset_y$ for all $y \in Y$ as described in Section 2.

We note that for $n = 4, 6$, the above lemma provides endomorphisms of S_n, but not a complete list. Using the above endomorphisms, however, we can determine $M_E(S_4)$. Because S_6 has outer automorphisms, we exclude the case $n = 6$ from further consideration.

In order for a function $f\colon G \to G$ to be in $M_E(G)$ we must have $f(0) = 0$, $f\varphi_x = \varphi_x f$ for all $x \in G$, and $f\emptyset_y = \emptyset_y f$ for all $y \in Y$. A criteria for this last condition is provided by Lemma 2.8 with $N = A_n$ in the statement of the result. Using Lemma 2.5 we discover a starting point for f to commute with the inner automorphisms, namely if $g \in G$, then $f(g) \in Z(C(g))$. We will need some definitions and a sequence of lemmas to find $Z(C(g))$ for $g \in G$.

Let $g \in G$. For notational convenience define the sets Fix $g = \{i \in \{1, \ldots, n\} \mid g(i) = i\}$ and Move $g = \{i \in \{1, 2, \ldots, n\} \mid g(i) \neq i\}$. Also define

$$D_g = \{h \in G \mid \text{for all} \ i = 1, 2, \ldots, n, \ \text{if} \ i \in \text{Move} \ g, \text{then} \ i \in \text{Fix} \ h\}$$

and

$$C^*(g) = \{h \in C(g) \mid \text{for all} \ i = 1, 2, \ldots, n, \ \text{if} \ i \in \text{Move} \ h, \text{then} \ i \in \text{Move} \ g\}.$$

As an example, let $g = (12) + (345) \in S_7$. Then $D_g = \{0, (67)\}$ and $C^*(g) = \{0, (12), (345), (354), (12) + (345), (12) + (354)\}$.

Let $g_1, \ldots, g_r$ be elements of G. We say the elements are *pairwise disjoint* if for every pair g_i, g_k the sets Move g_i and Move g_k are disjoint. In other words, g_j and g_k do not move any common element. We call a sum of pairwise disjoint elements a *pairwise disjoint sum*.

The next lemma consists of basic consequences of the preceding definitions. We include only portions of the proof.

Lemma 5.2: Let $g \in G$. Then

 (i) D_g and $C^*(g)$ are subgroups of $C(g)$;
 (ii) $D_g \cap C^*(g) = \{0\}$;
 (iii) $D_g \cong S_{n-k}$ where $k = |\text{Move} \ g|$;
 (iv) If $h \in G$ such that g and h are pairwise disjoint, then $C^*(g) \cap C^*(h) = \{0\}$;
 (v) If $h_1 \in D_g$ and $h_2 \in C^*(g)$, then $h_1 + h_2 = h_2 + h_1$.

Proof: (i) Clearly $0 \in D_g$. If $h \in D_g$ then $-h \in D_g$, for if $g(i) \neq i$, then $h(i) = i$ and $(-h)(i) = i$. If $h_1, h_2 \in D_g$, then $g(i) \neq i$ implies that $h_1(i) = h_2(i) = i$ and $(h_1 + h_2)(i) = h_1(i) = i$. So $h_1 + h_2 \in D_g$ and D_g is a subgroup of S_n. Now let $h \in D_g$. If $g(i) \neq i$, then $(g + g)(i) \neq g(i)$ since g is injective. So $(h + g)(i) = g(i)$ since $h \in D_g$. Also, since $g(i) \neq i$, then $h(i) = i$. So $(g + h)(i) = g(i)$. Thus $(h + g)(i) = (g + h)(i)$. If $g(i) = i$, applying h to both sides gives $(h + g)(i) = h(i)$. If $h(i) = i$, then $(h + g)(i) = h(i) = i = g(i) = (g + h)(i)$. So let $h(i) \neq i$. Suppose $(g + h)(i) \neq h(i)$. Then since $h \in D_g$, $(h + h)(i) = h(i)$ and $h(i) = i$, a contradiction. Hence $(g + h)(i) = h(i)$ and $(g + h)(i) = (h + g)(i)$. Therefore we have D_g is a subgroup of $C(g)$.

Clearly $0 \in C^*(g)$. If $h \in C^*(g)$, then $-h \in C^*(g)$, for if $(-h)(i) \neq i$, then $h(i) \neq i$ and $g(i) \neq i$. Furthermore, if $h_1, h_2 \in C^*(g)$, then $h_1(i) \neq i$ and $h_2(i) \neq i$ imply that $g(i) \neq i$. So if $(h_1 + h_2)(i) \neq i$, then either $h_1(i) \neq i$ or $h_2(i) \neq i$ and $g(i) \neq i$. Thus $C^*(g)$ is a subgroup of $C(g)$.

(v) Let $h_1 \in D_g$ and $h_2 \in C^*(g)$. If $h_1(i) = h_2(i) = i$, then $(h_1 + h_2)(i) = i = (h_2 + h_1)(i)$. If $h_1(i) \neq i$ and $h_2(i) = i$, then $(h_1 + h_2)(i) = h_1(i)$. Also, $h_1(i) \neq i$ implies that $(h_1 + h_1)(i) \neq h_1(i)$ and, therefore, that $(g + h_1)(i) = h_1(i)$ since $h_1 \in D_g$. But then $(h_2 + h_1)(i) = h_1(i)$ since $h_2 \in C^*(g)$. So $(h_2 + h_1)(i) = (h_1 + h_2)(i)$. If $h_1(i) = i$ and $h_2(i) \neq i$, then $(h_2 + h_2)(i) \neq h_2(i)$. Thus since $h_2 \in C^*(g)$, we deduce that $(g + h_2)(i) \neq h_2(i)$. But $h_1 \in D_g$ implies that $(h_1 + h_2)(i) = h_2(i)$. So $(h_1 + h_2)(i) = h_2(i) = (h_2 + h_1)(i)$. If $h_1(i) \neq i$ and $h_2(i) \neq i$, we arrive at a contradiction since $h_1(i) \neq i$ implies that $g(i) = i$ and $h_2(i) = i$. Since all cases have been considered, we have the result. ∎

Lemma 5.3: Let $g = g_1 + g_2 + \cdots + g_r$ be a pairwise disjoint sum in G where g_i is a pairwise disjoint sum of t_i, k_i-cycles ($k_i < k_{i+1}$) for every $i = 1, 2, \ldots, r$. Then $|C(g)| = k_1^{t_1} \cdot \; \cdots \; \cdot k_r^{t_r} t_1! t_2! \cdot \; \cdots \; \cdot t_r! \, (n - (t_1 k_1 + \cdots + t_r k_r))!$.

Proof: A straightforward counting argument gives that the number of elements in G with the same cycle structure as g is

$$\frac{n!}{k_1^{t_1} t_1!} \cdot \frac{1}{k_2^{t_2} t_2!} \cdot \; \cdots \; \cdot \frac{1}{k_r^{t_r} t_r!} \cdot \frac{1}{(n - (t_1 k_1 + \cdots + t_r k_r))!}.$$

Since this number is $\frac{|G|}{|C(g)|}$ and $|G| = n!$, we see that $|C(g)| = k_1^{t_1} \cdot \; \cdots \; \cdot k_r^{t_r} t_1! \cdot \; \cdots \; \cdot t_r!(n - (t_1 k_1 + \cdots + t_r k_r))!$ ∎

Lemma 5.4: Let $g \in S_n$. Then $C(g) = C^*(g) \oplus D_g$.

Proof: Lemma 5.2 (i) gives containment in one direction. Without a loss of generality assume that Move $g = \{1, 2, \ldots, k\}$. Regarding g as an element of S_k, therefore, we have

$$|C^*_{S_n}(g)| = |C_{S_k}(g)| = k_1^{t_1} \cdot \; \cdots \; \cdot k_r^{t_r} t_1! \cdot \; \cdots \; \cdot t_r!$$

where g is as in Lemma 5.3. Since $D_g \cong S_{n-k}$ by Lemma 5.2 (iii), $|D_g| = (n-k)! = (n - (t_1 k_1 + \cdots + t_r k_r))!$ Finally, by Lemma 5.2 (ii) we get

$$|C^*(g) + D_g| = |C^*(g)| \cdot |D_g| = k_1^{t_1} \cdot \cdots \cdot k_r^{t_r} t_1! \cdot \cdots \cdot t_r!(n - (t_1 k_1 + \cdots + t_r k_r))! = |C(g)|.$$

So $C^*(g) + D_g = C(g)$.

Let $x \in C(g)$. Then $x = c + d$ for some $c \in C^*(g)$ and $d \in D_g$. So for $h \in C^*(g)$, $x + h - x = (c + d) + h - (c + d) = c + d + h - d - c$. Since $h \in C^*(g)$ and $d \in D_g$, by Lemma 5.2 (v), we conclude that $d + h - d = h$. So $x + h - x = c + h - c \in C^*(g)$ and $C^*(g)$ is a normal subgroup of $C(g)$. A similar argument gives that D_g is also a subgroup of $C(g)$, hence the result. ∎

Lemma 5.5: For $g = g_1 + g_2 + \cdots + g_r \in S_n$ as in Lemma 5.3, $C(g) = C^*(g_1) \oplus C^*(g_2) \oplus \cdots \oplus C^*(g_r) \oplus D_g$.

Proof: Containment in one direction is clear. The proof of the above lemma gives $|C^*(g_i)| = k_i^{t_i} t_i!$. So by Lemma 5.2 (ii) and (iv) we conclude that

$$|C^*(g_1) + \cdots + C^*(g_r) + D_g| = |C^*(g_1)| \cdot \; \cdots \; \cdot |C^*(g_r)| \cdot |D_g|$$
$$= k_1^{t_1} t_1! \cdot \; \cdots \; \cdot k_r^{t_r} t_r!(n - (t_1 k_1 + \cdots + t_r k_r))! = |C(g)|.$$

The proof that each $C^*(g_i)$ is a normal subgroup of $C(g)$ is similar to the previous lemma. ∎

Lemma 5.6: Let $g \in S_n$ such that g is a k-cycle. Then $C(g) = \langle g \rangle \oplus D_g$ and $|C(g)| = (n - k)! k$.

Proof: The last statement follows from the proof of Lemma 5.4. Certainly $\langle g \rangle \oplus D_g \subseteq C(g)$. Since $D_g \cong S_{n-k}$, $|\langle g \rangle \oplus D_g| = |g| \cdot |D_g| = k \cdot (n-k)!$. Therefore $C(g) = \langle g \rangle \oplus D_g$. ∎

Lemma 5.7: Let $g_1, g_2 \in S_n$ such that g_1 and g_2 are pairwise disjoint. Then

$$(\langle g_1 \rangle + D_{g_1}) \cap (\langle g_2 \rangle + D_{g_2}) = \langle g_1 \rangle + \langle g_2 \rangle + (D_{g_1} \cap D_{g_2}).$$

Proof: Let $a + b \in (\langle g_1 \rangle + D_{g_1}) \cap (\langle g_2 \rangle + D_{g_2})$ such that $a \in \langle g_1 \rangle$ and $b \in D_{g_1}$. Since g_1 and g_2 are pairwise disjoint, then $a \in \langle g_1 \rangle \subseteq D_{g_2}$. Therefore $a + b \in \langle g_2 \rangle + D_{g_2}$ implies that $b \in \langle g_2 \rangle + D_{g_2}$ since a and b commute. So $b \in (\langle g_2 \rangle + D_{g_2}) \cap D_{g_1}$. Let $b = x_1 + x_2 = x_3$ such that $x_1 \in \langle g_2 \rangle$, $x_2 \in D_{g_2}$, and $x_3 \in D_{g_1}$. Then $x_2 = -x_1 + x_3 \in D_{g_1}$ since $x_1 \in \langle g_2 \rangle \subseteq D_{g_1}$.

We conclude that $b = x_1 + x_2 \in \langle g_2 \rangle + (D_{g_1} \cap D_{g_2})$ and $a + b \in \langle g_1 \rangle + \langle g_2 \rangle + (D_{g_1} \cap D_{g_2})$. The reverse inclusion is clear. ∎

Lemma 5.8: Let $g_1, g_2, \ldots, g_r \in S_n$ such that $g_1, g_2, \ldots, g_r$ are pairwise disjoint $(r \geq 2)$. Then

$$(\langle g_1 \rangle + D_{g_1}) \cap \cdots \cap (\langle g_r \rangle + D_{g_r}) = \langle g_1 \rangle + \cdots + \langle g_r \rangle + (D_{g_1} \cap \cdots \cap D_{g_r}).$$

Proof: A proof by induction on r is straightforward and similar to the proof of Lemma 5.7. ∎

Lemma 5.9: For $g = g_1 + \cdots + g_r \in S_n$ as in Lemma 5.3, $Z(C(g)) \subseteq \langle g_1 \rangle + \langle g_2 \rangle + \cdots + \langle g_r \rangle + D_g$.

Proof: First we consider g_1. Since g_1 is a pairwise disjoint sum of k_1-cycles, we can write $g_1 = (a_1 a_2 \ldots a_{k_1}) + (b_1 b_2 \ldots b_{k_1}) + \cdots + (m_1 m_2 \ldots m_{k_1})$. Here we assume that the number of cycles, m, is greater than 1. Let $x = (a_1 b_1 \ldots m_1 a_2 b_2 \ldots m_2 \ldots a_{k_1} b_{k_1} \ldots m_{k_1})$. Since $mx = g$ we conclude that $x \in C(g)$. Furthermore, since $(a_1 a_2 \ldots a_{k_1}) \in C(g)$ we have $Z(C(g)) \subseteq C(x) \cap C((a_1 a_2 \ldots a_{k_1}))$. Lemma 5.6 gives that $C(x) = \langle x \rangle + D_x$, so we wish to find $(\langle x \rangle + D_x) \cap C((a_1 a_2 \ldots a_{k_1}))$.

Let $w \in \langle x \rangle$ and $d \in D_x = D_{g_1}$ so that $w + d \in (\langle x \rangle + D_x) \cap C((a_1 a_2 \ldots a_{k_1}))$. But $d \in D_{g_1}$ implies that $d \in C((a_1 a_2 \ldots a_{k_1}))$. Therefore, $w \in C((a_1 a_2 \ldots a_{k_1}))$. Suppose $w(a_1) \notin \{a_1, a_2, \ldots, a_{k_1}\}$. Then the situation $w \in C((a_1 a_2 \ldots a_k))$ implies $[(a_1 a_2 \ldots a_k) + w](a_1) = [w + (a_1 a_2 \ldots a_k)](a_1)$ and $w(a_1) = w(a_2)$, a contradiction. So $w(a_1) \in \{a_1, \ldots, a_k\}$. Since $w \in \langle x \rangle$, we can write $w = (qm)x = q(mx) = qg_1$ for some integer q, i.e., $w + d \in \langle g_1 \rangle + D_{g_1}$. Note that for $m = 1$ we reach the same conclusion, so we have established that $Z(C(g)) \subseteq \langle g_1 \rangle + D_{g_1}$. Doing the same for $g_2, \ldots, g_r$ we conclude that

$$Z(C(g)) \subseteq (\langle g_1 \rangle + D_{g_1}) \cap \cdots \cap (\langle g_r \rangle + D_{g_r})$$
$$= \langle g_1 \rangle + \langle g_2 \rangle + \cdots + \langle g_r \rangle + (D_{g_1} \cap \cdots \cap D_{g_r})$$

by Lemma 5.8. But $D_{g_1} \cap \cdots \cap D_{g_r} = D_g$ and we have the result. ∎

Theorem 5.10: Let $g = g_1 + g_2 + \cdots + g_r \in S_n$ be as in Lemma 5.3.

(i) If $|\operatorname{Move} g| \neq n-2$, then

$$Z(C(g)) = \langle g_1 \rangle + \langle g_2 \rangle + \cdots + \langle g_r \rangle.$$

(ii) If $|\operatorname{Move} g| = n-2$, then

$$Z(C(g)) = \langle g_1 \rangle + \langle g_2 \rangle + \cdots + \langle g_r \rangle + D_g.$$

Proof: (i) If $|\operatorname{Move} g| > n-2$, then $D_g = \emptyset$ and $Z(C(g)) \subseteq \langle g_1 \rangle + \langle g_2 \rangle + \cdots + \langle g_r \rangle$ by Lemma 5.9. Assume $|\operatorname{Move} g| < n-2$. Let $h \in Z(C(g))$. Using Lemma 5.9 we conclude that $h = a + d$ where $a \in \langle g_1 \rangle + \cdots + \langle g_r \rangle$ and $d \in D_g$. So $h + x = x + h = a + d + x = x + a + d$ for all $x \in C(g)$. In particular for every $x \in D_g$ we have $x + a = a + x$ and thus $a + d + x = a + x + d$. Simplifying gives $d + x = x + d$ for every $x \in D_g$. But since $D_g \cong S_{n-k}$ where $k = |\operatorname{Move} g| < n-2$, D_g is centerless and $d = 0$. Hence $h = a \in \langle g_1 \rangle + \cdots + \langle g_r \rangle$ and $Z(C(g)) \subseteq \langle g_1 \rangle + \cdots + \langle g_r \rangle$.

For the reverse inclusion, let $g \in \langle g_1 \rangle + \cdots + \langle g_r \rangle$ and $z \in C(g)$. Then $z \in C^*(g_1) + \cdots + C^*(g_r) + D_g$ by Lemma 5.5. Clearly $g + z = z + g$ and we have the result.

(ii) By Lemma 5.9 we obtain inclusion one way and the proof above shows that $\langle g_1 \rangle + \cdots + \langle g_r \rangle \subseteq Z(C(g))$. Since $|\operatorname{Move} g| = n-2$, $D_g = \{0, (ab)\}$ for some transposition (ab) in S_n. We need to show that $(ab) \in Z(C(g))$. By Lemma 5.4, $C(g) = C^*(g) \oplus \langle (ab) \rangle$. Let $x \in C(g)$. If $x \in C^*(g) \cup \langle (ab) \rangle$, we clearly have $x + (ab) = (ab) + x$. If $x = y + (ab)$ for some $y \in C^*(g)$, then

$$x + (ab) = y + (ab) + (ab) = y = (ab) + y + (ab) = (ab) + x.$$

Since $x \in C(g)$ is arbitrary, $(ab) \in Z(C(g))$.   ■

Because the orbits of $G = S_n$ under $I = \operatorname{Inn} G$ are simply the various cycle structures of elements of G, to get a function $f \colon G \to G$ in $M_I(G)$ we first choose a representative from each cycle structure. Let $\{a_i\}$ be a set of such representatives. For each a_i we associate an arbitrary $b_i \in Z(C(a_i))$ via Lemma 2.5. Defining $f(-x + a_i + x) = -x + f(a_i) + x = -x + b_i + x$ and $f(0) = 0$ for every a_i and for every $x \in G$ gives $f \in M_I(G)$ by Betsch's Lemma. Furthermore by Betsch's Lemma, every function in $M_I(G)$ is constructible in this way.

From the comments following Lemma 5.1 we conclude that $f \in M_E(G)$ if and only if $f \in M_I(G)$ and f satisfies the conditions of Lemma 2.8 with $N = A_n$. In other words, we must construct f as above with the added restrictions of Lemma 2.8. Doing so yields our main characterization theorem stated without formal proof.

Theorem 5.11: Let $f \colon G \to G$ be a function. Then $f \in M_E(G)$ if and only if f can be constructed in the following manner:

(i) Choose a set of representatives $\{a_i\}$ of conjugacy classes consisting of even permutations under $I = \operatorname{Inn} G$ except for those classes contained in Y, the set of all elements of order two. Choose $b_i \in A_n \cap Z(C(a_i))$ for each i and define f on these classes as explained before the theorem.

(ii) Choose a set of representatives $\{a_i\}$ of conjugacy classes consisting of odd permutations under $I = \text{Inn } G$ except for those classes contained in Y. Choose $b_i \in A_n \cap Z(C(a_i))$ for every i or $b_i \in (G \backslash A_n) \cap Z(C(a_i))$ for every i. Define f as above on these classes with the following restrictions:

(a) If $\{a_i\} \subseteq A_n$, define $f|_Y \equiv 0$.
(b) If $\{a_i\} \in G \backslash A_n$, define $f|_Y \equiv id$.

With our characterization theorem we can now give our structural results for $M_E(G)$.

Lemma 5.12: $M_E(G)$ is not local for $n \geq 3$.

Proof: This lemma follows directly from Lemma 2.9. ∎

Lemma 5.13: $M_E(S_3) \cong Z_6$.

Proof: Let $f \in M_E(S_3)$. Using the notation of Theorem 5.11 we let $a_1 = (123)$. By Theorem 5.10, $Z(C(a_1)) = \{0, a_1, 2a_1 = (132)\}$, and $f(a_1) = b_1 \in \{0, a_1, 2a_1\}$. Furthermore for $y = (12) \in Y$, $f(y) \in \{0, (12)\}$. Since f is completely determined by its action on a_1 and y, there are at most six functions in $M_E(S_3)$. But $\langle id \rangle$ is a subnear-ring of $M_E(S_3)$ and $|\langle id \rangle| = 6$. We conclude that $M_E(S_3) = \langle id \rangle$ and establish the result. ∎

Lemma 5.14: $M_E(S_4) \cong Z_{12}$.

Proof: Let $f \in M_E(S_4)$. Although the maps listed in Lemma 5.1 are not a complete list of endomorphisms of S_4, since f commutes with each of these functions, f will be of the form in Theorem 5.11 with possibly more restrictions. Let $a_1 = (123)$ and $a_2 = (1234)$. By Theorem 5.10 we see that $Z(C(a_1)) = \{0, a_1, 2a_1 = (132)\}$ and $Z(C(a_2)) = \{0, a_2, 2a_2 = (13) + (24), 3a_2 = (1432)\}$. Since the action of f on Y is determined by $f(a_2)$, then f is completely determined on S_4 by $f(a_1)$ and $f(a_2)$. So there are at most twelve functions that commute with the endomorphisms in Lemma 5.1. But as in the previous lemma, $|\langle id \rangle| = 12$ so that $M_E(S_4) \cong Z_{12}$. ∎

Theorem 5.15: $M_E(S_n)$ is not simple for $n \geq 3$, $n \neq 6$.

Proof: The cases $n = 3, 4$, follow from the previous two lemmas. For $n \geq 5$, consider $\text{Ann}(A_n) = \{f \in M_F(S_n) \mid f(A_n) = 0\}$. It is straightforward to show that $\text{Ann}(A_n)$ is an ideal of $M_E(S_n)$ with $\text{Ann}(A_n) \neq M_E(S_n)$.

If a_i is a representative of a conjugacy class of S_n consisting of even permutations and not a subset of Y, we assign $f(a_i) = b_i = 0$ as its function value. If a_i is a representative of a conjugacy class of S_n consisting of odd permutations and not a subset of Y, then clearly $2a_i \in Z(C(a_i))$. Thus we can assign $f(a_i) = b_i = 2a_i$ as its function value. Since in this latter case the orbit representatives are all in A_n, we define $f|_Y = 0$. By Theorem 5.11 we have constructed a function $f \in M_E(S_n)$, namely $f(g) = \begin{cases} 0 & \text{if } g \in A_n \cup Y \\ 2g & \text{else} \end{cases}$. The function f is clearly in $\text{Ann}(A_n)$ and is nontrivial since $f((1234)) \neq 0$. Thus $\text{Ann}(A_n)$ is a nontrivial ideal and $M_E(S_n)$ is not simple. ∎

Theorem 5.16: For $n \geq 5$, $n \neq 6$, $M_E(S_n)$ is not a ring.

Proof: Define $f_1, f_2, f_3 \colon S_n \to S_n$ by

$$f_1(g) = \begin{cases} 0 & \text{if } g \in A_n \backslash Y \\ g & \text{else} \end{cases},$$

$$f_2(g) = \begin{cases} 0 & \text{if } g \in A_n \cup Y \\ 4g & \text{else} \end{cases}$$

and

$$f_3(g) = \begin{cases} g & \text{if } g \in Y \\ 0 & \text{if } g \in A_n \backslash Y \\ 3g & \text{else.} \end{cases}$$

As in the previous theorem each function is in $M_E(S_n)$ by Theorem 5.11. Furthermore

$$f_1[f_2((12) + (345)) + f_3((12) + (345))] = f_1[(345) + (12)] = (12) + (345)$$

while

$$f_1 f_2((12) + (345)) + f_1 f_3((12) + (345)) = f_1((345)) + f_1((12)) = 0 + (12) = (12).$$

So $f_1(f_2 + f_3) \neq f_1 f_2 + f_1 f_3$ and $M_E(S_n)$ is not a ring.   ∎

6. G a generalized quaternion group

As in the previous section we characterize the functions in $M_E(G)$, where G is a generalized quaternion group, by determining how to construct them. Before proceeding to the general case we first characterize $M_E(Q)$ where Q is the quaternion group of order eight.

Lemma 6.1: $M_E(Q) \cong Z_4$.

Proof: Pilz [9] lists all of the endomorphisms of Q. If we denote the generators of Q by a and b, then $Q = \{0, a, 2a, 3a, b, a+b, 2a+b, 3a+b\}$. Furthermore we see from the list that for each $q \in Q$, there exists $\psi_q \in \operatorname{End} Q$ such that $\psi_q(a) = q$. Thus for every $f \in M_E(Q)$, $f\psi_q(a) = \psi_q f(a)$ and $f(q) = \psi_q f(a)$. So f is completely determined by $f(a)$. Since $|\langle a \rangle| = 4$, we have $[Q : \langle a \rangle] = 2$ and $f(a) \in \langle a \rangle$ by Lemma 2.8. So $|M_E(Q)| \leq 4$. But $|\langle id \rangle| = 4$ gives that $M_E(Q) = \langle id \rangle$, hence the result.   ∎

We now generalize to Q_n, $n \geq 4$, the generalized quaternion group. Q_n is presented as $(a, b \mid 2^{n-1}a, b+a-b+a, 2^{n-2}a+2b)$. Elements of Q_n will be written in the form $xa + sb$ with $0 \leq x < 2^{n-1}$ and $0 \leq s \leq 1$.

For convenience let A, B, and C denote the normal subgroups of Q_n generated by a, b, and $a+b$, respectively. The next lemma provides results about the structure of Q_n. Because the details are either straightforward or can be found in [1] or [5], no proofs are included.

Lemma 6.2:

(i) $|Q_n| = 2^n$;

(ii) $A \cong B \cong C \cong Q_{n-1}$ and $|A| = |B| = |C| = 2^{n-1}$;

(iii) $B = \{xa + b \mid 0 \le x < 2^{n-1}$ and x is even$\} \cup \{xa \mid 0 \le x < 2^{n-1}$ and x is even$\}$;

(iv) $C = \{xa + b \mid 0 < x < 2^{n-1}$ and x is odd$\} \cup \{xa \mid 0 \le x < 2^{n-1}$ and x is even$\}$;

(v) $Q_n = A \cup B \cup C$;

(vi) For every integer x, $b + xa = -xa + b$;

(vii) $2^{n-2}a$ is the unique element of order two in Q_n;

(viii) $Z(Q_n) = \{0, 2^{n-2}a\}$;

(ix) Q_n has 2^{n-1} inner automorphisms, 2^{2n-3} automorphisms, and a total of $2^{2n-3} + 4$ endomorphisms.

Let φ_q denote the inner automorphism determined by q, i.e, $\varphi_q(g) = -q + g + q$ for every $g \in Q_n$. Also for every $0 < i < 2^{n-1}$ such that i is odd, let $\alpha_i \colon Q_n \to Q_n$ be defined by $\alpha_i(xa + sb) = (ix)a + sb$. Finally for $R \in \{A, B, C\}$ define the map $\emptyset_R \colon Q_n \to Q_n$ by $\emptyset_R(g) = \begin{cases} 0 & \text{if } g \in R \\ 2^{n-2}a & \text{if } g \notin R \end{cases}$. The proof that α_i is an automorphism for every odd i and $\emptyset_R$ is an endomorphism for $R \in \{A, B, C\}$ is direct. It is routine to show that if $\alpha_i \varphi_{q_1} = \alpha_j \varphi_{q_2}$, then $\alpha_i = \alpha_j$ and $\varphi_{q_1} = \varphi_{q_2}$. Since there are 2^{n-1} inner automorphisms and 2^{n-2} choices for $0 < i < 2^{n-1}$ with i odd, then there are $(2^{n-1})(2^{n-2}) = 2^{2n-3}$ different automorphisms of the form $\alpha_i \varphi_q$. All automorphisms, therefore, must be of this form by Lemma 6.2 (ix). Thus we have all of End G.

Lemma 6.3: The endomorphisms of Q_n are:

(i) 0, the zero map;

(ii) $\alpha_i \varphi_q$ for every $q \in Q_n$ and for every odd $0 < i < 2^{n-1}$;

(iii) $\emptyset_A, \emptyset_B$, and $\emptyset_C$.

Lemma 6.4: Let $D = \{0, a, b, a + b, 2a, 4a, 8a, \ldots, 2^{n-3}a\}$ and let $f \colon Q_n \to Q_n$ be a function. Then $f \in M_E(Q_n)$ if and only if $\gamma f(d) = f\gamma(d)$ for every $d \in D$ and for every $\gamma \in$ End Q_n.

Proof: The direct implication is obvious. Let $q \in Q_n$. It can be shown that $q = \beta(d)$ for some $\beta \in$ End Q_n and some $d \in D$. Thus for $\gamma \in$ End Q_n,

$$\gamma f(q) = \gamma f(\beta(d)) = \gamma(f\beta(d)) = \gamma(\beta f(d))$$
$$= (\gamma\beta)f(d) = f(\gamma\beta)(d) = f\gamma(\beta(d)) = f\gamma(q).$$

Since q and γ are arbitrary, $f \in M_E(Q_n)$. ∎

In view of Lemma 6.4 we now focus on determining restrictions on the images of elements in D under functions in $M_E(Q_n)$. This is done in a sequence of lemmas.

Lemma 6.5: Let $f \in M_E(Q_n)$. Then $f(b) \in \{0, b, 2^{n-2}a, 2^{n-2}a + b\}$.

Proof: Since $f \in M_E(Q_n)$, we have $\alpha_i f(b) = f\alpha_i(b)$ for every odd i. So $\alpha_i f(b) = f(\alpha_i(b)) = f(b)$ and $f(b)$ is fixed by α_i for every odd i. Let $f(b) = xa + yb$ and $i = 3$. Then $\alpha_3(xa + yb) = xa + yb$ and $(3x)a + yb = xa + yb$. So $(2x)a = 0$ and $2(xa) = 0$.

Since $2^{n-2}a$ is the unique element of order two in Q_n, either $x = 0$ or $x = 2^{n-2}$. Since $y \in \{0, 1\}$ we have the result.   ∎

Lemma 6.6: Let $f \in M_E(Q_n)$. Then

$$f(a + b) \in \{0, a + b, 2^{n-2}a, (2^{n-2} + 1)a + b\}.$$

Proof:

(i) Suppose $f(a + b) = xa + b$ for some $0 \le x < 2^{n-1}$. Since $f \in M_E(Q_n)$, then $f\alpha_3(a + b) = \alpha_3 f(a + b)$ and $f(3a + b) = \alpha_3(xa + b) = (3x)a + b$. Also, $f\varphi_{-a}(a + b) = \varphi_{-a}f(a + b)$ implies that $f(a + a + b - a) = a + (xa + b) - a$ and $f(3a + b) = (x + 2)a + b$. Hence $f(3a + b) = (3x)a + b = (x + 2)a + b$. Simplifying yields $(3x)a = (x + 2)a$ and $(2x - 2)a = 2(x - 1)a = 0$. Therefore $(x - 1)a \in \{0, 2^{n-2}a\}$. But this implies that $xa \in \{a, (2^{n-2} + 1)a\}$ and $f(a + b) \in \{a + b, (2^{n-2} + 1)a + b\}$.

(ii) Suppose $f(a+b) = xa$ for some $0 \le x < 2^{n-1}$. Then $f\alpha_3(a + b) = \alpha_3 f(a + b)$ and, as above, $f(3a + b) = (3x)a$. Also, $f\varphi_{-a}(a + b) = \varphi_{-a}f(a + b)$ gives that $f(3a + b) = xa$. So $xa = (3x)a$ and $(2x)a = 2(xa) = 0$. Thus $xa = f(a + b) \in \{0, 2^{n-2}a\}$.

In either case $f(a + b) \in \{0, a + b, 2^{n-2}a, (2^{n-2} + 1)a + b\}$.   ∎

Since $Z(Q_n) = \{0, 2^{n-2}a\}$, by Lemma 2.4, if $f \in M_E(Q_n)$, then $f(2^{n-2}a) \in \{0, 2^{n-2}a\}$. We consider each case separately.

Lemma 6.7: Let $f \in M_E(Q_n)$ with $f(2^{n-2}a) = 0$. Then $f(Q_n) \subseteq \{xa \mid x \text{ is even}\}$.

Proof: Let $q \in Q_n$.

(i) Assume $q \in \{xa \mid x \text{ is even}\}$. From Lemma 2.8 we have $f(A) \subseteq A$ and $f(B) \subseteq B$. Since $q \in A \cap B$ by Lemma 6.2 (iii), then $f(q) \in A \cap B = \{xa \mid x \text{ is even}\}$.

(ii) Assume $q \notin \{xa \mid x \text{ is even}\}$. Then by Lemma 6.2 (iii) and (iv) we see that $q \notin R_1 \cup R_2$ for R_1 and R_2 distinct elements of $\{A, B, C\}$. But by Lemma 2.8, $f(Q_n \backslash R_1) \subseteq R_1$ and $f(Q_n \backslash R_2) \subseteq R_2$. Since $q \in (Q_n \backslash R_1) \cup (Q_n \backslash R_2)$, $f(q) \in R_1 \cap R_2 = \{xa \mid x \text{ is even}\}$.

These two cases complete the proof.   ∎

Corollary 6.8: Let $f \in M_E(Q_n)$ with $f(2^{n-2}a) = 0$. Then $f(b) \in \{0, 2^{n-2}a\}$ and $f(a + b) \in \{0, 2^{n-2}a\}$.

Proof: Immediate from the previous three lemmas.   ∎

Lemma 6.9: Let $f \in M_E(Q_n)$ with $f(2^{n-2}a) = 2^{n-2}a$. Then $f(b) \in \{b, 2^{n-2}a+b\}$, $f(a+b) \in \{a+b, (2^{n-2}+1)a+b\}$, and $f(a) \in \{xa \mid 0 < x < 2^{n-1}$ and x is odd$\}$.

Proof: Since $b \in B \cap (Q_n \backslash A)$, then $f(b) \in B \cap (Q_n \backslash A) = \{xa + b \mid 0 \leq x < 2^{n-1}$ and x is even$\}$ by Lemma 2.8. But by Lemma 6.5, $f(b) \in \{b, 2^{n-2}a + b\}$.

Since $a+b \in C \cap (Q_n \backslash A)$, then $f(a+b) \in C \cap (Q_n \backslash A) = \{xa+b \mid 0 < x < 2^{n-1}$ and x is odd$\}$ by Lemma 2.8. By Lemma 6.6 we conclude that $f(a + b) \in \{a + b, (2^{n-2} + 1)a + b\}$.

Similarly, $a \in A \cap (Q_n \backslash B)$ and, thus, $f(a) \in A \cap (Q_n \backslash B)$. But this intersection is $\{xa \mid 0 < x < 2^{n-1}$ and x is odd$\}$.   ■

Lemma 6.10: Let $f \in M_E(Q_n)$ and $g \in \{xa \mid 0 \leq x < 2^{n-1}$ and x is even$\}$. Then $f(g) \in \{xa \mid 0 \leq x < 2^{n-1}$ and x is even$\}$.

Proof: Because $g \in A \cap B$, it follows that $f(g) \in A \cap B = \{xa \mid 0 \leq x < 2^{n-1}$ and x is even$\}$ by Lemma 2.8.   ■

Now that we have restrictions on the images of elements in D under functions in $M_E(Q_n)$, we wish to extend these restrictions to elements not in D. To simplify the statements of the results we say a function f *commutes with E at* $q \in Q_n$ if $\gamma f(q) = f\gamma(q)$ for every $\gamma \in \text{End } Q_n$. Because many of the proofs are computational in nature, we will omit the proofs for the next sequence of lemmas. The interested reader can find the details in [1].

Lemma 6.11: Let $f: Q_n \to Q_n$ be a function with $f(a) = xa$ where $0 \leq x < 2^{n-1}$. If x is even, then f commutes with E at a iff $f(0) = 0$ and $f(ia) = (ix)a$ for every odd i. If x is odd, we also include the restriction that $f(2^{n-2}a) = 2^{n-2}a$.

Lemma 6.12: Let $f: Q_n \to Q_n$ be a function. If $f(b) = 0$, then f commutes with E at b iff $f(2^{n-2}a) = f(0) = 0$ and $f(ia + b) = 0$ for every even i. If $f(b) = b$, we substitute $f(ia + b) = ia + b$ for every even i in the above statement. If $f(b) = 2^{n-2}a$, we substitute $f(ia + b) = 2^{n-2}a$ for every even i.

Lemma 6.13: Let $f: Q_n \to Q_n$ be a function with $f(b) = 2^{n-2}a + b$. Then f commutes with E at b iff $f(0) = 0$, $f(2^{n-2}a) = 2^{n-2}a$, and $f(ia + b) = (2^{n-2} + i)a + b$ for every even i.

Lemma 6.14: Let $f: Q_n \to Q_n$ be a function. If $f(a + b) = 0$, then f commutes with E at $a + b$ iff $f(2^{n-2}a) = f(0) = 0$ and $f(ia + b) = 0$ for every odd i. If $f(a + b) = 2^{n-2}a$, we substitute $f(ia + b) = 2^{n-2}a$ for every odd i in the above statement.

Lemma 6.15: Let $f: Q_n \to Q_n$ be a function. If $f(a+b) = a+b$, then f commutes with E at $a + b$ iff $f(0) = 0$, $f(2^{n-2}a) = 2^{n-2}a$, and $f(ia + b) = ia + b$ for every odd i. If $f(a + b) = (2^{n-2} + 1)a + b$, we substitute $f(ia + b) = (2^{n-2} + i)a + b$ for every odd i in the above statement.

Lemma 6.16: Let $j \in \{2, 2^2, 2^3, \ldots, 2^{n-3}\}$ and let $f: Q_n \to Q_n$ be a function with $f(ja) = xa$ where $0 \leq x < 2^{n-1}$ and x is even. Then f commutes with E at ja iff $f(0) = 0$ and $f((ij)a) = (ix)a$ for every odd i.

With the preceding six lemmas, we can completely determine all functions in $M_E(Q_n)$. To facilitate notation we write $f \in M_E(Q_n)$ as a 10-tuple $f = (g_1, g_2, \ldots, g_{10})$ where $g_1, g_2, \ldots, g_{10}$ is the image under f of 0, a, ia where i is odd, $2^{n-2}a, ja$ where $j \in \{2, 2^2, \ldots, 2^{n-3}\}$, $(ij)a$ where i is odd and $j \in \{2, 2^2, \ldots, 2^{n-3}\}$, b, $a + b$, $ia + b$ where i is odd, and $ia + b$ where i is even, respectively.

Theorem 6.17: Let $f: Q_n \to Q_n$ be a function. Then $f \in M_E(Q_n)$ iff f is of one of the following eight forms:

(i) $(0, xa, (ix)a, 0, y_ja, (iy_j)a, 0, 0, 0, 0)$ where x is even, $0 \leq y_j < 2^{n-1}$, and y_j is even for every j;

(ii) $(0, xa, (ix)a, 0, y_ja, (iy_j)a, 2^{n-2}a, 0, 0, 2^{n-2}a)$ with the same conditions as in (i);

(iii) $(0, xa, (ix)a, 0, y_ja, (iy_j)a, 0, 2^{n-2}a, 2^{n-2}a, 0)$ with the same conditions as in (i);

(iv) $(0, xa, (ix)a, 0, y_ja, (iy_j)a, 2^{n-2}a, 2^{n-2}a, 2^{n-2}a, 2^{n-2}a)$ with the same conditions as in (i);

(v) $(0, xa, (ix)a, 2^{n-2}a, y_ja, (iy_j)a, b, a + b, ia + b, ia + b)$ where x is odd, $0 \leq y_j < 2^{n-1}$, and y_j is even for every j;

(vi) $(0, xa, (ix)a, 2^{n-2}a, y_ja, (iy_j)a, 2^{n-2}a + b, a + b, ia + b, (2^{n-2} + i)a + b)$ with the same conditions as in (v);

(vii) $(0, xa, (ix)a, 2^{n-2}a, y_ja, (iy_j)a, b, (2^{n-2} + 1)a + b, (2^{n-2} + i)a + b, ia + b)$ with the same conditions as in (v);

(viii) $(0, xa, (ix)a, 2^{n-2}a, y_ja, (iy_j)a, 2^{n-2}a+b, (2^{n-2}+1)a+b, (2^{n-2}+i)a+b, (2^{n-2} + i)a + b)$ with the same conditions as in (v).

Corollary 6.18: $|M_E(Q_n)| = 2^{n^2 - 4n + 7}$.

Proof: In each of the eight cases in Theorem 6.17, different functions are obtained by varying $f(a)$ and $f(ja)$ for $j \in \{2, 2^2, \ldots, 2^{n-3}\}$. Since $|\{xa \mid 0 \leq x < 2^{n-1}$ and x is even$\}| = |\{xa \mid 0 \leq x < 2^{n-1}$ and x is odd$\}| = 2^{n-2}$, there are 2^{n-2} possible images for a and each ja. Thus there are $(2^{n-2})^{n-2}$ possible functions in each case. Because we have eight cases, we have $8(2^{n-2})^{n-2} = 2^{n^2 - 4n + 7}$ functions. ∎

Using our characterization theorem, we can now answer our three structure questions.

Theorem 6.19: Let G be a quaternion group. Then $M_E(G)$ is not simple.

Proof: If $G = Q$, then $M_E(G) \cong Z_4$ by Lemma 6.1 and is not simple. Assume $G = Q_n$ for some n.

It is straightforward to show that $\text{Ann}(Z(G))$ is a proper ideal of $M_E(G)$. Furthermore, the function $f = (0, a, ia, 0, 0, 0, 0, 0, 0, 0)$ is in $M_E(G)$ by Theorem 6.17 (i). Since $f \in \text{Ann}(Z(G))$, we have that $M_E(G)$ is not simple. ∎

Theorem 6.20: Let G be a quaternion group. The following are equivalent:

(i) $G = Q$;

(ii) $M_E(G)$ is local;

(iii) $M_E(G)$ is a ring.

Proof: If $G = Q$, $M_E(G)$ is a local ring by Lemma 6.1. To complete the proof we need to show that $M_E(Q_n)$ is neither local nor a ring. The former is easy since the function $f = (0, a, ia, 0, 0, 0, 0, 0, 0, 0)$ used in the proof of the previous theorem is an idempotent in $M_E(Q_n)$. Hence $M_E(Q_n)$ is not local.

Define $f_1 \colon Q_n \to Q_n$ by

$$f_1 = (0, a, ia, 2^{n-2}a, y_j a, (iy_j)a, b, a + b, ia + b, ia + b)$$

where $y_4 = 2$ and $y_j = j$ if $j \neq 4$. Then $f_1 \in M_E(Q_n)$ by Theorem 6.17 (v). Let $f_2 = f_3 = id$. Then

$$f_1(f_2 + f_3)(2a) = f_1(2a + 2a) = f_1(4a) = 2a$$

and

$$f_1 f_2(2a) + f_1 f_3(2a) = f_1(2a) + f_1(2a) = 2a + 2a = 4a.$$

So $f_1(f_2 + f_3) \neq f_1 f_2 + f_1 f_3$ and $M_E(Q_n)$ is not a ring. ∎

As an application of the results of this section we consider the class of groups known as the *Hamiltonian groups*, i.e., the nonabelian groups in which every subgroup is normal. A characterization theorem for these groups is given in [10].

Theorem 6.21: A finite group G is Hamiltonian iff $G = Q + A + B$ where Q is the quaternion group of order eight, A is an elementary abelian 2-group, and B is a finite abelian group with all elements of odd order.

Theorem 6.22: Let G be a finite Hamiltonian group. Then $M_E(G)$ is local iff $G = Z_2 \times Z_2 \times \cdots \times Z_2 \times Q$.

Proof: If $M_E(G)$ is local, then G is a p-group for some prime p by Lemma 2.9. Since G is nonabelian, using Theorem 6.21 we conclude that G is a 2-group, i.e., $G = Z_2 \times Z_2 \times \cdots \times Z_2 \times Q$.

If $G = Z_2 \times Z_2 \times \cdots \times Z_2 \times Q$, then $M_E(G) \hookrightarrow M_E(Q)$ by Lemma 2.2. Since $M_E(Q)$ is local by Lemma 6.1, we conclude that $M_E(G)$ is local. ∎

Acknowledgments

The author wishes to express his gratitude to Professor G. Pilz and Dr. P. Fuchs of Johannes Kepler Universität Linz for their hospitality and encouragement during the fall of 1992 when portions of this paper were completed.

References

1. G. A. Cannon, *Centralizer Near-Rings Determined by End G*, Doctoral Dissertation, Texas A&M University, to appear.

2. J. R. Clay, *Nearrings: Geneses and Applications*, Oxford Science Publ., Oxford, 1992.

3. Y. Fong and J. D. P. Meldrum, *Endomorphism Near-Rings of a Direct Sum of Isomorphic Finite Simple Nonabelian Groups*, *Near-Rings and Near-Fields* ed. G. Betsch, North-Holland, Amsterdam, 1987, 73-78.

4. D. Gorenstein, *Finite Groups*, Harper & Row, New York, 1968.

5. J. J. Malone, *Generalized Quaternion Groups and Distributively Generated Near-Rings*, Proc. Edinburgh Math. Soc., **18** (1973), 235-238.

6. C. J. Maxson, M. R. Pettet and K. C. Smith, *On Semisimple Rings that are Centralizer Near-Rings*, Pacific J. Math., **101** (1981), 451-461.

7. C. J. Maxson and K. C. Smith, *The Centralizer of a Set of Group Automorphisms*, Comm. Alg., **8** (1980), 211-230.

8. C. J. Maxson and K. C. Smith, *Centralizer Near-Rings Determined by Local Rings*, Houston J. Math., **11** (1985), 355-366.

9. G. Pilz, *Near-Rings*, North Holland/American Elsevier, Amsterdam, second, revised edition, 1983.

10. W. R. Scott, *Group Theory*, Dover, New York, 1987.

11. M. Suzuki, *Group Theory I*, Springer-Verlag, 1982.

ON CODES FROM RESIDUE CLASS RING
GENERATED FINITE FERRERO PAIRS

Roland A. Eggetsberger [‡]

Inst. f. Mathematik, Joh. Kepler Univ. Linz, Altenbergerstr. 69

A-4040 Linz, Austria

ABSTRACT

In the sequel we consider error-correcting codes constructed from the incidence matrix of a BIB-design, which is generated by a finite Ferrero pair based on a residue class ring. Codes from BIB-designs are already known for often having good properties concerning error-correction. Unfortunately many of the considered codes seem to have a lower quality at the first glance.

But via the simple trick of omitting a few of the codewords it is possible to highly improve the quality of these codes. This method also simplifies the calculation of the error-correcting properties and does not affect very much the size of the codes. Also other calculations, like encoding and decoding, do not become harder.

After an overview of the general method of constructing codes from finite Ferrero pairs we characterize the case of residue class ring generated ones and then we draw our attention on what the resulting codes look like.

1 Introduction

In the following we specify the basic relationships between finite Ferrero pairs, BIB-designs and coding theory (For more information see [2] and [8], [1] or [7], respectively).

At first we consider pairs of groups (N, Φ), where $(N, +)$ is finite and $(\Phi, \circ)$ is a non-trivial group of fixed point free automorphisms of $(N, +)$ (i.e. $(\forall \varphi \in \Phi \setminus \{id_N\})(\forall x \in N) \, [\varphi(x) = x \Rightarrow x = 0]$, where id_N is the identity map on N). Such a pair (N, Φ) is called a *finite Ferrero pair*. It is well known that we could then generate finite planar nearrings from a finite Ferrero pair and continue with the construction of BIB-designs and its codes. But because of the fact that different planar nearrings from one Ferrero pair yield the same BIB-designs [2;(4.24),(4.25)] we omit this step.

Furthermore we introduce BIB-designs (cf. [1]), which form the connection between finite Ferrero pairs and the resulting codes.

Let V be a finite non-empty set and $\mathcal{B} \subseteq 2^V$. Then the pair $\mathcal{D} = (V, \mathcal{B})$ is a so-called *incidence structure*. Additionally, $\mathcal{D}$ is called *BIB-design* (balanced

[‡]Supported by the Austrian Fonds zur Förderung der wissenschaftlichen Forschung, Projekt P9111-PHY

Y. Fong et al. (eds.), Near-Rings and Near-Fields, 113–122.
© 1995 *Kluwer Academic Publishers.*

incomplete block design) or $(v, k, \lambda)-$design $(v, k, \lambda \in \mathbf{N})$, if:

$$|V| = v \tag{1}$$
$$|B| = k \text{ for all } B \in \mathcal{B} \tag{2}$$
$$|\{B \in \mathcal{B} \mid p, q \in B\}| = \lambda \text{ for all } p, q \in V, p \neq q \tag{3}$$

and if $\mathcal{B} \neq \{B \subseteq V \mid |B| = k\}$ (The last condition only avoids that the block design is complete). The elements of V are called *points*, the ones of $\mathcal{B}$ *blocks* and $p \in B$ (for $p \in V, B \in \mathcal{B}$) is short for p *lies on* B. Furthermore the following parameters and properties are determined

$$|\{B \in \mathcal{B} \mid p \in B\}| = r \text{ for all } p \in V \tag{4}$$
$$r = \lambda \cdot (v - 1)/(k - 1) \tag{5}$$
$$b = v \cdot r/k \tag{6}$$

where $b = |\mathcal{B}|$. If an incidence structure only fulfils Eq. (1), (2), (4) and (6), then it is called *tactical configuration*. Now let $V = \{p_1, \ldots, p_v\}$ and $\mathcal{B} = \{B_1, \ldots, B_b\}$. Then the $v \times b-$matrix

$$A_{\mathcal{D}} = (a_{ij}) \text{ with } a_{ij} = \begin{cases} 1 & \text{if } p_i \in B_j \\ 0 & \text{otherwise} \end{cases} \tag{7}$$

is called *incidence matrix* of $\mathcal{D}$.

We also present the concept of an error-correcting code (cf. [7]). A *binary* $(n, M, d)-code$ $\mathcal{C}$ is a set of M vectors in $(\mathbf{Z}_2)^n$, such that d is minimal with respect to the property that two different vectors differ at least d entries. $d \ (= d_{min}(\mathcal{C}))$ is called *minimal distance* of the code $\mathcal{C}$. For $x, y \in (\mathbf{Z}_2)^n$ we define:

$$\bar{d}(x, y) := |\{i \mid x_i \neq y_i\}| \text{ the } \textit{Hamming-distance} \text{ between } x \text{ and } y \tag{8}$$
$$w(x) := |\{i \mid x_i \neq 0\}| \text{ the } \textit{(Hamming-) weight} \text{ of } x \tag{9}$$

The minimal distance d is a measure for the quality, because a code can detect up to $d - 1$ and correct up to $\lfloor (d - 1)/2 \rfloor$ errors ($\lfloor x \rfloor$ denotes the greatest integer less or equal to x).

2   Finite Ferrero Pairs from Residue Class Rings

Also finite Ferrero pairs are generated in certain ways. One of these is the following:

Definition 2.1 A finite Ferrero pair (N, Φ) is called *ring generated*, if there exist $(R, +, \cdot)$, a finite ring with identity 1, and $(\bar{\Phi}, \cdot)$, a non-trivial subgroup of $(\mathcal{U}(R), \cdot)$, the group of units of R, such that $(N, +) = (R, +)$ and

$$\Phi = \{\varphi_a \mid \varphi_a : R \to R, x \to x \cdot a, a \in \bar{\Phi}\} \tag{10}$$

Proposition 2.2 [2] *Let $(R, +, \cdot)$ be a finite ring with identity 1 and $(\bar{\Phi}, \cdot)$ be a non-trivial subgroup of $(\mathcal{U}(R), \cdot)$. Besides let $(N, +) = (R, +)$ and Φ be like in Eq. (10).*
Then (N, Φ) is a finite Ferrero pair, if we have $1 - a \in \mathcal{U}(R)$ for every $a \in \bar{\Phi} \setminus \{1\}$.

In the following, we use the meaning of Φ and $\bar{\Phi}$ as before.

Proposition 2.3 *Consider a residue class ring $(\mathbf{Z}_{p^m}, +, \cdot)$, where $p \in \mathbf{P} \setminus \{2\}$, $m \in \mathbf{N} \setminus \{1\}$ and let $(N, +) = (\mathbf{Z}_{p^m}, +)$. Furthermore let $(\bar{\Phi}, \cdot) < (\mathcal{U}(\mathbf{Z}_{p^m}), \cdot)$ with $(\bar{\Phi}, \cdot) \hookrightarrow (\mathbf{Z}_{p-1}, +)$ and $|\bar{\Phi}| > 1$.*
Then (N, Φ) is a finite Ferrero pair.

Proof. The case $(\bar{\Phi}, \cdot) \cong (\mathbf{Z}_{p-1}, +)$ is treated in [2]. Both conditions for $\bar{\Phi}$, namely inducing a non-trivial group of fixed point free automorphisms and having $1 - a \in \mathcal{U}(\mathbf{Z}_{p^m})$ for every element but 1, are also true for any non-trivial subgroup. □

Remark 2.4 A method of computing a generating element of $\bar{\Phi}$ can be found in [5].

Now we can generalize the result from Proposition 2.3 via the following construction.

Theorem 2.5 [2;(5.42)] *Let $(N_1, \Phi_1), (N_2, \Phi_2), \ldots, (N_t, \Phi_t)$ be finite Ferrero pairs, where the Φ_i are all isomorphic. Let $f_i : \Phi_i \to \Phi_{i+1}$ (for $1 \le i < t$) be isomorphisms. Then let $N = N_1 \oplus N_2 \oplus \ldots \oplus N_t$ and for $\varphi_1 \in \Phi_1$, define $\varphi : N \to N$ by*

$$\varphi(n_1, \ldots, n_t) = (\varphi_1(n_1), f_1(\varphi_1)(n_2), \ldots, f_{t-1}(\varphi_{t-1})(n_t)) \tag{11}$$

where $\varphi_i = f_{i-1}(\varphi_{i-1})$ for $1 < i \le t$ and $\Phi = \{\varphi \mid \varphi_1 \in \Phi_1\}$.
Then (N, Φ) is a finite Ferrero pair.

Theorem 2.6 *Consider a residue class ring $(\mathbf{Z}_s, +, \cdot)$, where $s = p_1^{m_1} \cdot \ldots \cdot p_t^{m_t}$ $(p_i \in \mathbf{P} \setminus \{2\}, m_i \in \mathbf{N})$ and let $r = \gcd(p_1 - 1, \ldots, p_t - 1)$. Furthermore let $(\bar{\Phi}, \cdot) < (\mathcal{U}(\mathbf{Z}_s), \cdot)$ and Φ be like in Eq. (10) (with $R = \mathbf{Z}_s$).*
Then $(\mathbf{Z}_s, \Phi)$ is a finite Ferrero pair iff $(\bar{\Phi}, \cdot) \hookrightarrow (\mathbf{Z}_r, +)$ and $|\bar{\Phi}| > 1$.

Proof.

⇒: Let $(\mathbf{Z}_s, \Phi)$ be a finite Ferrero pair. Let $i \in \{1, \ldots, t\}$ and π_i be the projection map $\pi_i : \mathbf{Z}_s \to \mathbf{Z}_{p_i^{m_i}}, z \to z \bmod p_i^{m_i}$. Since π_i is an epimorphism, we have $\pi_i(\mathcal{U}(\mathbf{Z}_s)) \subseteq \mathcal{U}(\mathbf{Z}_{p_i^{m_i}})$ and furthermore $\pi_i(\bar{\Phi}) < \mathcal{U}(\mathbf{Z}_{p_i^{m_i}})$. In addition we have to fulfil $1 - a \in \mathcal{U}(\mathbf{Z}_s)$ for $a \in \bar{\Phi} \setminus \{1\}$, so $\pi_i|_{\bar{\Phi}}$ has to be injective, because for any $a(\ne 1) \in \bar{\Phi}$ with $\pi_i(a) = 1$ we would get $\pi_i(1 - a) = \pi_i(1) - \pi_i(a) = 1 - 1 = 0$, which is not invertible. But also $\bar{\Phi}_i := \mathrm{Im}(\pi_i|_{\bar{\Phi}})$ fulfils $1 - b \in \mathcal{U}(\mathbf{Z}_{p_i^{m_i}})$ for $b \in \bar{\Phi}_i \setminus \{1\}$, so $|\bar{\Phi}_i| \mid (p_i - 1)$ and $\bar{\Phi} \cong \bar{\Phi}_i$. Therefore $|\bar{\Phi}| \mid \gcd(p_1 - 1, \ldots, p_t - 1)$, since i was arbitrary. Because $\bar{\Phi}_i$ is cyclic, so is $\bar{\Phi}$.

$\Leftarrow$: Use Theorem 2.5 with $N_i = \mathbf{Z}_{p_i^{m_i}}$ and Φ_i as in the first part of the proof.

$\square$

Remark 2.7 The result of Theorem 2.6 has been found independently in the context with other Ferrero pairs in [6].

3   The Resulting Codes

Now we are able to give a definition of the codes on which we focus our attention.

Definition and Theorem 3.1 [2, 4, 9] Let (N, Φ) be a finite Ferrero pair and

$$\mathcal{B}^* := \{\Phi(a) + b \mid a, b \in N, a \neq 0\} \tag{12}$$

Then $\mathcal{D}^* = (N, \mathcal{B}^*)$ is a BIB-design.

If we take the columns of the incidence matrix of this design, we receive $\mathcal{C}^{\mathcal{D}^*}$, the *column code* of $\mathcal{D}^*$. Then $\mathcal{C}^{\mathcal{D}^*}$ is an (n, M, d)−code with

$$n = |N| \tag{13}$$
$$M = n(n-1)/w \tag{14}$$
$$d = 2(w - \mu) \tag{15}$$

where $w := |\Phi|$ is the weight of any codeword and μ is determined as follows:

$$\mu := \max\{ |B_1 \cap B_2| \mid B_1, B_2 \in \mathcal{B}^*, B_1 \neq B_2\} \tag{16}$$

Now we focus on computing the parameter μ of the design in order to get information about the minimal distance of the codes. But at first a few words on a special case, where the resulting codes become pretty bad.

Proposition 3.2 *Let (N, Φ) be a finite Ferrero pair with $|\Phi| = w$. Let S be a Φ−invariant subgroup of N with $|S| = w + 1$.*
Then $\mu = w - 1$.

Proof. (cf. [3]) Let $S^* = S \setminus \{0\}$ and $x \in S^*$. Then both $B_x = \Phi(x) = S^*$ and $B_x - x = S \setminus \{x\}$ are blocks of the design $\mathcal{D}^*$. Therefore $B_x \cap (B_x - x) = S^* \setminus \{x\}$ and hence $|B_x \cap (B_x - x)| = |S| - 2 = w - 1$. On the other hand $|\Phi| = w$ and the intersection of two different blocks contains at most $w - 1$ elements, so $\mu = w - 1$. $\square$

Proposition 3.3 *Let (N, Φ) be a finite Ferrero pair, where $(N, +) = (\mathbf{Z}_s, +)$ with $s = p_1^{m_1} \cdot \ldots \cdot p_t^{m_t}$ $(p_i \in \mathbf{P} \setminus \{2\}, p_1 < p_2 < \ldots < p_t, m_i \in \mathbf{N})$ and $(\bar{\Phi}, \cdot) < (\mathcal{U}(\mathbf{Z}_s), \cdot)$ with $(\bar{\Phi}, \cdot) \cong (\mathbf{Z}_{p_1 - 1}, +)$.*
Then, for the corresponding column code, we have $\mu = p_1 - 2$ and therefore we receive a $(s, s\frac{s-1}{p_1-1}, 2)$−code.

Proof. Following proposition 3.2, it suffices to find a Φ−invariant subgroup of N of order p_1. Let $S := \frac{s}{p_1}\mathbf{Z}_s$. Then $|S| = p_1$ since $\frac{s}{p_1}\mathbf{Z}_s \cong \mathbf{Z}_{p_1}$, $\Phi(S) = S \cdot \bar{\Phi} = \frac{s}{p_1}\mathbf{Z}_s \cdot \bar{\Phi} = \frac{s}{p_1}\mathbf{Z}_s = S$ and furthermore $\frac{s}{p_1}\mathbf{Z}_s \leq \mathbf{Z}_s$. So we get $\mu = p_1 - 2$ and the parameters as above. $\square$

From the last propositions it seems to be obvious, which codewords should be omitted in order to improve the minimal distance of the code, namely the ones, which arise from the subgroup S, because these are the reason for $\mu = p_1 - 2$.

Definition and Proposition 3.4 Let (N, Φ) be a Ferrero pair, where $(N, +)$ $= (\mathbf{Z}_s, +)$ with $s = p_1^{m_1} \cdot \ldots \cdot p_t^{m_t}$ $(p_i \in \mathbf{P} \setminus \{2\}, p_1 < p_2 < \ldots < p_t, m_i \in \mathbf{N})$. Let $S := \frac{s}{p_1} \mathbf{Z}_s$. Furthermore take the set

$$\mathcal{B}_{red}^* := \{\Phi(a) + b \mid a, b \in N, a \notin S\} \tag{17}$$

Then $(N, \mathcal{B}_{red}^*)$ *is a tactical configuration.* The columns of its incidence matrix form the so-called *reduced code* $\mathcal{C}_{red}^{\mathcal{D}^*}$ of the column code $\mathcal{C}^{\mathcal{D}^*}$. *There we obtain an* $(n, M_{red}, d_{red})-$*code with*

$$n = s \tag{18}$$
$$M_{red} = n(n - p_1)/w \tag{19}$$
$$d_{red} = 2(w - \mu_{red}) \tag{20}$$

where $w = |\Phi|$ *as usual and*

$$\mu_{red} := \max\{ |B_1 \cap B_2| \mid B_1, B_2 \in \mathcal{B}_{red}^*, B_1 \neq B_2\} \tag{21}$$

Proof. We start with $(\mathbf{Z}_s, \mathcal{B}^*)$ and look at how the parameters change, when we throw away the blocks corresponding to S. At first let $b = 0$. Then for every point in S but 0 the number of incidence is reduced by 1. But b runs through all of $\mathbf{Z}_s$, so for every element in $\mathbf{Z}_s$ this number is reduced by $|S \setminus \{0\}| = p_1 - 1$. Therefore r is well-defined and we have $r = s - 1 - (p_1 - 1) = s - p_1$. For determining the number of eliminated blocks again let $b = 0$. Since the elements of $\bar{\Phi}$ are invertible, there are $\frac{p_1 - 1}{w}$ blocks of the form $\Phi(a)$ $(a \in S \setminus \{0\})$. Again we have to consider all translates $\Phi(a) + b$ for $b \in \mathbf{Z}_s$, so $s \cdot \frac{p_1 - 1}{w}$ blocks are taken away. Hence we have $s \cdot \frac{s-1}{w} - s \cdot \frac{p_1-1}{w} = s \cdot \frac{s-p_1}{w}$ blocks in $\mathcal{B}_{red}^*$ and therefore $(\mathbf{Z}_s, \mathcal{B}_{red}^*)$ is a tactical configuration via $v \cdot r = s \cdot (s - p_1) = s \cdot \frac{s-p_1}{w} \cdot w = b \cdot k$

Now also the parameters of the code are obvious, because $n = v = s$ and $M_{red} = b = s \cdot \frac{s-p_1}{w}$. The minimal distance d_{red} again depends on the maximal number of block intersections μ_{red}, but here only blocks of $\mathcal{B}_{red}^*$ are take into consideration. □

Proposition 3.5 *Let* $(\mathbf{Z}_s, \Phi)$ *be a finite Ferrero pair like in Definition and Proposition 3.4.*

Then, for computing the parameters μ *and* μ_{red} *of the design* $(\mathbf{Z}_s, \mathcal{B}^*)$ *and the tactical configuration* $(\mathbf{Z}_s, \mathcal{B}_{red}^*)$ *respectively, it suffices to compare only zeroblocks (i.e. blocks containing 0).*

Proof. The first fact follows immediately from the general proof for designs from finite Ferrero pairs (cf. [3]).

Now let $B_1, \ldots, B_b$ be the blocks of the tactical configuration $(\mathbf{Z}_s, \mathcal{B}_{red}^*)$. Let $\bar{B}_1, \bar{B}_2 \in \mathcal{B}_{red}^*$ such that

$$|\bar{B}_1 \cap \bar{B}_2| = \mu_{red} = \max\{ |B_1 \cap B_2| \mid B_1, B_2 \in \mathcal{B}_{red}^*, B_1 \neq B_2\} \tag{22}$$

Then we have $\bar{B}_1 \cap \bar{B}_2 \neq \emptyset$, since $\mu > 0$. So let $y \in \bar{B}_1 \cap \bar{B}_2$. Then we obtain $\bar{B}_1 - y, \bar{B}_2 - y \in \mathcal{B}^*_{red}$, and $0 \in \bar{B}_1 - y, \bar{B}_2 - y$. So $|(\bar{B}_1 - y) \cap (\bar{B}_2 - y)| = \mu_{red}$. $\quad\square$

Definition 3.6 Let (N, Φ) be a finite Ferrero pair, where $(N, +) = (\mathbf{Z}_s, +)$ with $s = p_1^{m_1} \cdot \ldots \cdot p_t^{m_t}$ $(p_i \in \mathbf{P} \setminus \{2\}, p_1 < p_2 < \ldots < p_t, m_i \in \mathbf{N})$.
For every q dividing s consider the set

$$\mathcal{B}^*(q) := \{\Phi(a) + b \mid a, b \in N, a \notin \frac{s}{q}\mathbf{Z}_s\} \qquad (23)$$

It is obvious that $\mathcal{B}^*(1) = \mathcal{B}^*$, $\mathcal{B}^*(p_1) = \mathcal{B}^*_{red}$ and $\mathcal{B}^*(s) = \emptyset$. Like in Definition and Proposition 3.4, we also specify the block intersection parameter $\mu(q)$ for $q \mid s, q \neq 1$, but now in a different way:

$$\mu(q) := \max\{|B_1 \cap B_2| \mid B_1, B_2 \in \bigcap_{\substack{u|q \\ u \neq q}} \mathcal{B}^*(u) \setminus \mathcal{B}^*(q), B_1 \neq B_2\} \qquad (24)$$

We could have also defined codes from these new block sets, but such a code would lose many of the codewords of the original column code and therefore have a lower quality. So we use the $\mu(q)$ in order to compute μ_{red} and μ.

Theorem 3.7 Let (N, Φ) be a finite Ferrero pair, where $(N, +) = (\mathbf{Z}_s, +)$ with $s = p_1^{m_1} \cdot \ldots \cdot p_t^{m_t}$ $(p_i \in \mathbf{P} \setminus \{2\}, p_1 < p_2 < \ldots < p_t, m_i \in \mathbf{N})$.
Then we have

$$\mu_{red} = \max\{\mu(q) \mid q \mid s, q > p_1\} \qquad (25)$$
$$\mu = \max\{\mu_{red}, \mu(p_1)\} \qquad (26)$$

Furthermore for computing $\mu(q)$ $(q \mid s, q \neq 1)$ it suffices to compare one arbitrary zeroblock $B \in \bigcap_{u|q, u \neq q} \mathcal{B}^*(u) \setminus \mathcal{B}^*(q)$ with all the other such blocks.

Proof. Let $1 \mid q_1 \mid q_2 \mid s$ $(1 \neq q_1 \neq q_2)$ and suppose that we have two zeroblocks $B_1 \in \bigcap_{u|q_1, u \neq q_1} \mathcal{B}^*(u) \setminus \mathcal{B}^*(q_1)$ and $B_2 \in \bigcap_{u|q_2, u \neq q_2} \mathcal{B}^*(u) \setminus \mathcal{B}^*(q_2)$.
Since $\frac{s}{q_1}\mathbf{Z}_s \leq \mathbf{Z}_s$ and $\frac{s}{q_2}\mathbf{Z}_s \leq \mathbf{Z}_s$ all elements of B_1 and B_2 are multiples of $\frac{s}{q_1}$ and $\frac{s}{q_2}$ respectively, but no element of B_2 except 0 can be divided by $\frac{s}{q_1}$, because $(\bar{\Phi}, \cdot) <$ $(\mathcal{U}(\mathbf{Z}_s), \cdot)$. Thus $B_1 \cap B_2 = \{0\}$ and we need not take such block intersections into consideration, when we compute the maximum. So only the $\mu(q)$ have a non-trivial contribute and therefore it suffices to take the maximum of these in order to compute μ_{red} and μ.
Furthermore all zeroblocks $B \in \bigcap_{u|q, u \neq q} \mathcal{B}^*(u) \setminus \mathcal{B}^*(q)$ can be divided by $\frac{s}{q}$ elementwise. The resulting blocks only consist of invertible elements and 0. Therefore it suffices to compare one fixed block of this set with the other ones similar as in Proposition 3.5. $\quad\square$

Example 3.8 We consider the case $t = 1$, $p_1 = 31$ and $m_1 = 2$. Clearly a resulting code has length $n = 961$. So let us have a look at the possible non-trivial weights:

- $w_1 = 5$, $w_2 = 6$: In both cases we already have $\mu = 2$. So the minimal distance cannot be improved by the construction of Definition and Proposition 3.4.

- $w_3 = 10$, $w_4 = 15$: Here the block intersection parameter is 4 and 8, respectively. The codes are relatively good, but for both of them the reduced version has $\mu_{red} = 2$ and this is better than before.

- $w_5 = 30$: For this case we know from Proposition 3.2 that $\mu = 29$ and therefore $d = 2$. But although this code seems not worth being looked at, the reduced code of it has $\mu_{red} = 4$ and therefore a minimal distance of 52.

4 Tables of Codes

Remark 4.1 In Tables 1, 2, 3 and 4 we present a listing of all possible codes up to the order of 1000 except the trivial ones with length 3 or less (cf. [3]). There n and w are as usual and γ denotes one generator of the group $(\bar{\Phi}, \cdot)$. Then certainly μ and d of the code $\mathcal{C}^{\mathcal{D}^{\bullet}}$ and the same parameters for the code $\mathcal{C}^{\mathcal{D}^{\bullet}}_{red}$, namely μ_{red} and d_{red} are given. When the reduced code does not yield any improvement, the entries for it are omitted.

The computation has been done in C on an Apple MacIntosh from the Department of Mathematics at the University of Linz.

Table 1: Column codes and reduced column codes from ring generated finite Ferrero pairs based on residue class rings

n		w	γ	μ	d	μ_{red}	d_{red}
25	$= \quad 5^2$	4	7	3	2	2	4
49	$= \quad 7^2$	6	19	5	2	2	8
65	$= \quad 5 \cdot 13$	4	8	3	2	2	4
85	$= \quad 5 \cdot 17$	4	13	3	2	2	4
91	$= \quad 7 \cdot 13$	6	10	5	2	3	6
121	$= \quad 11^2$	5	3	3	4	2	6
		10	40	9	2	3	14
125	$= \quad 5^3$	4	57	3	2	2	4
133	$= \quad 7 \cdot 19$	6	12	5	2	3	6

R. A. Eggetsberger

Table 2: Column codes and reduced column codes from ring generated finite Ferrero pairs based on residue class rings (ctd.)

n			w	γ	μ	d	μ_{red}	d_{red}
145	=	$5 \cdot 29$	4	12	3	2	2	4
169	=	13^2	4	70	2	4		
			6	23	3	6	2	8
			12	19	11	2	3	18
185	=	$5 \cdot 37$	4	43	3	2	2	4
205	=	$5 \cdot 41$	4	32	3	2	2	4
217	=	$7 \cdot 31$	6	26	5	2	2	8
221	=	$13 \cdot 17$	4	21	2	4		
247	=	$13 \cdot 19$	6	69	3	6		
259	=	$7 \cdot 37$	6	101	5	2	2	8
265	=	$5 \cdot 53$	4	23	3	2	2	4
289	=	17^2	4	38	2	4		
			8	110	4	8	2	12
			16	40	15	2	4	24
301	=	$7 \cdot 43$	6	80	5	2	2	8
305	=	$5 \cdot 61$	4	72	3	2	2	4
325	=	$5^2 \cdot 13$	4	18	3	2	2	4
341	=	$11 \cdot 31$	5	4	3	4	2	6
			10	29	9	2	4	12
343	=	7^3	6	19	5	2	2	8
361	=	19^2	6	69	3	6	2	8
			9	28	5	8	2	14
			18	116	17	2	3	30
365	=	$5 \cdot 73$	4	27	3	2	2	4
377	=	$13 \cdot 29$	4	70	2	4		
403	=	$13 \cdot 31$	6	88	3	6	2	8
425	=	$5^2 \cdot 17$	4	132	3	2	2	4
427	=	$7 \cdot 61$	6	75	5	2	2	8
445	=	$5 \cdot 89$	4	123	3	2	2	4
451	=	$11 \cdot 41$	5	16	3	4	2	6
			10	72	9	2	4	12
469	=	$7 \cdot 67$	6	38	5	2	2	8
481	=	$13 \cdot 37$	4	31	2	4		
			6	101	3	6	2	8
			12	45	11	2	5	14
485	=	$5 \cdot 97$	4	22	3	2	2	4
493	=	$17 \cdot 29$	4	157	2	4		
505	=	$5 \cdot 101$	4	192	3	2	2	4
511	=	$7 \cdot 73$	6	82	5	2	2	8

Table 3: Column codes and reduced column codes from ring generated finite Ferrero pairs based on residue class rings (ctd.)

n			w	γ	μ	d	μ_{red}	d_{red}
529	=	23^2	11	118	6	10	2	18
			22	28	21	2	4	36
533	=	$13 \cdot 41$	4	73	2	4		
545	=	$5 \cdot 109$	4	33	3	2	2	4
553	=	$7 \cdot 79$	6	24	5	2	2	8
559	=	$13 \cdot 43$	6	166	3	6	2	8
565	=	$5 \cdot 113$	4	98	3	2	2	4
589	=	$19 \cdot 31$	6	88	3	6	2	8
625	=	5^4	4	182	3	2	2	4
629	=	$17 \cdot 37$	4	191	2	4		
637	=	$7^2 \cdot 13$	6	166	5	2	3	6
671	=	$11 \cdot 61$	5	9	3	4	2	6
			10	41	9	2	4	12
679	=	$7 \cdot 97$	6	159	5	2	2	8
685	=	$5 \cdot 137$	4	37	3	2	2	4
689	=	$13 \cdot 53$	4	83	2	4		
697	=	$17 \cdot 41$	4	132	2	4		
			8	161	4	8	3	10
703	=	$19 \cdot 37$	6	27	3	6	2	8
			9	9	5	8	4	10
			18	3	17	2	9	18
721	=	$7 \cdot 103$	6	47	5	2	2	8
725	=	$5^2 \cdot 29$	4	157	3	2	2	4
745	=	$5 \cdot 149$	4	193	3	2	2	4
763	=	$7 \cdot 109$	6	173	5	2	2	8
781	=	$11 \cdot 71$	5	5	3	4	2	6
			10	17	9	2	3	14
785	=	$5 \cdot 157$	4	28	3	2	2	4
793	=	$13 \cdot 61$	4	255	2	4		
			6	75	3	6	2	8
			12	32	11	2	4	16
817	=	$19 \cdot 43$	6	50	3	6	2	8
841	=	29^2	4	41	2	4		
			7	190	3	8	2	10
			14	63	7	14	2	24
			28	14	27	2	6	44
845	=	$5 \cdot 13^2$	4	268	3	2	2	4
865	=	$5 \cdot 173$	4	93	3	2	2	4
871	=	$13 \cdot 67$	6	30	3	6	2	8

Table 4: Column codes and reduced column codes from ring generated finite Ferrero pairs based on residue class rings (ctd.)

n			w	γ	μ	d	μ_{red}	d_{red}
889	$=$	$7 \cdot 127$	6	108	5	2	2	8
901	$=$	$17 \cdot 53$	4	30	2	4		
905	$=$	$5 \cdot 181$	4	162	3	2	2	4
925	$=$	$5^2 \cdot 37$	4	43	3	2	2	4
931	$=$	$7^2 \cdot 19$	6	31	5	2	3	6
949	$=$	$13 \cdot 73$	4	265	2	4		
			6	82	3	6	2	8
			12	24	11	2	3	18
961	$=$	31^2	5	374	2	6		
			6	440	2	8		
			10	333	4	12	2	16
			15	235	8	14	2	26
			30	115	29	2	4	52
965	$=$	$5 \cdot 193$	4	112	3	2	2	4
973	$=$	$7 \cdot 139$	6	236	5	2	2	8
985	$=$	$5 \cdot 197$	4	182	3	2	2	4

References

1. Th. Beth, D. Jungnickel and H. Lenz, *Design Theory* (Bibl. Inst., Mannheim, 1985).
2. J. R. Clay, *Near-Rings: Geneses and Applications* (Oxford Univ. Press, 1992).
3. R. A. Eggetsberger, *Some Topics in Frobenius Groups, BIB-Designs and Coding Theory* (Contributions to General Algebra 8, Hölder-Pichler-Tempsky/Teubner, Wien/Stuttgart, 1992), P. 45–56.
4. P. Fuchs, G. Hofer and G. Pilz, *Codes from Planar Near-Rings* (IEEE Trans. on Information Theory 36, 1990) P. 647–651.
5. H. Griffin, *Elementary Theory of Numbers* (McGraw-Hill Book Company, Inc. New York, Toronto, London, 1954).
6. W.-F. Ke and H. Kiechle, *Characterization of Some Finite Ferrero Pairs* (to appear).
7. F. J. MacWilliams and N. J. A. Sloane, *The Theory of Error-Correcting Codes I, II* (North-Holland, Amsterdam, 1977).
8. G. Pilz, *Near-Rings, 2nd edition* (North-Holland, Amsterdam, 1983).
9. G. Pilz, *Codes, Block Designs, Frobenius Groups and Near-Rings* (Combinatorics 90, Elsevier Science Publishers B.V. 1992) P. 471–476.

ON MINIMAL VARIETIES OF NEAR-RINGS

Y. Fong

Department of Mathematics, National Cheng Kung University
Tainan 701, Taiwan, R.O.C.

K. Kaarli

Department of Mathematics, Tartu University
EE2400 Tartu, Estonia

W.-F. Ke

Department of Mathematics, National Cheng Kung University
Tainan 701, Taiwan, R.O.C.

ABSTRACT

It is well known that every minimal variety of associative rings is generated by a finite ring of prime order, in particular it is locally finite. In this paper we focus at locally finite minimal varieties of near-rings. They are exactly the varieties generated by finite strictly simple near-rings. We prove that every finite strictly simple near-ring is either a near-ring with the so-called trivial multiplication on a group of prime order or a finite planar near-ring whose additive group is elementary abelian. We describe the multiplicative subgroups of Galois fields which lead to strictly simple Ferrero near-rings and prove that in this way one obtains all finite strictly simple near-rings satisfying the identity $xyz = yxz$. In particular, this proves that the finite, strictly simple near-rings with non-prime order are abundant.

Introduction

Throughout the paper, N is a left near-ring, i.e., N is a non-empty set with two binary operations: addition and multiplication, such that $\langle N; + \rangle$ is a group, $\langle N; \cdot \rangle$ is a semigroup and the left distributivity law $x(y + z) = xy + xz$ is satisfied. All rings are associative.

The left distributive law implies that the left multiplication operators l_a, ($a \in N$) defined via $l_a(x) = ax$ are endomorphisms of $\langle N; + \rangle$. Hence, if 0 is the neutral element of the group $\langle N; + \rangle$ then the identity $x0 = 0$ holds in N. However, $0x$ does not equal to 0, in general. Given a near-ring N, the subnear-rings $N_c = \{x \in N \mid 0x = x\}$ and $N_0 = \{x \in N \mid 0x = 0\}$ are said to be the *constant* and the *0-symmetric* parts of N, respectively. If $N = N_c$ then N is a *constant near-ring* and if $N = N_0$ then N is a *0-symmetric near-ring*. The basics of near-ring theory can be found in the books [1,5,8].

This research was done during the second author's stay in National Cheng-Kung University, Tainan. Support by the grant No.VRP92034 from the National Science Council of ROC is gratefully acknowledged.

Y. Fong et al. (eds.), Near-Rings and Near-Fields, 123–131.
© 1995 Kluwer Academic Publishers.

Obviously near-rings form a variety of Ω-groups with Ω consisting of a single operation: the binary multiplication. A variety is called *locally finite* if all its finitely generated members are finite. A variety V is called *minimal* if it has exactly 2 subvarieties: the V itself and the variety of one-element algebras. It is well known that a variety generated by a single finite algebra is locally finite. While the converse is not true, in general, a minimal locally finite variety is obviously generated by any of its finite members.

It is known that a variety of rings is minimal if and only if it is generated by a ring whose additive group is of prime order. Hence for every prime number p, there are exactly two minimal varieties of rings: one generated by the field of integers modulo p, and the other generated by the ring with zero multiplication on the group of order p.

It is an open question whether there exist minimal varieties of near-rings which are not locally finite. The equivalent question is whether every minimal variety of near-rings does contain a finite member. In this paper we focus on locally finite minimal varieties of near-rings.

Clearly, every locally finite minimal variety of near-rings is generated by a finite simple near-ring which has no proper non-zero subnear-rings. In what follows we refer to those near-rings as to *strictly simple*. The converse is also true. Since the variety of near-rings is congruence permutable, it follows from a general fact of universal algebra [9] that every finite strictly simple near-ring generates a minimal variety.

Hence, the problem of how to describe those locally finite minimal varieties of near-rings reduces to characterizing finite strictly simple near-rings.

1. Ferrero near-rings

In what follows we need a well-known construction from near-ring theory which we present in a little bit generalized form. Let us consider a pair (Γ, G) where G is a group written additively and Γ is a group of automorphisms of G. Then G is a disjoint union of Γ-orbits. Let E be a subset of G satisfying the following two conditions:
(1) if $a \in G$, $\gamma \in \Gamma \setminus \{1\}$ and $\gamma a = a$ then $a \notin E$;
(2) if $a \in G$ then $|E \cap \Gamma a| \leq 1$.
In other words, E does not contain fixed points of any $\gamma \neq 1$ from Γ (in particular, $0 \notin E$) and there is at most one element from each Γ-orbit of G. It is easy to understand that the elements of the orbits Γe, $e \in E$, cannot be fixed by some $\gamma \in \Gamma \setminus \{1\}$.

Then the multiplication on G is defined as follows: $(\gamma e)a = \gamma a$ for all $\gamma \in \Gamma$, $e \in E$ and $a \in G$; all other products are zero. What we get in this way is a 0-symmetric near-ring. We denote it by $F(\Gamma, G, E)$. Also denote $\Gamma E = \{\gamma e \mid \gamma \in \Gamma, e \in E\}$ and $G^0 = G \setminus \Gamma E$.

This construction is basically due to Ferrero [2], so we shall refer to the near-rings $F(\Gamma, G, E)$ as *Ferrero near-rings*. In fact originally it was assumed that Γ is a group of fixed-point-free automorphisms of G. However, what we need to get the construction to work is that the set E does not contain the fixed points of Γ. It is straightforward from the definition that the members of E are exactly the left unity elements of $F(\Gamma, G, E)$. Moreover, E is also the set of all idempotents of $F(\Gamma, G, E)$.

The class of near-rings $F(\Gamma, G, E)$ contains all planar near-rings which have important applications in combinatorics [1].

If $E = \emptyset$ then $F(\Gamma, G, E)$ is a near-ring with zero multiplication.

If $|\Gamma| = 1$ then $F(\Gamma, G, E)$ is the so-called near-ring with *trivial multiplication* [3].

If $G^0 = \{0\}$ and $|E| = 1$ then $F(\Gamma, G, E)$ is a near-field.

If G is an additive group of a (skew) field K and Γ is a subgroup of the multiplicative group K^* of K then $F(\Gamma, G, E)$ is said to be a field-generated Ferrero near-ring. The near-field-generated Ferrero near-rings are defined similarly.

Clearly, if $N = F(\Gamma, G, E)$, then Γ is exactly the set of non-zero left multiplication operators l_a for N. Thus, all non-zero left multiplication operators of N are automorphisms of G. It is an easy exercise to show that in the case of finite N the converse is also true. Hence we have the following lemma and an obvious corollary.

Lemma 1.1 *For a finite 0-symmetric near-ring N, all non-zero l_a with $a \in N$ are automorphisms of the additive group of N if and only if N is isomorphic to some Ferrero near-ring.*

Corollary 1.2 *A subnear-ring of a finite Ferrero near-ring is again a Ferrero near-ring.*

It will turn out that all finite strictly simple near-rings are Ferrero near-rings. Therefore more information is needed about subnear-rings of Ferrero near-rings. In particular we need conditions which ensure that a given subgroup of G is a subnear-ring of $F(\Gamma, G, E)$.

Proposition 1.3 *Let $N = F(\Gamma, G, E)$ be finite. A subgroup G_1 of G is a subnear-ring of N if and only if the following conditions are satisfied:*

(1) *G_1 has enough idempotents, i.e., for $\gamma \in \Gamma$ and $e \in E$, if $\gamma e \in G_1$ then $e \in G_1$;*
(2) *if $\gamma \in \Gamma$ and $e \in E \cap G_1$ then $\gamma e \in G_1$ implies $\gamma G_1 \subseteq G_1$.*

Then this subnear-ring is $F(\Gamma_1, G_1, E_1)$ where $E_1 = E \cap G_1$ and $\Gamma_1 = \{\gamma \in \Gamma \mid \gamma G_1 = G_1\}$.

Proof. It follows directly from the multiplication rule of $F(\Gamma, G, E)$ that the conditions (1) and (2) are sufficient. Let now N_1 be a subnear-ring of N with additive group G_1. By Corollary 1.2, N_1 is a Ferrero near-ring; let $N_1 = F(\Gamma_1, G_1, E_1)$. Since E is exactly the set of all left unity elements of $F(\Gamma, G, E)$, we have $E_1 = E \cap G_1$. Take an arbitrary $a \in G_1 \backslash G^0$ and let $a = \gamma e$ with $\gamma \in \Gamma$ and $e \in E$. Since $aN_1 \neq 0$ and N_1 is a Ferrero near-ring, $N_1 a$ must contain a left unity element of N_1. However, $N_1 a \subseteq \Gamma a = \Gamma e$ and e is the only idempotent contained in Γe. Hence, $e \in N_1 a \subseteq N_1$.

Obviously,

$$\Gamma_1 \subseteq \{\gamma \in \Gamma \mid \gamma G_1 = G_1\} \subseteq \{\gamma \in \Gamma \mid \gamma e \in G_1\}$$

for every $e \in E_1$. Conversely, if $\gamma \in \Gamma$, $e \in E_1$ and $\gamma e \in G_1$ then $\gamma = l_{\gamma e}$ implying $\gamma \in \Gamma_1$.

The proof is complete.

Corollary 1.4 *If $F(\Gamma, G, E)$ is a Ferrero near-ring then every Γ-invariant subgroup of G is a subnear-ring.*

2. Strictly simple near-rings

Now assume that N is an arbitrary strictly simple near-ring. Since any near-ring has both constant and 0-symmetric subnear-rings, N has to be either constant or 0-symmetric.

The structure of a constant near-ring is completely determined by its additive structure. All of its additive subgroups are subnear-rings. Thus, a strictly simple constant near-ring N cannot have proper non-zero additive subgroups. It follows that N is of prime order. Hence, all constant minimal varieties of near-rings are locally finite. There is exactly one such varieties for every prime number p and it is determined by the identities:

$$x + y = y + x, \qquad px = 0, \qquad xy = y.$$

Let now N be a strictly simple 0-symmetric near-ring and first assume that it is nilpotent. It is well known that every nilpotent near-ring has a non-zero subnear-ring with zero multiplication. Since N is strictly simple, then N itself must have the zero multiplication. Thus again the structure of N is completely determined by its additive structure. It follows that N must be of prime order. Consequently, all nilpotent minimal varieties of near-rings are locally finite. There is exactly one such varieties for every prime number p and it is determined by the identities:

$$x + y = y + x, \qquad px = 0, \qquad xy = 0$$

Assume now that N is a finite strictly simple 0-symmetric near-ring and $N^2 \neq 0$. Note that if there exist minimal varieties of near-rings which are not locally finite, there is no obvious reason why they should be generated by strictly simple near-rings. Therefore we are not interested in infinite strictly simple near-rings here. It is possible that they do not exist at all. The next theorem shows that quite a lot can be said about the structure of finite strictly simple 0-symmetric near-rings.

Theorem 2.1 *A finite strictly simple 0-symmetric near-ring is isomorphic to some Ferrero near-ring $F(\Gamma, G, E)$ such that G is an elementary abelian group and Γ is a fixed-point-free automorphism group of G.*

Proof. Let N be a finite strictly simple 0-symmetric near-ring. It is convenient to have a different notation for the additive group of N: let it be G.

If N has zero multiplication then clearly N is isomorphic to $F(\Gamma, G, E)$ with $\Gamma = \{1\}$ and $E = \emptyset$. Thus assume that there is an $a \in N$ such that $aN \neq 0$. Since N is strictly simple, $aN = N$. On the other hand, the right ideal $(0 : a)_N$ does not equal to the whole N. Since right ideals of 0-symmetric near-rings are subnear-rings, we have $(0 : a)_N = 0$. Hence the left multiplication operator $l_a : N \to N$ is an automorphism of G and by Lemma 1.1, N is a Ferrero near-ring.

Let now $\gamma \in \Gamma$ and let H be the set of all fixed points of γ. Obviously H is a subgroup of G and it follows from the definition of the near-ring $F(\Gamma, G, E)$ that $H \subseteq G^0$. Then H is a subnear-ring with the zero multiplication of N. Since N is strictly simple, we have $H = \{0\}$.

It remains to prove that G is elementary abelian. If Γ is a unity group then Corollary 1.4 implies that every subgroup of G is a subnear-ring of $F(\Gamma, G, E)$. Hence N has to be of prime order.

Let now $|\Gamma| \geq 2$. By Thompson's theorem, we get that G is a nilpotent group. However, it is well known that a finite nilpotent group has a nontrivial fully invariant subgroup, unless it is elementary abelian. Thus our claim follows from Corollary 1.4.

This completes the proof.

Theorem 2.1 says that finite strictly simple near-rings come from linear algebra. We may consider G as a vector space over the prime field Z_p and Γ as a group of non-singular fixed-point-free linear transformations of G. Now we show that in the case of commutative Γ, all finite strictly simple near-rings $F(\Gamma, G, E)$ are field-generated. Note that the commutativity of Γ actually means that the Ferrero near-ring $F(\Gamma, G, E)$ satisfies the identity $xyz = yxz$.

Theorem 2.2 *Every finite strictly simple 0-symmetric near-ring satisfying the identity $xyz = yxz$ is isomorphic to a field-generated Ferrero near-ring.*

Proof. Let N be a finite strictly simple 0-symmetric near-ring satisfying the identity $xyz = yxz$. If N has zero multiplication then obviously it is isomorphic to a Ferrero near-ring generated by a prime field Z_p (one has to take $E = \emptyset$).

Let now $N^2 \neq 0$. By Theorem 2.1, N is isomorphic to a Ferrero near-ring $F(\Gamma, G, E)$ where G is an n-dimensional vector space over a prime field Z_p, and Γ is a subgroup of the general linear group $\mathrm{GL}(n, Z_p)$. Since the identity $xyz = yxz$ is satisfied in N, the group Γ has to be commutative. Let R be a subring of the matrix ring $\mathrm{M}_n(Z_p)$ generated by Γ. Obviously R is commutative and G is a left R-module. It follows from Corollary 1.4 that G is an irreducible R-module. Hence $R = \mathrm{GF}(p^n)$.

Now fixing $0 \neq a \in G$ we have a 1-1 correspondence $r \longleftrightarrow ar$ between R and G. Hence the group G becomes a Galois field if we define the multiplication $*$ via $(r_1 a) * (r_2 a) = (r_1 r_2)a$ where $r_1, r_2 \in R$. In particular, for $\gamma \in \Gamma$ and $r \in R$, we have $\gamma(ra) = (\gamma r)a = (\gamma a) * (ra)$ showing that the action of Γ is induced by the multiplication of the Galois field.

The proof is complete.

Our next goal is to characterize those multiplicative subgroups of Galois fields which lead to strictly simple near-rings. Since the multiplicative groups of Galois fields are cyclic, the solution is purely number-theoretical. Note that, given a Galois field $K = \mathrm{GF}(p^n)$ with $n \geq 2$ and a subgroup $\Gamma \leq K^*$, it is easy to choose a subset $E \subseteq K$ such that the related Ferrero near-ring is not strictly simple. Thus the reasonable question is: what are the pairs (K, Γ) for which there exists a subset $E \subseteq K$ such that the corresponding Ferrero near-ring is strictly simple?

We start with a lemma which will simplify the proof of strict simplicity of Ferrero near-rings derived from Galois fields.

Lemma 2.3 *Let* $N = F(\Gamma, G, E)$ *be a Ferrero near-ring generated by a Galois field* K. *If* N *is not strictly simple then there exists a proper subfield* L *of* K *and a 1-dimensional* L-*subspace of* G *which is a subnear-ring of* N.

Proof. Assume that M is a non-zero proper subnear-ring of N and denote its additive group by G_1. Then G_1 satisfies the two conditions of Proposition 1.3 and $M = F(\Gamma_1, G_1, E_1)$ where $E_1 = E \cap G_1$ and $\Gamma_1 = \{\gamma \in \Gamma \mid \gamma G_1 = G_1\}$. Let L be the subring of K generated by Γ_1. Since K is a finite field, L is actually a subfield of K. Obviously G_1 is an L-subspace of G. Since G_1 is a proper subgroup of G, the L is a proper subfield of K.

Take an arbitrary L-subspace G_2 of G_1 and prove that it also satisfies the two conditions from Proposition 1.3.

(1) Let $e \in E$, $\gamma \in \Gamma$ and $\gamma e \in G_2$. Then $\gamma e \in G_1$ and applying Proposition 1.3 we have $e \in G_1$ and $\gamma \in \Gamma_1$. Hence $\gamma^{-1} \in L$ and $e = \gamma^{-1}(\gamma e) \in LG_2 \subseteq G_2$.

(2) Now suppose that $\gamma \in \Gamma$, $e \in G_2 \cap E$ and $\gamma e \in G_2$. Since $G_2 \subseteq G_1$ and M is a subnear-ring of N, we have $\gamma \in \Gamma_1 \subseteq L$. Hence, $\gamma G_2 \subseteq LG_2 \subseteq G_2$.

This completes the proof.

Theorem 2.4 *Let* $N = F(\Gamma, G, E)$ *be a Ferrero near-ring generated by a Galois field* $\mathrm{GF}(p^n)$ *with* $E \neq \emptyset$, $n \geq 2$. *Put* $k = |\Gamma|$. *If* N *is strictly simple then the following conditions are satisfied:*

(1) $p - 1$ *does not divide* k;

(2) k *divides none of the integers* $p^m - 1$ *where* m *is a proper divisor of* n.

Conversely, if the two conditions are satisfied then E *can be chosen so that* N *is strictly simple.*

Proof. Assume first that N is strictly simple. Since the group K^* is cyclic, $p - 1 | k$ implies that Γ contains the multiplicative group of the prime field $\mathrm{GF}(p^n)(p)$. Let G_1 be a cyclic subgroup of G generated by any of elements $e \in E$. Then it is easy to check that both conditions of Proposition 1.3 are satisfied which contradicts the strict simplicity of N.

Now assume that $k | p^m - 1$ where m is one of the proper divisors of n. Then Γ is contained in the subfield $L = \mathrm{GF}(p^n)(p^m)$ of K yielding that $\langle L, + \rangle$ is a proper non-zero Γ-invariant subgroup of G. This contradicts Corollary 1.4.

It remains to prove the sufficiency of our conditions. Let a be any generator of K, i.e., $a \in K$ and $K^* = \{a^i \mid i = 1, \ldots, p^n - 1\}$. Take $E = \{a^i \mid i = 1, \ldots, k'\}$ where $k' = (p^n - 1)/k$ and prove that the Ferrero near-ring $F(\Gamma, G, E)$ is strictly simple. In view of Proposition 1.3 and Lemma 2.3 it is sufficient to prove that given any proper subfield L of K, none of the 1-dimensional L-subspaces of G contains enough idempotents.

Let $L = \mathrm{GF}(p^n)(p^m)$, v be the greatest common divisor of k and $p^m - 1$ and $k = vl$, $p^m - 1 = vt$. Pick an arbitrary 1-dimensional L-subspace H of G. It is easy to see that H^* (the set of non-zero elements of H) is exactly a coset of K^* by L^*. Hence the intersections of H with non-zero Γ-orbits of G are exactly cosets of K^* by $L^* \cap \Gamma$. Since $|L^* \cap \Gamma| = v$, exactly t non-zero Γ-orbits intersect H. Thus we must prove that H contains at most $t - 1$ members of E.

Observe that L^* is the kernel of the group endomorphism of K^* defined by $x \to x^{p^m - 1}$. Hence the elements $x \in H$ are determined by the condition that $x^{p^m - 1}$ is equal to a constant. Suppose that $a^{i(p^m - 1)} = a^{j(p^m - 1)}$. Then $a^{(i-j)(p^m - 1)} = 1$ implying $p^n - 1 | (i-j)(p^m - 1)$ which is equivalent to $u | i - j$ where $u = (p^n - 1)/(p^m - 1)$. Thus, H can contain enough idempotents only provided that $k' \geq u(t - 1) + 1$ and we shall be done if we are able to prove the inequality $k' \leq u(t - 1)$.

Directly from the definitions, $vlk' = kk' = p^n - 1 = (p^m - 1)t = vut$, hence $lk' = ut$ and the equality $k' \leq u(t - 1)$ is equivalent to $t/l \leq t - 1$. Clearly the latter inequality is valid if both t and l are greater than 1. But this is exactly what we have in view of our hypothesis.

The theorem is proved.

3. Examples and concluding remarks

Due to Theorem 2.1, it makes sense to speak about the characteristic of a finite strictly simple near-ring. Obviously, there exist finite strictly simple near-rings for every prime characteristic p: the prime field $\mathrm{GF}(p^n)(p)$, the ring with zero multiplication on Z_p and the constant near-ring on Z_p are three strictly simple near-rings of order p.

In the case $p = 2$ these three near-rings form the complete list of finite strictly simple near-rings of characteristic p. Obviously, there are no more near-rings of order 2. Suppose there is a strictly simple near-ring N of order 2^n, $n \geq 2$. Then it cannot be constant or with zero multiplication and hence it is a Ferrero near-ring with $E \neq \emptyset$. Now, if $e \in E$ then $\{0, e\}$ is obviously a subnear-ring of N contradicting the fact that N is strictly simple.

If $p \neq 2$ then there exist always other strictly simple near-rings of characteristic p except those three mentioned above. In particular, since in this case the multiplicative group of $\mathrm{GF}(p^n)(p)$ has at least one proper subgroup, we find them already among field-generated Ferrero near-rings of order p. Note that in fact all Ferrero near-rings of order p are field-generated because the automorphism group of Z_p is commutative.

Applying Theorem 2.4, it is easy to find many strictly simple near-rings whose order is not a prime. Actually almost all prime powers p^n with $p \neq 2$ may occur as the orders of strictly simple field-generated near-rings. There are exceptions like 9, because every divisor $d \neq 1$ of 8 has 2 as a divisor. On the other hand, if $p \geq 5$ and $n = 2$ then taking $k = p + 1$ we get the two conditions of Theorem 2.4

satisfied.

Also, if $n \geq 2$ is odd then the conditions of Theorem 2.4 will be satisfied if we choose $k = (p^n - 1)/2$.

Thus, Theorem 2.4 provides lot of examples of strictly simple near-rings but we must remember that they all satisfy the identity $xyz = yxz$. Now we show that there exist finite strictly simple near-rings which do not satisfy this identity. One might hope that they all are near-field-generated but this turns out to be not the case. It is easy to see, that if $F(\Gamma, G, E)$ is a near-field-generated Ferrero near-ring then all $\gamma \in \Gamma$, considered as linear transformations of the vector space G, have no eigenvalues unless they are scalars. However, in the following example originally presented in [6,7], all $\gamma \in \Gamma$ do have eigenvalues though only two of them are scalars.

Let G be the elementary abelian group of order 29^2 and Γ consist of the following four matrices and their negatives:

$$\begin{pmatrix} 1 & 0 \\ 0 & 1 \end{pmatrix}, \begin{pmatrix} 0 & 1 \\ -1 & 0 \end{pmatrix}, \begin{pmatrix} 12 & 0 \\ 0 & 17 \end{pmatrix}, \begin{pmatrix} 0 & 12 \\ -17 & 0 \end{pmatrix}.$$

Then Γ is a quaternion group acting fixed-point-freely on G. We show that a set $E \subseteq G$ may be chosen so that the resulting Ferrero near-ring $F(\Gamma, G, E)$ is strictly simple. Beside $\{0\}$ and G, there are 30 '1-dimensional' subgroups in G; let $\mathcal{A}$ stand for the set of all these subgroups. In view of Proposition 1.3, we shall be done if we can choose the set E so that every $H \in \mathcal{A}$ contains fewer members of E than the number of non-zero Γ-orbits it intersects.

The group Γ acts on $\mathcal{A}$ and under this action $\mathcal{A}$ has three 2-element orbits and six 4-element orbits. Note that the 6 subgroups belonging to the 2-element orbits are exactly the eigenspaces of the 6 non-scalar members of Γ. Let $\{H_1, H_2\}$ be one of the 2-element orbits of $\mathcal{A}$. For example, the subgroups H_1 and H_2 may be generated by $(1,0)$ and $(0,1)$, respectively. Then 56 non-zero elements of $H_1 \cup H_2$ are divided into 7 Γ-orbits of G. Moreover, each of those 7 orbits has 4 common elements with both H_1 and H_2. Hence, we can pick the members of E from those 7 orbits so that some of them are in H_1 and some are not. Under such choice neither H_1 nor H_2 will be a subnear-ring of $F(\Gamma, G, E)$.

Any of the 4-element orbits $\{H_1, H_2, H_3, H_4\}$ of $\mathcal{A}$ can be handled in a rather similar way. There are 14 Γ-orbits which cover the union of non-zero elements of these 4 subgroups. It is easy to check that every orbit and every subgroup have exactly 2 elements in common now. Hence it is again not difficult to pick the members of E from those 14 orbits so that they are not contained all together in one of the 4 subgroups.

To conclude the paper we add some words about the structure of minimal varieties of near-rings. As mentioned above, for every prime number p, there exist two minimal varieties all of whose members are completely determined with their additive structure. These are varieties of elementary abelian groups with constant or zero multiplication, respectively. Using the language of Ω-groups or universal algebra, these are minimal abelian varieties of near-rings. Their structure completely coincides with the structure of the corresponding variety of elementary abelian groups. Hence they are congruence permutable (as all near-ring varieties) but not congruence distributive.

It follows from results of [3] that all other locally finite minimal varieties of near-rings are congruence distributive. Moreover, all strictly simple 0-symmetric near-rings with non-zero multiplication are quasiprimal universal algebras and therefore the varieties which are generated by these near-rings are discriminator.

References

[1] J. R. Clay. *Nearrings: Geneses and Applications,* Oxford University Press, 1992.

[2] G. Ferrero. Classificazione e construzione degli stems *p*-singolari, *Ist. Lombardo Accad. Sci. Lett. Rend. A* **102** (1968), 597–613.

[3] Y. Fong, K. Kaarli and W.-F. Ke. On arithmetical varieties of near-rings, in preparation.

[4] J. J. Malone. Near-rings with trivial multiplications, *American Mathematical Monthly* **74** (1967), 1111–1112.

[5] J. D. P. Meldrum. *Near-rings and their links with groups,* Pitman, Boston-London-Melbourne, 1985.

[6] M. C. Modisett. *A characterization of the circularity of certain balanced incomplete block designs,* Ph. D. Dissertation, Univ. of Arizona, Tucson 1988.

[7] M. C. Modisett. A characterization of the circularity of balanced incomplete block designs, *Utilitas Math.* **35** (1989), 83-94.

[8] G. Pilz. *Near-rings,* North-Holland, Amsterdam-New York-xford, 1983.

[9] J. D. H. Smith. *Mal'cev varieties,* Lecture Notes in Mathematics 554, Springer-Verlag, Berlin-Heidelberg-New York, 1976.

SYNTACTIC NEARRINGS

Y. Fong & W.-F. Ke

Department of Mathematics, National Cheng Kung University
Tainan, Taiwan, R. O. C.

C.-S. Wang

Department of Mathematics, Chinese Air Force Academy
Kang San, Taiwan, R. O. C.

ABSTRACT

Let G be a finite group written additively and $\mathcal{S} = (G, G, \delta)$ a group semiautomaton. In this paper we investigate the subnearring $N_{(\sigma,\tau)}(\mathcal{S})$ of $M(G)$, referred to as the syntactic nearring of $\mathcal{S}$, with $\delta(x, y) = x\sigma + y\tau$ for $\sigma, \tau \in M(G)$. By definition, $N_{(\sigma,\tau)}(\mathcal{S})$ is the subnearring of $M(G)$ generated by $M_{\mathcal{S}} = \{f_a : G \to G, a \in G \mid x f_a = \delta(x, a)\ \forall x \in G\}$ under pointwise addition and composition of mappings. We first prove the key result that if $(G, +)$ is a finite group, $\sigma \in M(G)$ and $m, n \in \mathbf{Z}$, then $N_{(m,\sigma)}(\mathcal{S}) = N_{(n,\sigma)}(\mathcal{S})$. Finally, we apply the result to the case where $G = S_3$, the symmetric group of degree 3, and give a detailed study for the case $\mathcal{S} = (S_3, S_3, \delta)$ with $\delta(x, y) = mx + ny$ for various syntactic nearrings.

§1 Preliminary

A *nearring* is a triple $(N, +, \cdot)$ satisfying the conditions

(i) $(N, +)$ is a group which may not be abelian;
(ii) $(N, \cdot)$ is a semigroup;
(iii) $\forall x, y, z \in N,\ x(y + z) = xy + xz$.

In fact, condition (iii) makes N a *left* nearring. If we replace (iii) by

(iii') $\forall x, y, z \in N,\ (x + y)z = xz + yz$,

then we obtain a *right* nearring N. We shall only consider left nearrings in the sequel.

Of course, all rings are nearrings. The converse is not true, however. Let $(G, +)$ be a (not necessarily abelian) group and denote by $M(G) = \{f \mid f : G \to G\}$ the set of all mappings from G into itself. Then $M(G)$, together with pointwise addition and composition of mappings, forms an honest nearring. Readers should refer to [1], [11] or [12] for additional information on nearrings.

In order to capture the originality and motivation of this work, readers are asked to study those references from [1] to [8]. Recall that a *semiautomata* is an ordered triple $\mathcal{S} = (Q, A, \delta)$, where Q and A are nonempty sets, and $\delta : Q \times A \to Q$ is a function. The set Q is called the *set of states*, A the *set of inputs*, and δ the *transition function*. If a group structure is imposed on Q, then $\mathcal{S}$ is called a *group semiautomata*.

Y. Fong et al. (eds.), Near-Rings and Near-Fields, 133–139.

Let $(Q, +)$ be a group and $\mathcal{S} = (Q, A, \delta)$ a group semiautomata. Let $\bar{A}$ be the free semigroup generated by A; thus, every element in $\bar{A}$ is a string of letters from A. There is an identity in $\bar{A}$, i.e., the empty string Λ. Now the transition function δ can be extended to a function $\bar{\delta} : Q \times \bar{A} \to Q$ via $\bar{\delta}(\omega, \Lambda) = \omega$, $\bar{\delta}(\omega, a) = \delta(\omega, a)$, and $\bar{\delta}(\omega, a_1 a_2 \cdots a_n) = \delta(\bar{\delta}(\omega, a_1 a_2 \cdots a_{n-1}), a_n)$ for $n > 1$, where $a, a_1, a_2, \ldots, a_n \in A$, and $\omega \in Q$.

For each $x \in \bar{A}$, define $f_x : Q \to Q$ via $\omega f_x = \bar{\delta}(\omega, x)$ for all $\omega \in Q$. Let $M_{\mathcal{S}} = \{f_x \mid x \in \bar{A}\}$. Then $(M_{\mathcal{S}}, \circ)$ is a monoid with the property that $f_x \circ f_y = f_{xy}$ for all $x, y \in \bar{A}$. Note that $f_\Lambda = id_Q$ is the identity of $M_{\mathcal{S}}$. Since $(Q, +)$ is a group, $M_{\mathcal{S}} \subseteq M(Q)$. The subnearring $N(\mathcal{S})$ of $M(Q)$ generated by $M_{\mathcal{S}}$ is called the *syntactic nearring* of $\mathcal{S}$ on Q (cf. [8], [9]). It is easy to see that

$$N(\mathcal{S}) = \langle \{f_a \mid a \in A\} \cup \{id_Q\}, +, \circ \rangle.$$

It is proved in [7] that if $\delta(x, y) = x+y$ for all $x, y \in Q$, then $N(\mathcal{S}) = E(Q)+M_c(Q)$, where $E(Q) = \{\sum_{\text{finite}} f_i \mid f_i \in End\, Q\}$ is the endomorphism nearring of Q and $M_c(Q) = \{\theta_a : Q \to Q, a \in Q \mid x\theta_a = a \,\forall x \in G\}$ the constant nearring of Q.

§2 General properties

Let $(G, +)$ be a finite group and $\sigma, \tau \in M(G)$. If we define a mapping $\delta_{(\sigma, \tau)}$ from $G \times G$ to G via

$$\delta_{(\sigma, \tau)}(x, y) = x\sigma + y\tau \quad \forall x, y \in G,$$

then $\mathcal{S}_{(\sigma, \tau)} = (G, G, \delta_{(\sigma, \tau)})$ is a group semiautomata. For simplicity, we denote the set $M_{\mathcal{S}_{(\sigma, \tau)}}$ by $M_{(\sigma, \tau)}$ and the nearring $N(\mathcal{S}_{(\sigma, \tau)})$ by $N_{(\sigma, \tau)}$. The mapping $f_a : G \to G$ such that $x f_a = \delta_{(\sigma, \tau)}(x, a) = x\sigma + a\tau$ for all $x \in G$ will be written as $f_{a(\sigma, \tau)}$ in order to be distinguished from the mapping $f_a \in M_{(\sigma', \tau')}$ which takes an $x \in G$ to $\delta_{(\sigma', \tau')}(x, a) = x\sigma' + a\tau'$. It is obvious that if $(\sigma, \tau) \neq (\sigma', \tau')$, where $\sigma, \sigma', \tau, \tau' \in M(G)$, then $N_{(\sigma, \tau)}$ and $N_{(\sigma', \tau')}$ may not be the same in general. However, in some situation, they may be identical.

(2.1) Theorem   *Let $(G, +)$ be a finite group and let $\sigma \in M(G)$ be fixed. If $\rho_a, \rho_b \in M(G)$ are the inner automorphisms of G induced by the conjugate of $a, b \subset G$, respectively, then $N_{(\sigma, \rho_a)} = N_{(\sigma, \rho_b)}$.*

Proof. It suffices to show that $M_{(\sigma, \rho_a)} = M_{(\sigma, \rho_b)}$. Let $x, y \in G$, then

$$
\begin{aligned}
x f_{y(\sigma, \rho_a)} &= x\sigma + y\rho_a = x\sigma + (-a + y + a) \\
&= x\sigma + (-b) + (b - a + y + a - b) + b \\
&= x\sigma + (b - a + y + a - b)\rho_b \\
&= x f_{\bar{y}(\sigma, \rho_b)},
\end{aligned}
$$

where $\bar{y} = b - a + y + a - b \in G$. So $f_{y(\sigma, \rho_a)} = f_{\bar{y}(\sigma, \rho_b)} \in M_{(\sigma, \rho_b)}$. This shows that $M_{(\sigma, \rho_a)} \subseteq M_{(\sigma, \rho_b)}$. A similar argument shows the converse. Therefore, $M_{(\sigma, \rho_a)} = M_{(\sigma, \rho_b)}$ and the result follows. ∎

For an integer $n \in \mathbf{Z}$, there is a mapping $\bar{n} : G \to G; x \mapsto nx$. We shall identify $\bar{n}$ with n, so n is the mapping from G to G which sends every $x \in G$ to nx. The following result gives more examples.

(2.2) Theorem   *Let $(G, +)$ be a finite group and $\tau \in M(G)$. If $m, n \in \mathbf{Z}$, then $N_{(m,\tau)} = N_{(n,\tau)}$.*

Proof. Let $|G| = s$. We notice that $N_{(qs+r,\tau)} = N_{(r,\tau)}$ for any $q, r \in \mathbf{Z}$ with $0 \le r < s$. For if $x, a \in G$, then

$$x f_{a(qs+r,\tau)} = (qs+r)x + a\tau = (qs)x + rx + a\tau$$
$$= q(sx) + rx + a\tau = rx + a\tau = x f_{a(r,\tau)}.$$

Therefore, $M_{(qs+r,\tau)} = M_{(r,\tau)}$; hence $N_{(qs+r,\tau)} = N_{(r,\tau)}$. It remains to show that $N_{(0,\tau)} = N_{(1,\tau)} = \cdots = N_{(s-1,\tau)}$.

Let $k \in \{0, 1, \ldots, s-1\}$. For any $x, a \in G$, we have

$$x f_{a(k+1,\tau)} = (k+1)x + a\tau = x + (kx + a\tau)$$
$$= x\, id_G + x f_{a(k,\tau)} = x(id_G + f_{a(k,\tau)}).$$

Therefore, $f_{a(k+1,\tau)} \in N_{(k,\tau)} = \langle M_{(k,\tau)} \cup \{id_G\} \rangle$, and so $M_{(k+1,\tau)} \subseteq N_{(k,\tau)}$. It follows that

$$N_{(s,\tau)} \le N_{(s-1,\tau)} \le \cdots \le N_{(1,\tau)} \le N_{(0,\tau)}.$$

Since $N_{(0,\tau)} = N_{(s,\tau)}$, the result follows. ∎

(2.3) Corollary   *Let $(G, +)$ be a finite group and $k \in \mathbf{Z}$. Then $N_{(m,k)} = N_{(n,k)}$ for all $m, n \in \mathbf{Z}$.* ∎

§3 Syntactic nearrings $N_{(m,k)}$ on S_3

Consider $G = S_3$ and we shall use addition as the group operation of $S_3 = \{0, a, b, c, d, e\}$, where $0 = (1)$, $a = (12)$, $b = (13)$, $c = (23)$, $d = (132)$ and $e = (123)$. Also, a mapping in $M(S_3)$ will be represented by a 6-tuple of the images of S_3. For example, $(0acbed)$ represents the mapping which fixes 0 and a, and sends b to c, c to b, d to e and e to d.

In the light of (2.3), if $m, k \in \mathbf{Z}$, then $N_{(m,k)} = N_{(1,k)}$. So for $k \in \mathbf{Z}$, let

$$N_k = N_{(1,k)} \quad \text{and} \quad M_k = M_{(1,k)}.$$

We observe that

(1) if $k \in \{0, 1, 2, 3, 4, 5\}$, then $f_{0(1,k)} = id_{S_3}$;
(2) if $w \in \{a, b, c\}$ and $k \in \{1, 3, 5\}$, then $kw = w$; hence $f_{w(1,1)} = f_{w(1,3)} = f_{w(1,5)}$;
(3) if $w \in \{a, b, c\}$ and $k \in \{2, 4\}$, then $kw = 0$; hence $f_{w(1,2)} = f_{w(1,4)}$.

First of all, we claim that $N_1 = N_3 = N_5$. To see this, it suffices to show that $f_{d(1,k)}$ and $f_{e(1,k)}$ can be expressed as sums of $f_{w(1,k)}$ with $w \in \{0, a, b, c\}$. The result will then follow from (1) and (2) of the above observations.

Let $x \in S_3$. Then:

(a)
$$xf_{d(1,1)} = x + d = x + (a + c) = x + (a + 6x + c)$$
$$= (x + a) + 5x + (x + c) = xf_{a(1,1)} + x(5f_{0(1,1)}) + xf_{c(1,1)}$$
$$= x(f_{a(1,1)} + 5f_{0(1,1)} + f_{c(1,1)}),$$

so $f_{d(1,1)} = f_{a(1,1)} + 5f_{0(1,1)} + f_{c(1,1)}$. Similarly, $f_{e(1,1)} = f_{a(1,1)} + 5f_{0(1,1)} + f_{b(1,1)}$.

(b)
$$xf_{d(1,3)} = x + 3d = x = xf_{0(1,3)};$$

and

$$xf_{e(1,3)} = x + 3e = x = xf_{0(1,3)}.$$

So $f_{d(1,3)} = f_{e(1,3)} = f_{0(1,3)}$.

(c)
$$xf_{d(1,5)} = x + 5d = x + e = x + (a + b) = x + (a + 6x + b)$$
$$= (x + a) + 5x + (x + b) = xf_{a(1,1)} + x(5f_{0(1,1)}) + xf_{b(1,1)}$$
$$= x(f_{a(1,5)} + 5f_{0(1,5)} + f_{b(1,5)}),$$

so $f_{d(1,5)} = f_{a(1,5)} + 5f_{0(1,5)} + f_{b(1,5)}$. Similarly, $f_{e(1,5)} = f_{a(1,5)} + 5f_{0(1,5)} + f_{c(1,5)}$.

Therefore, $f_{d(1,k)}$ and $f_{e(1,k)}$ can be expressed as sums of $f_{w(1,k)}$ with $w \in \{0, a, b, c\}$. This concludes our claim.

Secondly, we note that if $\{f_{d(1,2)}, f_{e(1,2)}\} = \{f_{d(1,4)}, f_{e(1,4)}\}$, then as a consequence of (1) and (3) of the above observations, we will have $N_2 = N_4$.

For any $x \in G$, we have

$$xf_{d(1,2)} = x + 2d = x + e = x + 4e = xf_{e(1,4)}$$

and

$$xf_{e(1,2)} = x + 2e = x + 4d = x + 4d = xf_{d(1,4)}.$$

Therefore, $f_{d(1,2)} - f_{e(1,4)}$ and $f_{e(1,2)} - f_{d(1,4)}$ as desired, and so $N_2 = N_4$.

Hence, to determine the structure of N_k, $k = 0, 1, \ldots, 5$, we only need to discuss N_0, N_1 and N_2.

Case 1. N_0.

Since $f_{w(1,0)} = id_{S_3}$ for any $w \in S_3$, we easily obtain $N_0 = \langle id_{S_3}, + \rangle \cong \mathbf{Z}_6$, the ring of integers modulo 6.

Case 2. N_1.

It is proved in [1] or [7] that if the transition function is given by $\delta(x, y) = x + y$ for all $x, y \in G$, then $N(\mathcal{S}) = E(G) + M_c(G)$. Since the transition function of $\mathcal{S}_{(1,1)}$ is given by $\delta_{(1,1)}(x, y) = x + y$ for all $x, y \in S_3$, so we have $N_1 = E(S_3) + M_c(S_3)$.

Case 3. N_2.

Let

$$E(S_3|A_3) = \langle \{\rho_t \mid t \in A_3\}, + \rangle = \langle \{\rho_0, \rho_d, \rho_e\}, + \rangle,$$

and

$$M_c(S_3|A_3) = \{\theta_t \mid t \in A_3\}.$$

Then it is clear that $E(S_3|A_3)$ is a subnearring of the d. g. nearring $E(S_3)$ and $M_c(S_3|A_3)$ is a subnearring of the constant nearring $M_c(S_3)$. Our final goal is to show that $N_2 = E(S_3|A_3) + M_c(S_3|A_3)$.

We argue that $E(S_3|A_3) + M_c(S_3|A_3)$ is a proper subnearring of N_1. It is clear that $E(S_3|A_3) + M_c(S_3|A_3) \subseteq N_1$ since N_1 is a nearring. If $s, t \in A_3$ and $x \in S_3$, then

$$x(\theta_s + \rho_t - \theta_s) = x\theta_s + x\rho_t - x\theta_s = s + (-t + x + t) - s$$
$$= -(t - s) + x + (t - x) = x\rho_{t-s}$$

and

$$x(\theta_s \rho_t) = (x\theta_s)\rho_t = s\rho_t = x\theta_{s\rho_t};$$

thus,

$$\theta_s + \rho_t - \theta_s = \rho_{t-s} \quad \text{and} \quad \theta_s \rho_t = \theta_{s\rho_t}.$$

Now, let $\alpha, \beta = \sum_{j=1}^{n} \rho_{y_j} \in E(S_3|A_3)$ and $\theta_s, \theta_t \in M_c(S_3|A_3)$, where $n \in \mathbf{N}$, $y_j \in A_3$ for $1 \le j \le n$, and $s, t \in A_3$. Then

(a) $(\alpha + \theta_s) - (\beta + \theta_t) = \alpha + (\theta_s - \theta_t - \sum_{j=1}^{n} \rho_{y_j})$

$$= \alpha + (\theta_s - \theta_t - \sum_{j=1}^{n} \rho_{y_j} + \theta_t - \theta_s) + \theta_s - \theta_t$$

$$= \alpha - \sum_{j=1}^{n} (\theta_{s-t} + \rho_{y_j} - \theta_{s-t}) + \theta_{s-t}$$

$$= (\alpha - \sum_{j=1}^{n} \rho_{y_j - (s-t)}) + \theta_{s-t}$$

$$\in E(S_3|A_3) + M_c(S_3|A_3);$$

(b) $(\alpha + \theta_s)(\beta + \theta_t) = (\alpha + \theta_s)\beta + (\alpha + \theta_s)\theta_t$

$$= (\alpha + \theta_s) \sum_{j=1}^{n} \rho_{y_j} + \theta_t$$

$$= \sum_{j=1}^{n} ((\alpha + \theta_s)\rho_{y_j}) + \theta_t$$

$$= \sum_{j=1}^{n} (\alpha\rho_{y_j} + \theta_s\rho_{y_j}) + \theta_t$$

$$= \sum_{j=1}^{n} (\alpha \rho_{y_j} + \theta_{s \rho_{y_j}}) + \theta_t$$

$$= \sum_{j=1}^{n-1} (\alpha \rho_{y_j} + \theta_{s \rho_{y_j}}) + (\alpha \rho_{y_n} + \theta_{s \rho_{y_n}} + \theta_t) = (*).$$

From (a), we see that $(E(S_3|A_3) + M_c(S_3|A_3), +)$ is a subgroup of N_1 containing $E(S_3|A_3)$ as a normal subgroup. Therefore, $(*)$ is an element of $E(S_3|A_3) + M_c(S_3|A_3)$, and so (b) implies that $(E(S_3|A_3) + M_c(S_3|A_3), \circ)$ is a semigroup. Hence $E(S_3|A_3) + M_c(S_3|A_3)$ is indeed a subnearring of N_1.

Now we want to make sure that $E(S_3|A_3) + M_c(S_3|A_3)$ is exactly the nearring N_2. First of all we have $\rho_0 = id_{S_3} = f_{0(1,2)} \in N_2$. Let $x \in S_3$. Then

$$x \rho_d = -d + x + d = 2d + x + 2e = 5x + (x + 2d) + (x + 2e)$$
$$= x(5id_{S_3}) + x f_{d(1,2)} + x f_{e(1,2)} = x(5id_{S_3} + f_{d(1,2)} + f_{e(1,2)}),$$

so $\rho_d = 5id_{S_3} + f_{d(1,2)} + f_{e(1,2)} \in N_2$. Similarly, $\rho_e = 5id_{S_3} + f_{e(1,2)} + f_{d(1,2)} \in N_2$. This shows that $E(S_3|A_3) \subseteq N_2$. It is also easy to verify that

$$\theta_0 = 6id_{S_3}, \quad \theta_d = 5id_{S_3} + f_{e(1,2)}, \quad \text{and} \quad \theta_e = 5id_{S_3} + f_{d(1,2)},$$

and so $M_c(S_3|A_3) \subseteq N_2$. Thus we conclude that $E(S_3|A_3) + M_c(S_3|A_3) \leq N_2$.
On the other hand, if $x, y \in S_3$, then

$$x f_{y(1,2)} = x + 2y = \begin{cases} x, & \text{if } 2y = 0; \\ x + e, & \text{if } y = d; \\ x + d, & \text{if } y = e; \end{cases}$$

$$= \begin{cases} x\rho_0, & \text{if } 2y = 0; \\ x(\rho_0 + \theta_e), & \text{if } y = d; \\ x(\rho_0 + \theta_d), & \text{if } y = e. \end{cases}$$

Therefore, $M_{(1,2)} \subseteq E(S_3|A_3) + M_c(S_3|A_3)$, and so $N_2 \leq E(S_3|A_3) + M_c(S_3|A_3)$ since $E(S_3|A_3) + M_c(S_3|A_3)$ is itself a nearring. Hence $N_2 = E(S_3|A_3) + M_c(S_3|A_3)$ as desired.

Now we know from Case 1 that $|N_0| = 6$. It is showed in [10] that $|E(S_3)| = 54$. Therefore, from Case 2, we have $|N_1| = |E(S_3)| \cdot |M_c(S_3)| = 54 \cdot 6 = 324$. Also, it follows from Case 3 that $|N_2| = |E(S_3|A_3)| \cdot |M_c(S_3|A_3)|$. Certainly, $|M_c(S_3|A_3)| = |A_3| = 3$, but what is $|E(S_3|A_3)|$? Here we notice that

$$\rho_0 = (0abcde) = (0abc00) + (0000de),$$
$$\rho_d = (0bcade) = (0bca00) + (0000de),$$
$$\rho_e = (0cabde) = (0cab00) + (0000de),$$

and each of $(0abc00), (0bca00), (0cab00)$ commutes with $(0000de)$. Thus, if we set

$$A = \langle \{(0abc00), (0bca00), (0cab00)\}, + \rangle \leq (E(S_3|A_3), +)$$
$$B = \langle (0000de), + \rangle \leq (E(S_3|A_3), +).$$

Then $A \cap B = \{(000000)\}$ and

$$E(S_3|A_3) = \langle\{\rho_0, \rho_d, \rho_e\}, +\rangle = A + B.$$

A quick calculation shows that

$$A = \{(0abc00), (0bca00), (0cab00), (0ddd00), (0eee00), (000000)\},$$
$$B = \{(0000de), (0000ed), (000000)\}.$$

Therefore, $|E(S_3|A_3)| = |A| \cdot |B| = 6 \cdot 3 = 18$, and so $|N_2| = 18 \cdot 3 = 54$.
Summarizing the above discussions, we have

$$|N_k| = \begin{cases} 324, & \text{if } k \equiv 1 \pmod 2; \\ 54, & \text{if } k \equiv 2 \text{ or } 4 \pmod 6; \\ 6, & \text{if } k \equiv 0 \pmod 6. \end{cases}$$

Acknowledgement: The first author wishes to express his appreciation to the National Science Council of R. O. C. This work was supported in part by NSC grant no. NCSC81-0208-M006-07.

References

[1] J. R. Clay. *Nearrings: Geneses and Applications.* Oxford Univ. Press, Oxford, 1992.

[2] J. R. Clay and Y. Fong. On syntactic near-rings of even dihedral groups. *Results in Mathematics* **23** (1993), 23–44.

[3] Y. Fong. On near-rings and automata. *Proc. National Sci. Counc. R.O.C. (A)* **12** (1988), 240–246.

[4] Y. Fong. On the structure of abelian syntactic near-rings. In *Proceedings of the First International Symposium on Algebraic Structure and Number Theory, Hong Kong, 1987,* 114–123. Word Scientific, 1990.

[5] Y. Fong and J. R. Clay. Computer programs for investigating syntactic near-rings of finite group-semiautomata. *Bull. Math. Academia Sinica* **12** (1988), 295–304.

[6] Y. Fong, F.-K. Huang, and W.-F. Ke. Syntactic near-rings associated with group semiautomata. *PU. M. A. Ser. A* **2** (1992), 187–204.

[7] Y. Fong and W.-F. Ke. Syntactic near-rings of finite group-semiautomata. In *Proceedings of the Conference on Ordered Structures and Algebra of Computer Languages, Hong Kong 1991,* 31–39. Word Scientific, Singapore, 1993.

[8] G. Hofer and G. Pilz. Group automata and near-rings. In *Proc. Conf. Univ. Algebra, Klagenfurt (Austria),* 153–162, 1982.

[9] M. Holcombe. The syntactic near-rings of a linear sequential machine. *Proc. Edin. Math. Soc.* **26** (1983), 12–25.

[10] C. G. Lyons and J. J. Malone. Endomorphism near-rings. *Proc. Edin. Math. Soc.* **17** (1970), 71–78.

[11] J. D. P. Meldrum. *Near-rings and Their Links with Groups.* No. 134 in Research Note Series. Pitman Publ. Co., 1985.

[12] G. Pilz. *Near-rings.* North-Holland/American Elsevier, second, revised edition, 1983.

ON SUFFICIENT CONDITIONS FOR NEAR-RINGS
TO BE ISOMORPHIC

R. L. Fray

Department of Mathematics, University of the Western Cape
Private Bag X17, BELLVILLE 7535, South Africa

ABSTRACT

In some special cases it is possible for two algebraic structures, each having a set of generators, to be isomorphic. This is shown by the provision of sufficient conditions for two near-rings, each being generated by a nonempty subset, to be isomorphic. Examples in group near-rings serve to illustrate this.

1. INTRODUCTION

The group near-ring of a group G over a near-ring R is defined in [3] as a subnear-ring of $M(R^G)$, the near-ring of all mappings of the group R^G into itself. The authors show that there are two possible choices for the definition of the group near-ring constructed from R and G. These two subnear-rings of $M(R^G)$ are both generated by certain subsets of $M(R^G)$ and are shown to be isomorphic. This example begs the question whether sufficient conditions can be given for two near-rings, each being generated by non-empty subsets, to be isomorphic. We provide sufficient conditions for this to be the case and show that this facilitates alternative proofs of results in [3] concerning the fact that the two candidates for the definition of the group near-ring are isomorphic as well as the fact that in the case of distributively generated near-rings, the group near-ring constructed in [3] is a homomorphic image of the group near-ring defined in Meldrum [4].

2. NOTATION

Let R^G be the collection of all maps from a group $(G, \cdot)$ into a near-ring R with identity. Special maps from R^G to R^G are given by $[r, g]$ and $\langle r, g \rangle$, with $r \in R$, $g \in G$ and actions $([r, g]\mu)(h) = r\mu(hg)$ and $(\langle r, g \rangle \mu)(h) = r\mu(g^{-1}h)$, for all $\mu \in R^G$, $h \in G$. The subnear-rings of $M(R^G)$ generated by the sets $\{[r, g] : r \in R, g \in G\}$ and $\{\langle r, g \rangle : r \in R, g \in G\}$ are denoted by $R[G]$ and $R\langle G \rangle$, respectively. $R\langle G \rangle$ is the group near-ring constructed from the group G and the near-ring R. It is clear that R^G is a faithful $R\langle G \rangle$-module as well as a faithful $R[G]$-module.

Denote by $R^{(G)}$ the subgroup of $(R^G, +)$ consisting of all functions with finite support, that is, $R^{(G)} = \{\mu \in R^G : \text{supp}\,\mu \text{ is finite}\}$, where $\text{supp}\,\mu := \{x \in G : \mu(x) \neq 0\}$ is the support of μ. It is shown in [3] that $R^{(G)}$ is a faithful $R\langle G \rangle$-module.

Y. Fong et al. (eds.), *Near-Rings and Near-Fields*, 141–144.

3. PRELIMINARIES

Let S be a non-empty subset of the right near-ring $(N, +, \cdot)$. For an element A of $N(S)$, the subnear-ring of N generated by S, the author describes in [1] how A is constructed starting from elements of S. Recall that a generating sequence of length m on S is a sequence A_1, A_2, ..., A_m where each A_i for all $1 \leq i \leq n$ has either of the following two forms:

(i) A_i is an element of S,

(ii) $A_i = t_i$, where $t_i \in N(S)$ and $t_i = t_k *_i t_l$ with $1 \leq k, l < i$, $*_i \in \{+, -, \cdot\}$.

Given any $A \in N(S)$, it is possible to construct a generating sequence for A. Let $A_1, \ldots, A_m$ be such a generating sequence on S with $A_m = A$ for some integer $m \geq 1$. Then we say A is the *result* of the generating sequence.

The length of a generating sequence of minimal length for A will be called the *complexity* of A and denoted $c(A)$.

It is clear that $c(A) = 1$ if and only if $A \in S$ and if $c(A) \geq 1$ then $A = B + C$ or $A = BC$ where $B, C \in N(S)$ with $c(B), c(C) < c(A)$.

For a distributively generated near-ring (hereafter written as dg), (N, S), the notion of a generating sequence on S is significantly simplified. We redefine the complexity of an element A, which we also denote by $c(A)$, as the smallest positive integer m such that $A = \sum_{i=1}^{m} \sigma_i s_i$ for some $s_i \in S$, $\sigma_i = \pm 1$. It is immediately obvious that if $c(A) \geq 2$ then $A = A_1 + A_2$ with $c(A_1), c(A_2) < c(A)$ and $A_1, A_2 \in N$.

Now let N_1 and N_2 be right near-rings with non-empty subsets S_1 and S_2, respectively. Suppose that $\phi : S_1 \to S_2$ is a surjection and let $A_1, \ldots, A_m, m \geq 1$, be a generating sequence on S_1. Consider the generating sequence B_1, B_2, B_3, ..., B_m on S_2 where

$$B_i = \begin{cases} \phi(t_i), & \text{if } A_i = t_i \in S_i \\ t_k' *_i' t_l', & \text{if } A_i = t_k *_i t_l, \text{ with } 1 \leq k, l < i \end{cases}$$

with $t_k' *_i' t_l' \in N(S_2)$ where t_k' and t_l' are elements of $N(S_2)$ obtained from t_k and t_l, respectively, by replacing every occurrence of $s \in S_1$ by $\phi(s)$ and $*_i'$ is the corresponding operation in N_2.

This defines a mapping Φ from the set of all generating sequences on S_1 onto the set of all generating sequences on S_2. Denote by $T = (A_1, \ldots, A_m)$ the generating sequence A_1, ..., A_m of length m, $m \geq 1$ on S_1 and by $\Phi(T)$ the generating sequence $T' = (B_1, \ldots, B_m)$ on S_2 obtained from T in the way described above. Note that the result of T, denoted $r(T)$, is A_m.

The following result in [1] will be needed in the next section.

Theorem 3.1 *Let N_1 and N_2 be right near-rings with faithful left modules H_1 and H_2, respectively. Suppose $S_i \subseteq N_i$ are such that $N_i = N(S_i)$ for $i = 1, 2$ and let $\phi : S_1 \to S_2$ be a surjection. Let $\theta : H_1 \to H_2$ be a (group) epimorphism such that for every generating sequence T_1 on S_1 we have $\theta(r(T_1)h_1) = r(\Phi(T_1))\theta(h)$ for all $h_1 \in H_1$. Then $\tilde{\Phi} : r(T_1) \to r(\Phi(T_1))$ is an epimorphism from N_1 onto N_2. If in addition θ is injective then $N_1 \cong N_2$.*

4. EXAMPLES

It was shown in [3] that $R[G] \cong R\langle G \rangle$. We show that Theorem 3.1 yields an alternative proof for this result.

Let $S_1 = \{[r,g] : r \in R, g \in G\}$ and $S_2 = \{\langle r,g \rangle : r \in R, g \in G\}$.

Recall that $N(S_1) = R[G]$ and $N(S_2) = R\langle G \rangle$. Let $\phi : S_1 \to S_2$ be the surjection defined by $\phi : [r,g] \to \langle r,g \rangle$. It is easy to check that the mapping $\theta : R^G \to R^G$ defined by $(\theta\mu)(h) = \mu(h^{-1})$ for all $h \in G$, $\mu \in R^G$ is an automorphism of $(R^G, +)$.

It only remains to show that $\theta(r(T_1)\mu) = r(\Phi(T_1))\theta(\mu)$, for all generating sequences T_1 on S_1, $\mu \in R^G$. This is easily done by induction on the length of $r(T_1)$. We have thus proved

Theorem 4.1 $R[G] \cong R\langle G \rangle$.

Let (R,S) be a dg near-ring, G a group. It is shown in [2] by using induction on the complexity of an element A in $R\langle G \rangle$, that $R\langle G \rangle$ is dg by the subset $\langle S, G \rangle = \{\langle s,g \rangle : s \in S, g \in G\}$. We state this as

Theorem 4.2 *If (R,S) is a dg near-ring then*

$$R\langle G \rangle = \left\{ \sum_{i=1}^{m} \sigma_i \langle s_i, g_i \rangle : m \in \mathbb{N}, \sigma_i = \pm 1, s_i \in S, g_i \in G \right\}.$$

Furthermore, let H be the free (R,S) product of the set of (R,S) modules $\{Rg : g \in G\}$, where the action is defined by left multiplication. Then the group dg near-ring $(R(G), SG)$ is defined in [4] as the subnear-ring of $M(H)$ generated by SG, where $sg \cdot rg' = sr \cdot gg'$ for $s \in S$, $r \in R$, $g, g' \in G$.

As a second application of theorem 3.1 we show that $(R\langle G \rangle, S\langle G \rangle)$ is a homomorphic image of $(R(G), SG)$.

Let $S_1 = \{sg : s \in S, g \in G\}$ and $S_2 = \{\langle s,g \rangle : s \in S, g \in G\}$ and consider the surjection $\phi : S_1 \to S_2$ defined by $\phi(sg) = \langle s,g \rangle$. Define a mapping $\theta : Rg \to R^{(G)}$ by $\theta(rg) = \mu$ where $\mu : g \mapsto r$, $\mu(k) = 0$ if $k \neq g$. Then θ is an (R,S)-homomorphism.

By identifying Rg with its image in H and using the definition of a free product of the set of (R,S)-modules $\{Rg : g \in G\}$ it is possible to define a unique (R,S) homomorphism from H into $R^{(G)}$ which we also denote by θ. Now θ is onto, for if $\mu \in R^{(G)}$ and $\text{supp}\,\mu = \{g_1, \ldots, g_n\}$ with $\mu(g_i) = r_i$ for $i = 1, 2, \ldots, n$ and $\mu(k) = 0$ if $k \in \text{supp}\,\mu$, then let $\mu_{r_i}(g_i) = r_i$ and $\mu_{r_i}(x) = 0$ if $x \neq g_i$ for $i = 1, 2, \ldots, n$. Therefore $\theta(\sum_{i=1}^{n} r_i g_i) = \sum_{i=1}^{n} \theta(r_i g_i) = \sum_{i=1}^{n} \mu_{r_i} = \mu$.

It is an easy matter to show that $\theta((sg)(rg')) = \langle s,g \rangle \theta(rg')$. It therefore follows that

$$\theta\left((sg)\left(\sum_{i=1}^{n} r_i g_i\right)\right) = \langle s,g \rangle \theta\left(\sum_{i=1}^{n} r_i g_i\right)$$

and hence, in the notation of theorem 3.1, $\theta(r(T_1)h_1) = r(\Phi(T_1))\theta(h_1)$, for all $h_1 \in H_1 = H$ and for every generating T_1 on S_1 of length one. For any generating sequence T_1 on $S_1 = SG$, $r(T_1)$ is of the form $r(T_1) = r(T_1') + r(T_1'')$ for some generating sequences T', T'' on S_1. The proof by induction that $\theta(r(T_1)h_1) = r(\Phi(T_1))\theta(h_1)$, for all $h_1 \in H$ and for every generating sequence T_1 on S_1 can now easily be completed.

By theorem 3.1, $\tilde{\Phi} : (R(G), SG) \to (R\langle G\rangle, S\langle G\rangle)$ defined by $\tilde{\Phi}(r(T_1)) = r(\Phi(T_1))$, i.e., $\tilde{\Phi}(\sum_{i=1}^{n} \sigma_i s_i g_i) = \sum_{i=1}^{n} \sigma_i \langle s_i g_i\rangle$ is an epimorphism from $(R(G), SG)$ onto $(R\langle G\rangle, S\langle G\rangle)$ with $\ker \tilde{\Phi} = (\ker \theta : H)$ where $\ker \theta$ is the normal subgroup of H generated by the set of commutators $\{[r_1 g_1, r_2 g_2] : r_i \in R, g_i \in G, g_1 \neq g_2, i = 1, 2\}$.

By theorem 3.1 we have thus proved

Theorem 4.3 $(R(G), SG)/(\ker \theta : H) \cong (R\langle G\rangle, S\langle G\rangle)$.

References

[1] R. L. Fray, On group distributively generated near-rings, *J. Austral. Math. Soc. (Series A)* **52** (1992), 40–56.

[2] R. L. Fray, Dissertation, University of Stellenbosch, South Africa, 1989.

[3] L. R. Le Riche, J. D. P. Meldrum and A. P. J. van der Walt, On group near-rings, *Arch. math.* **52** (1989), 132–139.

[4] J. D. P. Meldrum, The group distributively generated near-ring. *Proc. London Math. Soc. (3)* **32** (1976), 323–346.

SIMPLICITY OF SOME NONZERO-SYMMETRIC CENTRALIZER NEAR-RINGS

Lucyna Kabza

Department of Mathematics, University of St. Thomas
3800 Montrose Boulevard, Houston, TX 77006, USA

1. Introduction

Let G be a group written additively with 0 and S a semigroup of endomorphisms of G. The set of functions

$$M_S(G) = \{f : G \to G \mid f \circ \alpha = \alpha \circ f, \forall \alpha \in S\}$$

is a centralizer near-ring under pointwise addition and function composition. Centralizer near-rings are very general, since every near-ring with identity is isomorphic to a centralizer near-ring for some pair (S, G) [9]. The structure of this type of near-rings has been an object of interest for some time and has been studied extensively by Maxson, Meldrum, Oswald, Smith, and Zeller. The simplicity of

$$M_A^0(G) = \{f \in M_A(G) \mid f(0) = 0\},$$

where A is a group of automorphisms was investigated in [3], [7], [10], and [11]. In [7] Maxson and Smith completely determined the simplicity of $M_A^0(G)$, when G is a finite group and A is a group of automorphisms, in terms of the stabilizers of elements of G^*.

Theorem 1. *Let G be a finite group and A a group of automorphisms of G. Then $M_A^0(G)$ is simple if and only if all stabilizer subgroups, $\mathrm{stab}(v)$ for v in G^*, are conjugate.*

In [11] the authors gave sufficient conditions for a near-ring $M_A^0(G)$ to have a proper ideal. They also showed that if A is fixed point free (f.p.f.) and $|A| < |G|$, then $M_A^0(G)$ is simple. The case when A is f.p.f. was further investigated in [3] where the authors gave necessary conditions on (G, A) for $M_A^0(G)$ to be simple. They also showed that if $M_A^0(G)$ is not simple then it has at most one proper ideal.

The study of the structure of a centralizer near-ring $M_S^0(G)$ with S a semigroup of endomorphisms was initiated by Maxson and Smith in [8], and continued by Maxson and Oswald in [6], where the two cases $G = Sx$ for some $x \in G$, and S a union of groups were investigated.

All the cases mentioned above involve various centralizer near-rings that are zero-symmetric. In this paper the simplicity of the following nonzero-symmetric centralizer near-rings is investigated:

- $M_A(G)$, where A is a group of automorphisms,
- $M_S(G)$, where S is an inverse semigroup of endomorphisms,

Y. Fong et al. (eds.), Near Rings and Near Fields, 145 151.
© *1995 Kluwer Academic Publishers.*

with emphasis on the relation between the simplicity of the near-ring $M_A(G)$ and the simplicity of its zero-symmetric subnear-ring, $M_A^0(G)$.

2. Simplicity of $M_A(G)$, A a group of automorphisms

Let A be a group of automorphisms of a group G, and consider the centralizer near-rings associated with the pair (G, A), namely:

$$N_0 = M_A^0(G) = \{f \in M_0(G) \mid f \circ \alpha = \alpha \circ f, \forall \alpha \in A\}$$

and

$$N = M_A(G) = \{f \in M(G) \mid f \circ \alpha = \alpha \circ f, \forall \alpha \in A\}.$$

If $A = \{id\}$, then $N = M(G)$ and $N_0 = M_0(G)$, and the simplicity of these two near-rings has been investigated by Berman and Silverman in [1]. Hence we assume for the rest of this paper that the cardinality of A, A a group of automorphisms, and $S \setminus \{0\}$, S a semigroup of endomorphisms, is at least 2. As was mentioned above the simplicity of $M_A^0(G)$ was determined completely when G is finite in [7]. Here we investigate the simplicity of the nonzero-symmetric near-ring $M_A(G)$ for an arbitrary group G. We find in fact that the simplicity of $M_A(G)$ coincides exactly with that of $M_A^0(G)$.

Theorem 2. *Let A be a group of automorphisms of a group G. Then $M_A(G)$ is simple if and only if $M_A^0(G)$ if simple.*

Proof. Since every near-ring is a group semi-direct sum of a zero-symmetric and a constant near-ring, we have $M_A(G) = M_A^0(G) + M_A^c(G)$. Let $W = \{w \in G \mid \alpha(w) = w, \forall \alpha \in A\}$. Under the addition in G and multiplication defined by $w \cdot v = w$, W is a near-ring. We show that $M_A^c(G) \cong W$. Let $w \in W$ and let c_w be the function in $M(G)$ that takes every element of G to w. We claim that c_w is an element of $M_A^c(G)$. First we need to show that c_w is in $M_A(G)$. Let $x \in G$, $\alpha \in A$, then $c_w(\alpha(x)) = w = \alpha(w) = \alpha(c_w(x))$. Now let $g \in M_A(G)$. Then $c_w(g(x)) = w = c_w(x)$ for every $x \in G$. Hence $c_w \in M_A^c(G)$. Let f be in $M_A^c(G)$. Then for every $g \in M_A(G)$, $f \circ g = f$. Hence $f(c_w(x)) = f(x)$. But, for every $x \in G$, $f(c_w(x)) = f(w) = f(x)$. Hence f is also constant on G, say $f(x) = v$ for every $x \in G$. Moreover if $\alpha \in A$, then $\alpha(f(x)) = \alpha(v) = f(\alpha(x)) = v$, so $v \in W$. Now if one defines $\phi : W \to M_A^c(G)$ by $v \mapsto c_v$, then ϕ is a bijection and it is straightforward to verify that ϕ is a near-ring homomorphism. To establish our result we look at various possibilities for W.

(i) If A is t.p.f., that is, for every $\alpha \in A$ and $0 \neq x \in G$ we have $\alpha(x) \neq x$, then $W = \{0\}$ and $M_A(G) = M_A^0(G)$. Hence the result follows.

(ii) If A is not f.p.f. then either $W = \{0\}$ and $M_A(G) = M_A^0(G)$, and the result follows, or $W \neq \{0\}$ and $M_A(G) = M_A^0(G) + M_A^c(G)$. We show that in the latter case both $M_A(G)$ and $M_A^0(G)$ are not simple, namely $l_N(W) = \{f \in M_A(G) \mid f(W) = 0\} = l_{N_0}(W)$ is a nontrivial ideal in $M_A(G)$ and $M_A^0(G)$ respectively. First we note that W is N-invariant. If $f \in M_A(G)$ then for every $\alpha \in A$ and for every $w \in W$ we have $f(\alpha(w)) = f(w) = \alpha f(w)$, hence $f(w) \in W$. Let $f : G \to G$ be defined by

$$f(x) = \begin{cases} 0, & \text{if } x \in W \\ x, & \text{if } x \notin W. \end{cases}$$

Then $f \in M_A(G)$. To see this let $\alpha \in A$ and $x \in G$. Then either $x \in W$ and $f(\alpha(x)) = f(x) = 0 = \alpha(f(x))$, or $x \notin W$ and then $\alpha(x) \notin W$ since α is injective (A is a group of automorphisms). Consequently $f(\alpha(x)) = \alpha(x) = \alpha(f(x))$. Hence f is an element of $M_A(G)$. Moreover $f \in l_N(W) = l_{N_0}(W)$. Since $W \neq G$, $f \neq 0$. Also $id \notin l_N(W) = l_{N_0}(W)$ since $W \neq \{0\}$. It follows that $l_N(W) = l_{N_0}(W)$ is a nontrivial ideal and $M_A(G)$ and $M_A^0(G)$ are not simple. This completes the proof. ■

Corollary 3. *Let G be a group and A a group of automorphisms of G. If $M_A^0(G) \subset M_A(G)$ then they are both non-simple.*

Using Theorem 1 we get

Corollary 4. *If A is a group of automorphisms of a finite group G, then $M_A(G)$ is simple if and only if $M_A(G) = M_A^0(G)$ and the stabilizers of G determined by A are conjugate.*

Now we give an example to show that the situation $M_A^0(G) \subset M_A(G)$ can indeed occur.

Example 1. Let $G = \mathbf{Z}_5 \times \mathbf{Z}_5$ and $A = \{id, \alpha_1, \alpha_2, \alpha_3\}$, where

$$\alpha_1 = \begin{bmatrix} 1 & 0 \\ 0 & 2 \end{bmatrix}, \alpha_2 = \begin{bmatrix} 1 & 0 \\ 0 & 3 \end{bmatrix}, \alpha_3 = \begin{bmatrix} 1 & 0 \\ 0 & 4 \end{bmatrix}.$$

Then $W = \left\{ \begin{bmatrix} 0 \\ 0 \end{bmatrix}, \begin{bmatrix} 1 \\ 0 \end{bmatrix}, \begin{bmatrix} 2 \\ 0 \end{bmatrix}, \begin{bmatrix} 3 \\ 0 \end{bmatrix}, \begin{bmatrix} 4 \\ 0 \end{bmatrix} \right\} \neq \left\{ \begin{bmatrix} 0 \\ 0 \end{bmatrix} \right\}$. Hence $M_A^c(G) \neq \{0\}$ and $M_A^0(G) \subset M_A(G)$. From Corollary 3 it follows that $M_A(G)$ and $M_A^0(G)$ are both nonsimple.

Next we show that the two near-rings can be nonsimple even though $M_A(G) = M_A^0(G)$.

Example 2. Let G he as before, $G = \mathbf{Z}_5 \times \mathbf{Z}_5$, and $A = \left\{ \begin{bmatrix} a & 0 \\ 0 & b \end{bmatrix}, a, b \in \mathbf{Z}_5 \right\}$. It is straightforward to verify that $W = \left\{ \begin{bmatrix} 0 \\ 0 \end{bmatrix} \right\}$. Hence $M_A(G) = M_A^0(G)$. We observe that stab $\left\{ \begin{bmatrix} 1 \\ 0 \end{bmatrix} \right\} = \left\{ \begin{bmatrix} 1 & 0 \\ 0 & b \end{bmatrix}, a, b \in \mathbf{Z}_5 \right\}$, and stab $\left\{ \begin{bmatrix} 2 \\ 1 \end{bmatrix} \right\} = \left\{ \begin{bmatrix} 1 & 0 \\ 0 & 1 \end{bmatrix} \right\}$. Hence the stabilizer subgroups cannot be conjugate and $M_A(G) = M_A^0(G)$ is not simple.

3. Simplicity of $M_S(G)$, S a semigroup of endomorphisms

Next we investigate the simplicity of the near-rings $M_S(G)$ and $M_S^0(G)$ where S is a semigroup of endomorphisms of a group G. We get the following necessary condition for the simplicity of $M_S(G)$ and $M_S^0(G)$.

Lemma 5. *If S is a semigroup of endomorphisms of a group G and $M_S^0(G)$ or $M_S(G)$ is simple, then no idempotent e, $0 \neq e \neq id$, of S can be central in S.*

Proof. Let e be a nontrivial idempotent in S that is central. Then $Ker\,e$ and $Im\,e$ are nontrivial subgroups of G and $G = Ker\,e + Im\,e$. Moreover, $Ker\,e$ and $Im\,e$ are S-invariant. Since if $x \in Ker\,e$, $y \in Im\,e$ and $\alpha \in S$, then $e(\alpha(x)) = \alpha(e(x)) = \alpha(0) = 0$ and $e(\alpha(y)) = \alpha(e(y)) = \alpha(y)$. Hence $\alpha(x) \in Ker\,e$ and $\alpha(y) \in Im\,e$. Also $Im\,e$ is N-invariant since if $x \in Im\,e$, then $f(x) = f(e(x)) = e(f(x))$ for every $f \in N$ and $f(x) \in Im\,e$. Consider $I = l_N(Im\,e) = l_{N_0}(Im\,e)$. Clearly I is an ideal since $Im\,e$ is N-invariant and hence N_0-invariant. To see that $I \neq \{0\}$ define $f : G \to G$ by $f(x) = x'$ where $x = x' + x''$, $x' \in Ker\,e$, $x'' \in Im\,e$. Then $f \in M_S(G)$. For if $x \in G$, $\alpha \in S$ then $\alpha(x') \in Ker\,e$, $\alpha(x'') \in Im\,e$ and $f(\alpha(x)) = f(\alpha(x' + x'')) = f(\alpha(x') + \alpha(x'')) = \alpha(x') = \alpha(f(x))$. Furthermore if $x \in Im\,e$ then $f(x) = 0$ and $f \in l_N(Im\,e) = l_{N_0}(Im\,e)$. Since $0 \neq e \neq id$, $\{0\} \neq Im\,e \neq G$, we have $f \not\equiv 0$ and $id \notin l_N(Im\,e)$. This shows that I is a nontrivial ideal in N and N_0, and the result follows. ∎

We recall that a Clifford semigroup is defined to be a regular semigroup with central idempotents, where regular means that every element has an inverse, i.e., for every $\alpha \in S$ there is an $\alpha^{-1} \in S$, such that $\alpha\alpha^{-1}\alpha = \alpha$ and $\alpha^{-1}\alpha\alpha^{-1} = \alpha^{-1}$. If a Clifford semigroup is not a group, then it has nontrivial idempotents, namely, $\alpha\alpha^{-1}$ is an idempotent for every $\alpha \in S$.

Corollary 6. *If S is an abelian or a Clifford semigroup with a nontrivial idempotent then $M_S(G)$ and $M_S^0(G)$ are not simple.*

Corollary 7. *Let G be a finite group and S an abelian semigroup, such that not all the elements are nilpotent. Then $M_S^0(G)$ is simple if and only if S is a group of f.p.f. automorphisms. In this case $M_S^0(G) = M_S(G)$, hence $M_S(G)$ is also simple.*

Proof. ($\Rightarrow$) If S is a finite abelian semigroup and not all the elements are nilpotent, then it has a nontrivial idempotent, since we are assuming the cardinality of $S \setminus \{0\}$ to be at least 2. Then it follows from Lemma 5, that $M_S^0(G)$ is not simple unless S is a group of automorphisms of G. From Theorem 1 all stabilizer subgroups of S are conjugate, i.e., $\forall v, w \in G^*$, $\exists \alpha \in S$ such that $stab(v) = \alpha\,stab(w)\alpha^{-1} = stab(w)$, since S is abelian. Hence all stabilizer subgroups are equal. This implies that for every $\alpha \in stab(v)$, $\alpha = id$. Hence $stab(v) = \{id\}$ and S is f.p.f.

($\Leftarrow$) If S is a f.p.f. group of automorphisms then clearly all stabilizers are conjugate and by Theorem 1, $M_S^0(G)$ is simple. When S is f.p.f., $M_S^c(G) = \{0\}$, hence $M_S(G) = M_S^0(G)$. ∎

4. Simplicity of $M_S(G)$, S an inverse semigroup of endomorphisms of G.

Now we consider $M_S(G)$ and $M_S^0(G)$ where S is a nonabelian semigroup of endomorphisms of a group G. In particular, we investigate in this section the nonzero-symmetric near-ring $M_S(G)$, where S is a finite inverse semigroup of endomorphisms of G. Recall that a semigroup is an inverse semigroup if for every element $\alpha \in S$ there is a unique $\alpha^{-1} \in S$ such that $\alpha\alpha^{-1}\alpha = \alpha$ and $\alpha^{-1}\alpha\alpha^{-1} = \alpha^{-1}$. Another characterization of an inverse semigroup is the following: S is regular and all idempotents commute with each other. More information can be found in [4] and [2]. As before,

$$M_S(G) = M_S^0(G) + M_S^c(G),$$

where $M_S^c(G) \cong W = \{x \in G \mid \alpha(x) = x \,\forall\alpha \in S\}$. Since we investigate the nonzero-symmetric near-ring $M_S(G)$, we assume for the rest of this section that $W \neq \{0\}$, and the zero function is not in S. Notice that for every $\alpha \in S$, $\alpha \neq 0$ (since $0 \notin S$) and $\alpha\alpha^{-1}$ and $\alpha^{-1}\alpha$ are nonzero idempotents. First we have the following necessary condition for the simplicity of $M_S(G)$.

Lemma 8. *If $M_S(G)$ or $M_S^0(G)$ is simple then for every $x \in G$ there is an $\alpha \in S$ such that $\alpha(x) \in W$.*

Proof. (i) If $W = G$ then the result follows.

(ii) If $W \neq G$ then assume that there is $0 \neq x_0 \in G$ such that for ever $\alpha \in S$, $\alpha(x_0) \notin W$. We claim that $I = l_N(W) = l_{N_0}(W)$ is a proper ideal in N and N_0 respectively. Similarly as in the proof to Theorem 2, it is straightforward to verify that W is N-invariant. Next, we define $f : G \to G$ by

$$f(x) = \begin{cases} 0 & \text{if } x \in W \text{ or if } \exists\alpha \in S \text{ such that } \alpha(x) \in W \\ x & \text{otherwise.} \end{cases}$$

To see that $f \in M_S(G)$, let $\beta \in S$ and $x \in G$. If $x \in W$ then $\beta(x) = x$. Hence $f(\beta(x)) = f(x) = 0 = \beta(0) = \beta(f(x))$. If $x \notin W$ and for some $\alpha \in S$, $\alpha(x) \in W$, then

$$(\alpha^{-1}\alpha)(\beta^{-1}\beta)(x) = (\beta^{-1}\beta)(\alpha^{-1}\alpha)(x) = (\beta^{-1}\beta\alpha^{-1})(\alpha(x)),$$

since idempotents commute. But $\alpha(x) \in W$, which in turn implies

$$(\beta^{-1}\beta\alpha^{-1})(\alpha(x)) = \alpha(x) \in W.$$

Let $\gamma = \alpha^{-1}\alpha\beta^{-1}$. Then γ has the property that $\gamma(\beta(x)) \in W$. It follows that $f(\beta(x)) = 0 = \beta(0) = \beta(f(x))$.

(iii) If $x \notin W$ and for every $\alpha \in S$, $\alpha(x) \notin W$, then for every $\alpha \in S$,

$$\alpha(\beta(x)) \notin W \quad \text{and} \quad f(\beta(x)) = \beta(x) = \beta(f(x)).$$

This shows that $f \in M_S(G)$ and $f \in l_N(W) = l_{N_0}(W)$. Moreover, since $f(x_0) = x_0 \neq 0$, $f \not\equiv 0$. We assumed that $W \neq \{0\}$ which implies that $id \notin l_N(W)$. Hence $I = l_N(W) = l_{N_0}(W)$ is a nontrivial ideal in $M_S(G)$ and $M_S^0(G)$ respectively, which contradicts the simplicity of $M_S(G)$ and $M_S^0(G)$. The result now follows. ∎

For the rest of this section, we assume that for every $x \in G$, there is an $\alpha \in S$ such that $\alpha(x) \in W$. Before we state and prove our final result, we first state some properties of elements of an inverse semigroup of endomorphisms. Details and more information can be found in [2], [5].

Lemma 9. *For every* $\alpha \in S$, $Ker\,\alpha = Ker(\alpha^{-1}\alpha)$ *and* $Im\,\alpha = Im(\alpha\alpha^{-1})$.

We note that $\alpha^{-1}\alpha$ and $\alpha\alpha^{-1}$ are idempotents.

Lemma 10. *If* e *and* f *are idempotents in* S *then* (i) $Ker\,ef = Ker\,e + Ker\,f$, (ii) $Im\,ef = Im\,e \cap Im\,f$.

Since S is finite let $e_1, e_2, \ldots, e_n$ be the non-zero idempotents of S, and let $e = e_1 \cdot e_2 \cdot \ldots \cdot e_n$. Then e is a primitive idempotent, i.e., if e' is any idempotent in S, then either $ee' = e'e = 0$, or $ee' = e'e = e$. From Lemma 10 we have $Ker\,e = Ker\,e_1 + Ker\,e_2 + \cdots + Ker\,e_n$ and $Im\,e = Im\,e_1 \cap Im\,e_2 \cap \cdots \cap Im\,e_n$. Moreover from Lemma 9, for every $\alpha \in S$ there is an i, j, $1 < i, j \leq n$, such that $Ker\,\alpha = Ker\,e_i$ and $Im\,\alpha = Im\,e_j$. Hence $Im\,e = \cap_{i=1}^{n} Im\,e_i = \cap_{\alpha \in S} Im\,\alpha \supseteq W \neq \{0\}$, and $Ker\,\alpha = Ker\,e_i \subset Ker\,e$. As before $G = Ker\,e + Im\,e$.

Lemma 11. *Let* $x \notin Ker\,e$, *say* $x = x_1 + x_2$, *where* $x_1 \in Ker\,e$, *and* $0 \neq x_2 \in Im\,e$. *Then there is some* $\beta \in S$ *such that* $0 \neq \beta(x) \in W$.

Proof. Since $x \notin Ker\,e$ then $e(x) = e(x_1 + x_2) = e(x_2) = x_2 \neq 0$. By our working assumption, there is an $\alpha \in S$ such that $\alpha(x_2) \in W$. If $\alpha(x_2) = 0$ then $x_2 \in Ker\,\alpha = Ker\,e_i \subseteq Ker\,e$ which contradicts the choice of x. Hence $\alpha(x_2) \neq 0$, and $\alpha e(x) = \alpha(x_2) \in W$. Let $\beta = \alpha e$ and the result follows. ∎

Lemma 12. *Let* $x \in Ker\,e$. *If* $\alpha(x) \in W$ *then* $\alpha(x) = 0$.

Proof. Let $\alpha \in S$ such that $\alpha(x) \in W$. Then $\alpha^{-1}(\alpha(x)) = \alpha(x) \in W$. For some i, $1 \leq i \leq n$, $\alpha^{-1}\alpha = e_i$, hence $e_i(x) = \alpha^{-1}\alpha(x) = \alpha(x) \in W$. Since $e = e_1 \cdot e_2 \cdot \ldots \cdot e_n$ is a primitive idempotent we have $ee_i(x) = e(x) = 0$, ($x \in Ker\,e$). On the other hand $ee_i(x) = e(\alpha^{-1}\alpha(x)) = (e\alpha^{-1})(\alpha(x)) = \alpha(x) \in W$. It follows that $\alpha(x) = 0$. ∎

Now we define $f_0 : G \to G$ by

$$f_0 = \begin{cases} 0 & \text{if } \exists\alpha \subset S, \text{ such that } \alpha(x) \neq 0 \text{ and } \alpha(x) \in W \\ x & \text{if } \forall\alpha \in S \text{ such that } \alpha(x) \in W, \ \alpha(x) = 0. \end{cases}$$

Lemma 13. *The function* f_0 *is an element of* $M_S(G)$.

Proof. Let $\beta \in S$ and $x \in G$.

Case (i) If there is an $\alpha \in S$ such that $0 \neq \alpha(x) \in W$, then $f_0(x) = 0$. Moreover

$$(\alpha^{-1}\alpha)(\beta^{-1}\beta)(x) = (\beta^{-1}\beta)(\alpha^{-1}\alpha)(x) = (\beta^{-1}\beta\alpha^{-1})(\alpha(x)) = \alpha(x)$$

since $\alpha(x) \in W$. Further, since $\alpha(x) \neq 0$ then $\alpha(x) = (\alpha^{-1}\alpha\beta^{-1})(\beta(x)) \neq 0$. This implies that $f_0(\beta(x)) = 0$. Hence $f_0(\beta(x)) = 0 = \beta(0) = \beta(f_0(x))$.

Case (ii) If for every $\alpha \in S$, $\alpha(x) \in W$ implies that $\alpha(x) = 0$, then the same holds for $\beta(x)$. It follows that $f_0(\beta(x)) = \beta(x) = \beta(f_0(x))$.

Hence for every $\beta \in S$ and every $x \in G$ we have $f_0(\beta(x)) = \beta(f_0(x))$ and the result follows. ∎

Lemma 14. *If* $Ker\, e \neq \{0\}$ *then* $l_N(Im\, e) = l_{N_0}(Im\, e)$ *is a nontrivial ideal in* $M_S(G)$ *and* $M_S^0(G)$.

Proof. We first show that $Im\, e$ is N-invariant. Let $x \in Im\, e = \cap_{\alpha \in S} Im\, \alpha$. Then for every $\alpha \in S$, $x \in Im\, \alpha$. Hence there is a $y_\alpha \in G$ such that $x = \alpha(y_\alpha)$. Now if $f \in N$ then $f(x) = f(\alpha(y_\alpha)) = \alpha(f(y_\alpha)) \in Im\, \alpha$. This implies that for every $\alpha \in S$, $f(x) \in Im\, \alpha$. Hence $f(x) \in Im\, e = \cap_{\alpha \in S} Im\, \alpha$, and $Im\, e$ is N-invariant. Next, we see from Lemma 11 that f_0 as defined above is in $l_N(Im\, e) = l_{N_0}(Im\, e)$. Moreover $f \not\equiv 0$ since $Ker\, e \neq \{0\}$ and f_0 acts as an identity on $Ker\, e$ by Lemma 12. Further $id \notin l_N(Im\, e) = l_{N_0}(Im\, e)$ since $Im\, e \neq \{0\}$. ∎

Now we are ready to state and prove our final result.

Theorem 15. *Let* G *be a group and* S *a finite inverse semigroup of endomorphisms of* G *such that* $W \neq \{0\}$. *Then* $M_S(G)$ *and* $M_S^0(G)$ *are both nonsimple.*

Proof. Since $W \neq \{0\}$, $M_S^0(G) \subset M_S(G)$. If S is a group then the result follows from Corollary 3. If S is not a group then from Lemma 14, $l_N(Im\, e) = l_{N_0}(Im\, e)$ is a nontrivial ideal in $M_S(G)$ and $M_S^0(G)$ respectively. Hence $M_S(G)$ and $M_S^0(G)$ are both nonsimple. ∎

References

[1] G. Berman and R. J. Silverman, Simplicity of near-rings of transformations, *Proc. Amer. Math. Soc.* **10** (1959). 456–459.

[2] A. H. Clifford and G. B. Prestson, *The algebraic theory of semigroups*, Math. Surveys No. 7, Amer. Math. Soc., Providence RI, Vol. I (1961), Vol. II (1967).

[3] P. Fuchs, C. J. Maxson, M. R. Pettet and K. C. Smith, Centralizer near-rings determined by fixed point free automorphism groups. *Proc. Roy. Soc. Edinburgh Sect. A* **107** (1987), 327–337.

[4] J. M. Howie, *An introduction to semigroup theory*, Academic Press, London, 1976.

[5] S. J. Mahmood, J. D. Meldrum and L. O'Carroll, Inverse semigroups and near-rings, *J. London Math. Soc.* **23** (1981), 45–60.

[6] C. J. Maxson and A. Oswald, On the centralizer of a semigroup of group endomorphisms, *Semigroup Forum* **28** (1984), 30–45.

[7] C. J. Maxson and K. C. Smith, The centralizer of a set of group automorphisms, *Comm. Algebra* **8** (1980), 211–230.

[8] C. J. Maxson and K. C. Smith, The centralizer of a group endomorphism, *J. Algebra* **57** (1979), 441–448,

[9] C. J. Maxson and K. C. Smith, Near-ring centralizers, *Proc. 9th USL-Math. Conf.*, University of Southwestern Louisiana, Lafayette, 1979.

[10] C. J. Maxson and K. C. Smith, The centralizer of a group automorphism, *J. Algebra* **54** (1978), 27–41.

[11] J. D. P. Meldrum and M. Zeller, The simplicity of near-rings of mappings, *Proc. Roy. Soc. Edinburgh Sect. A* **90** (1981), 185–193.

CHARACTERIZATION OF SOME FINITE FERRERO PAIRS

Wen-Fong Ke

Department of Mathematics, National Cheng Kung University
Tainan 701, Taiwan, R. O. C.

Hubert Kiechle*

Mathematisches Institut, Technische Universität München
D-80290 München, Germany

Introduction

Every planar nearring gives rise to a Ferrero pair [2, §4] and isomorphic nearrings yield equivalent Ferrero pairs. From a given Ferrero pair one can construct different (i.e., nonisomorphic) planar nearrings. However, the geometries (cf. [2, §§6,7]) derived from these nearrings will be isomorphic. The case of 2-designs has been studied in [6] and some sort of a converse turned out to be true, also.

Clay [2] gives a lot of methods to construct Ferrero pairs (N, Φ). One purpose of this paper is to give a unifying description of these methods in some special cases. Specifically, we study the cases where N is elementary abelian or cyclic, and Φ is abelian.[1] Here, we can give a complete characterization, i.e., we have a simple description of all the possible Ferrero pairs and their equivalences or isomorphisms. All of these Ferrero pairs are *ring-generated* (cf. [2, (4.18)]), where the rings are actually direct products of fields, or $\mathbf{Z}_n$, respectively. However, we use the language of vector spaces, and cyclic groups, respectively, which seems more appropriate in our contexts.

In some instances, one or more of the $\Phi_0 a, a \in N$, are subgroups of N, where $\Phi_0 := \Phi \cup \{0\}$. Again, we give a characterization, generalizing approaches in [2, §§7.3,7.4]. The interest in this situation stems from geometric investigations (cf. [2, §7], [3]). We will not talk about these in the sequel.

In two supplementary statements (2.3) and (3.2.3) the *circular* Ferrero pairs are characterized in each case. They are easy consequences of theorems in [2, §5] and our results. We refer the reader to [2] for the definition of circularity and for what it is all about.

After a preliminary section, we cover the elementary abelian case in §2 and the cyclic case in §3.

---

* This work is one result of a Feodor-Lynen Fellowship. I'm very grateful both to the Alexander von Humboldt Foundation and to Prof. James R. Clay for making it possible.

[1] Φ is always abelian if N is cyclic, and Φ abelian implies Φ cyclic (cf. [8, (5.2.e)]).

Y. Fong et al. (eds.), Near-Rings and Near-Fields, 153–160.
© 1995 *Kluwer Academic Publishers.*

1. Definitions and basic results

Let $(N, +)$ be a group and Φ a subgroup of Aut N. Let $\mathbf{1}$ denote the identity map on N. We call (N, Φ) a *Ferrero pair* if the map $-\mathbf{1} + \phi : N \to N$ is bijective for every $\phi \in \Phi \setminus \{\mathbf{1}\}$. If N is finite, we call (N, Φ) a *finite* Ferrero pair. Since injectivity implies bijectivity for finite sets, finite Ferrero pairs are just *fixed point free* representations of Φ on the group N. Let (N, Φ) and (N', Φ') be Ferrero pairs. An isomorphism $\sigma : N \to N'$ is called an *equivalence* or an *isomorphism* between these Ferrero pairs if $\Phi' = \sigma \Phi \sigma^{-1}$. (This makes the representations $\Phi \to \text{Aut } N$ and $\Phi' \to \text{Aut } N'$ equivalent.)

For our description, we shall need a way to break Ferrero pairs into pieces. We do that by describing some sort of a product construction. However, our construction is not unique, so it is not a product in any category. Let $(N_1, \Phi), \ldots, (N_s, \Phi)$, $s \in \mathbf{N}$, be Ferrero pairs and put $N := N_1 \times \ldots \times N_s$. Note that the same group Φ is acting on each N_i. According to Clay [2, (5.42) p. 77], we have

(1.1) *Let* $\alpha_1, \ldots, \alpha_s \in \text{Aut } \Phi$. *For every* $\phi \in \Phi$ *define the action of* $\phi^{(\alpha_1, \ldots, \alpha_s)}$ *on* N *by*

$$\phi^{(\alpha_1, \ldots, \alpha_s)}(x_1, \ldots, x_s) := \left(\phi^{\alpha_1}(x_1), \ldots, \phi^{\alpha_s}(x_s)\right), \quad (x_1, \ldots, x_s) \in N,$$

and let $\Phi^{(\alpha_1, \ldots, \alpha_s)} := \{\phi^{(\alpha_1, \ldots, \alpha_s)}; \phi \in \Phi\}$, *then* $\left(N, \Phi^{(\alpha_1, \ldots, \alpha_s)}\right)$ *is a Ferrero pair.* ∎

(1.2) *Let* $\alpha_1, \ldots, \alpha_s, \beta_1, \ldots, \beta_s \in \text{Aut } \Phi$.

(1) *If* $\alpha_i^{-1} \alpha_i \hat{\sigma}_i = \beta_i^{-1} \beta_i$ *with an inner automorphism* $\hat{\sigma}_i$ *induced by* $\sigma_i \in \text{Aut } N_i$ *for every* $i \in \{2, \ldots, s\}$, *then*

$$\mathbf{1} \times \sigma_2 \times \ldots \times \sigma_s : \left(N, \Phi^{(\alpha_1, \ldots, \alpha_s)}\right) \to \left(N, \Phi^{(\beta_1, \ldots, \beta_s)}\right)$$

is an equivalence.

(2) *Let* $\sigma_1 \times \ldots \times \sigma_s : \left(N, \Phi^{(\alpha_1, \ldots, \alpha_s)}\right) \to \left(N, \Phi^{(\beta_1, \ldots, \beta_s)}\right)$ *be an equivalence, so* $\sigma_i \in$ *Aut* N_i *for* $i \in \{1, 2, \ldots, s\}$, *then* $\alpha_i^{-1} \alpha_i = \hat{\sigma}_1 \beta_i^{-1} \beta_i \hat{\sigma}_i^{-1}$ *for every* $i \in \{2, \ldots, s\}$, *where* $\hat{\sigma}_i$ *denotes the inner automorphism induced by* σ_i. *In particular, if* $\text{Aut } N =$ *Aut* $N_1 \times \ldots \times \text{Aut } N_s$, *then every equivalence is of this form.*

Proof. (1) For an arbitrary $\phi \in \Phi$ let $\phi^{\alpha_1} = \psi = \chi^{\beta_1}$, where $\psi, \chi \in \Phi$. Then we have

$$(\mathbf{1} \times \sigma_2 \times \ldots \times \sigma_s)\phi^{(\alpha_1, \ldots, \alpha_s)}(\mathbf{1} \times \sigma_2 \times \ldots \times \sigma_s)^{-1}$$
$$= (\phi^{\alpha_1}, \sigma_2 \phi^{\alpha_2} \sigma_2^{-1}, \ldots, \sigma_s \phi^{\alpha_s} \sigma_s^{-1})$$
$$= (\psi, \psi^{\alpha_1^{-1} \alpha_2 \sigma_2}, \ldots, \psi^{\alpha_1^{-1} \alpha_s \hat{\sigma}_s})$$
$$= (\chi^{\beta_1}, \chi^{\beta_2}, \ldots, \chi^{\beta_s}).$$

Therefore $\mathbf{1} \times \sigma_2 \times \ldots \times \sigma_s$ defines an equivalence.

(2) For every $\phi \in \Phi$ we find a $\psi \in \Phi$ such that

$$(\sigma_1 \times \ldots \times \sigma_s)\phi^{(\alpha_1, \ldots, \alpha_s)}(\sigma_1 \times \ldots \times \sigma_s)^{-1} = (\sigma_1 \phi^{\alpha_1} \sigma_1^{-1}, \ldots, \sigma_s \phi^{\alpha_s} \sigma_s^{-1})$$
$$= (\psi^{\beta_1}, \ldots, \psi^{\beta_s}),$$

and so we get
$$\phi^{\alpha_i \hat{\sigma}_i} = \psi^{\beta_i} \quad \text{for every} \quad i \in \{1, \dots, s\}.$$

Thus for all $i \in \{2, \dots, s\}$ we have $\phi^{\alpha_1 \hat{\sigma}_1 \beta_1^{-1}} = \psi = \phi^{\alpha_i \hat{\sigma}_i \beta_i^{-1}}$. Since this is independent of ψ, we have $\hat{\sigma}_1 \beta_1^{-1} \beta_i = \alpha_1^{-1} \alpha_i \hat{\sigma}_i$. This shows the result. ∎

Remark. Choosing $\hat{\sigma}_i = 1$ and $\alpha_1 = 1$ in (1.2.1) gives a "normalization" of the Ferrero pair $\left(N, \Phi^{(\beta_1, \dots, \beta_s)}\right)$. The statements [2, (14.23) and (14.31)] are special cases of this observation.

In this paper we will focus on the following specialization of the above construction. Let $\Phi = \langle \phi \rangle$ be a finite cyclic group of order k acting fixed point free on each N_i. For $d_i, i \in \{1, \dots, s\}$, all relatively prime to k, we will write $\Phi^{(d_1, \dots, d_s)}$ for the group obtained by the automorphisms $\phi \mapsto \phi^{d_i}$ of Φ as in (1.1). In this situation, Clay's [2, (14.45)] is our (1.1).

The following lemma can be found in [6, (1.5.1)] or [1, (40.1) p. 205], and will be useful later.

(1.3) *Let (N, Φ) be a finite Ferrero pair, then $|\Phi| \big| |N| - 1$.* ∎

We will write $\mathbf{Z}_n^\times$ for the automorphism group of $\mathbf{Z}_n = \mathbf{Z}/n\mathbf{Z}$, which is identified with the unit group of the ring. Then $\varphi(n) := |\mathbf{Z}_n^\times|$ becomes the Euler function. The multiplicative group of a field F will be denoted by F^*.

2. The elementary abelian case

Let $N_i = F$ for some finite field F of characteristic p, i.e., $N = F^s$ is a vector space over F, and let $\phi \in F$ be a primitive k-th root of unity. Then ϕ defines a fixed point free automorphism on the additive group of F by left multiplication. Throughout this section, we will identify ϕ with this automorphism.

(2.1) *Let $\Phi = \langle \phi \rangle$ and $c_i, d_i \in \mathbf{N}, i \in \{1, \dots, s\}$. In each of the following cases $\left(N, \Phi^{(c_1, \dots, c_s)}\right)$ and $\left(N, \Phi^{(d_1, \dots, d_s)}\right)$ are equivalent Ferrero pairs:*
(i) *There exists a permutation π of $\{1, \dots, s\}$ and $u \in \mathbf{N}$, such that $c_{\pi(i)} = u d_i$ for every $i \in \{1, \dots, s\}$.*
(ii) *For every $i \in \{1, \dots, s\}$ there exist ν_i and $u \in \mathbf{N}$, such that $c_i p^{\nu_i} = u d_i$.*

Proof. With respect to the standard basis of F^s, the maps $\phi^{(c_1, \dots, c_s)}$ and $\phi^{(d_1, \dots, d_s)}$ are represented by the diagonal matrices $\text{diag}(\phi^{c_1}, \dots, \phi^{c_s})$ and $\text{diag}(\phi^{d_1}, \dots, \phi^{d_s})$, respectively. For case (i), let M be the permutation matrix given by π, then

$$M \, \text{diag}(\phi^{c_1}, \dots, \phi^{c_s}) M^{-1} = \text{diag}(\phi^{d_1}, \dots, \phi^{d_s})^u,$$

so M gives the desired equivalence.

For (ii) we consider the map $\eta : N \to N; (x_1, \dots, x_s) \mapsto (x_1^{p^{\nu_1}}, \dots, x_s^{p^{\nu_s}})$. This is clearly in Aut N. By looking at the action on a typical element in N, one readily verifies that

$$\eta \phi^{(c_1, \dots, c_s)} \eta^{-1} = \phi^{(c_1 p^{\nu_1}, \dots, c_s p^{\nu_s})} = \left(\phi^{(d_1, \dots, d_s)}\right)^u.$$

Therefore η does it in this case.   ∎

Note that (ii) is actually a special case of (1.2.1), and generalizes [2, (14.17)].

The following example shows that the various $(N, \Phi^{(d_1,\dots,d_s)})$'s are not all equivalent. Let $s = 2$ and $F = \mathbf{Z}_p$, the prime field with p elements. For a divisor $k > 2$ of $p-1$ choose a primitive k-th root of unity ϕ. Let d be relatively prime to k, then by considering the eigenvalues, one easily sees that the only diagonal matrices conjugate to $\mathrm{diag}(\phi, \phi^d)$ are $\mathrm{diag}(\phi, \phi^d)$ and $\mathrm{diag}(\phi^d, \phi)$. So if for $d' \in \mathbf{N}$, relatively prime to k,

$$d' \not\equiv d \quad \text{and} \quad dd' \not\equiv 1 \mod k,$$

then $\phi^{(1,d)}$ and $\left(\phi^{(1,d')}\right)^u$ are not conjugate in $\mathrm{GL}(2, \mathbf{Z}_p) = \mathrm{Aut}\, N$ for any $u \in \mathbf{N}$. The Ferrero pairs $(N, \Phi^{(1,d)})$ and $(N, \Phi^{(1,d')})$ are therefore not equivalent. The parameters $p = 5, k = 4, d = 1, d' = 2$ constitute the "minimal" example. It is conjectured that if two Ferrero pairs (N, Φ) and (N, Φ') are not equivalent, then the geometric structures derived from them (cf. [2, §§6,7]) are not isomorphic, except for the trivial case $k = |N| - 1$. Some evidence for this is contained in [6]. Furthermore, a class of examples, where the conjecture is true, can be found in [5]. In fact, this class of examples contains the case when $k = p - 1, d = 1, d' \not\equiv 1$ mod k in the above construction.

(2.2) Theorem. *Let (N, Φ) be a finite Ferrero pair, where N is an elementary abelian p-group and Φ is abelian of order k. Let F be the k-th cyclotomic field over $\mathbf{Z}_p$, i.e., the smallest extension field of $\mathbf{Z}_p$ such that $k \,|\, |F^*|$. Then $N \cong F^s$ for some $s \in \mathbf{N}$ and there are a primitive k-th root of unity $\gamma \in F$ and $d_2, \dots, d_s \in \mathbf{N}$ such that (N, Φ) and $\left(F^s, \langle \gamma^{(1,d_2,\dots,d_s)} \rangle\right)$ are equivalent Ferrero pairs. In particular, Φ acts F-linear.*

Proof.[2] By [8, (I.5.2.e) p. 19] $\Phi = \langle \phi \rangle$ is cyclic. Let $\mathbf{Z}_p[t]$ act on N by

$$h(t)(x) := h(\phi)(x) \quad \text{for all} \quad h \in \mathbf{Z}_p[t] \quad \text{and all} \quad x \in N.$$

This makes N a $\mathbf{Z}_p[t]$-module.

Let us first assume that N is an irreducible $\mathbf{Z}_p[t]$-module. Then the minimal polynomial g of ϕ over $\mathbf{Z}_p$ is irreducible, hence $F := \mathbf{Z}_p[t]/(g)$ is a field and N is an irreducible F-vector space. This means $N = Fb$ for some $b \in N$. Furthermore, Φ acts F-linear since Φ is commutative. By the definition of g, the class $\gamma := t + (g)$ is a primitive k-th root of unity (it has multiplicative order $k = |\phi|$) and generates F, so F is the k-th cyclotomic field.

Now $\phi(xb) = (t + (g))xb = \gamma xb$ for all $x \in F$. Hence the map $F \to N; x \mapsto xb$ is an equivalence of the Ferrero pairs $(F, \langle \gamma \rangle)$ and (N, Φ) in this case.

To handle the general case, we use Maschke's Theorem [1, (12.9) p. 40] to decompose N into a direct sum of irreducible $\mathbf{Z}_p[t]$-modules and apply the above argument to each summand. The easy details are left to the reader.   ∎

---

[2]   We like to thank Emanuel Kolb for some helpful suggestions.

From this and [2, (5.43) p. 79 and (5.21) p. 69] we get the immediate corollary

(2.3) *Let* (N, Φ) *be a finite Ferrero pair, where* N *is an elementary abelian p-group and* Φ *is abelian of order* k. *Then* (N, Φ) *is circular if and only if there exists a finite extension field* F *of* $\mathbf{Z}_p$ *and a primitive* k-th *root of unity* γ *in* F, *such that* $(F, \langle \gamma \rangle)$ *is circular.* ∎

We can now give a simple criterion for equivalence of our Ferrero pairs.

(2.4) Theorem. *Let* (N, Φ) *and* (N', Φ') *be Ferrero pairs with elementary abelian p-groups* N *and* N' *and cyclic* Φ *and* Φ'. *Then* (N, Φ) *and* (N', Φ') *are equivalent if and only if there are generators* ϕ *and* ϕ' *of* Φ *and* Φ', *respectively, which, viewed as* $\mathbf{Z}_p$-*linear maps, have the same characteristic polynomial.*

Proof. If the characteristic polynomials are the same, then, in particular, they have the same degree. Thus N and N' have the same dimension and are therefore isomorphic $\mathbf{Z}_p$-vector spaces. Thus we may assume $N = N'$ for both directions!

Now let (N, Φ) and (N, Φ') be equivalent. This means that Φ and Φ' are conjugate. Choosing a generator ϕ of Φ, and letting ϕ' be the image of ϕ under this conjugation, ϕ and ϕ' have the same characteristic polynomial.

For the converse, let f be the common characteristic polynomial of ϕ and ϕ'. We apply (2.2) to (N, Φ), identifying whenever appropriate: Let F be the k-th cyclotomic field over $\mathbf{Z}_p$, where $k := |\Phi|$, then $N = F^s$ for some $s \in \mathbf{N}$ and $\phi = \gamma^{(d_1, \ldots, d_s)}$ with a primitive k-th root of unity γ, and integers $d_1, \ldots, d_s$ relatively prime to k. Clearly ϕ is represented by the diagonal matrix $\mathrm{diag}(\gamma^{d_1}, \ldots, \gamma^{d_s})$ over F. Using the Cayley-Hamilton theorem [7, XV.4.1. p. 545], we obtain

$$\mathrm{diag}(f(\gamma^{d_1}), \ldots, f(\gamma^{d_s})) \overset{!}{=} f(\phi) = 0.$$

Thus all the γ^{d_i} are roots of f.

Since ϕ acts on the direct sum $F \oplus \ldots \oplus F$ (s copies), leaving each summand invariant, we have $f = f_1 \cdot \ldots \cdot f_s$, where f_i is the characteristic polynomial of ϕ acting on the i-th summand of F^s (cf. [7, XV §4 p. 547]). This action is merely left multiplication by γ^{d_i} on F. So f_i is the minimal polynomial of γ^{d_i}, because $F = \mathbf{Z}_p(\gamma^{d_i})$. In particular, all the roots of f in F are primitive k-th roots of unity.

Similarly, all the eigenvalues of ϕ' are roots of f, hence have order k, and are thus in F. So by (2.2) ϕ' is diagonalizable over F and has order k. Since diagonalizing does not change the conjugacy class of ϕ', we may assume that ϕ' is represented by a diagonal matrix as well. Therefore $\phi' = \gamma^{(c_1, \ldots, c_s)}$, for appropriate integers c_i that are relatively prime to k. By (2.1.i) and the preceding paragraph, we may assume that $f_i(\gamma^{c_i}) = 0$ for every $i \in \{1, \ldots, s\}$ (choose a suitable ordering). Hence, the transitivity of the Galois group on the roots of an irreducible polynomial (cf. [4, Thm. 4.14 p. 259]) ensures that there is a field automorphism of F mapping γ^{d_i} to γ^{c_i}, i.e., $d_i p^{\nu_i} = c_i$ for some $\nu_i \in \mathbf{N}$. By (2.1.ii), (N, Φ) and (N, Φ') are equivalent. ∎

In the sequel, it will be convenient to write Φ_0 for $\Phi \cup \{0\}$ and extend the multi-plication by 0 and the action of 0 in the obvious way.

(2.5) *Let* (N, Φ) *be a finite Ferrero pair, where* N *is an elementary abelian p-group and* Φ *is abelian of order* k. *For* F *as in (2.2), if there exists* $a \in N^*$ *such that* $\Phi_0 a$ *is a subgroup of* N, *then* $\Phi \cong F^*$.

Proof. Carrying the addition on $\Phi_0 a$ over to Φ_0 gives a nearfield. Since Φ is abelian, Φ_0 is actually a field. Now F is the smallest field to contain a subgroup of order $|\Phi|$ in the multiplicative group, so we must have $\Phi \cong F^*$. ∎

We shall now describe a class of examples. Let F be a finite field of order $q = p^f, p$ prime, and let U be the subgroup of $\mathbf{Z}_{q-1}^\times$ generated by the residue class of p. This group has obviously order f, and is naturally isomorphic to Aut F. Choose $c_1, \ldots, c_r \in \{1, \ldots, q-1\}$, $r \leq \frac{\varphi(q-1)}{f}$, such that $c_1 U, \ldots, c_r U$ are different cosets of U in $\mathbf{Z}_{q-1}^\times$. In particular, c_i and $q-1$ are relatively prime. Let $s, s_1, \ldots, s_r \in \mathbf{N}$, such that $s = s_1 + \ldots + s_r$. Then we have $F^s = F^{s_1} \times \ldots \times F^{s_r}$. Let ϕ be a fixed generator of F^*. Then ϕ^{c_i} is also a generator of F^*, and the different ϕ^{c_i}'s are not conjugate, i.e., there is no automorphism of F which maps one to another.

Define the action of $\phi^{(c_1, \ldots, c_r)}$ on $F^{s_1} \times \ldots \times F^{s_r}$ as in (1.1). Then $\phi^{(c_1, \ldots, c_r)}$ is represented by a diagonal matrix $\text{diag}(\phi^{c_1}, \ldots, \phi^{c_1}, \ldots \ldots, \phi^{c_r}, \ldots, \phi^{c_r})$, with s_i entries ϕ^{c_i} at the appropriate place. Write $\Phi := \langle \phi^{(c_1, \ldots, c_r)} \rangle$, then (F^s, Φ) is a Ferrero pair by (1.1).

By (2.4) the parameters $r, s, s_1, \ldots, s_r$ and $c_1 U, \ldots, c_r U$ determine the equivalence of the Ferrero pair up to a permutation of the indices, i.e., "essentially" different parameters give nonequivalent Ferrero pairs. This is, because ϕ^{c_i} and ϕ^{c_j} have different minimum polynomials over $\mathbf{Z}_p$ for different i and j. From (2.2) it is clear that every Ferrero pair satisfying the assumptions of (2.5) has an equivalent Ferrero pair in the above described class of examples.

(2.6) Theorem. $\Phi_0 a$ *is a subgroup of* F^s *if and only if* $a \in F^{s_i}$ *for some* $i \in \{1, \ldots, r\}$, *where* F^{s_i} *is naturally embedded in* $F^s = F^{s_1} \times \ldots \times F^{s_r}$.

Proof. The "$\Longleftarrow$"-part is obvious.

Let $a = (a_1, \ldots, a_s) \in F^s$ be such that $\Phi_0 a$ is a subgroup of F^s. The case $a = 0$ is trivial, so let $a \neq 0$. In the following argument, there is no loss of generality to assume that $a_1 \neq 0$. For any $u, v \in \mathbf{N}$ there is a $w \in \mathbf{N}$ such that

$$\left(\phi^{(c_1, \ldots, c_r)}\right)^u a + \left(\phi^{(c_1, \ldots, c_r)}\right)^v a = \left(\phi^{(c_1, \ldots, c_r)}\right)^w a$$

$$\text{or} \quad \left(\phi^{(c_1, \ldots, c_r)}\right)^u a + \left(\phi^{(c_1, \ldots, c_r)}\right)^v a = 0.$$

Looking at the first component, we find

$$\phi^{c_1 u} + \phi^{c_1 v} = \phi^{c_1 w} \quad \text{or} \quad \phi^{c_1 u} + \phi^{c_1 v} = 0, \quad \text{respectively.}$$

For any $i \in \{s_1 + 1, \ldots, s\}$ with $a_i \neq 0$ there is a $j \in \{2, \ldots, r\}$ such that $(\phi^{c_j u} + \phi^{c_j v}) a_i = \phi^{c_j w} a_i$ or $(\phi^{c_j u} + \phi^{c_j v}) a_i = 0$, respectively. It follows that

$$\phi^{c_j u} + \phi^{c_j v} = \phi^{c_j w} \quad \text{or} \quad \phi^{c_j u} + \phi^{c_j v} = 0, \quad \text{respectively.}$$

Let $c_1' \in \mathbf{N}$ be the inverse of $c_1 \mod (q-1)$, i.e., $c_1 c_1' \equiv 1 \mod q-1$. Putting the last two displayed equations together, we find in both cases

$$(\phi^{uc_1})^{c_1' c_j} + (\phi^{vc_1})^{c_1' c_j} = (\phi^{uc_1} + \phi^{vc_1})^{c_1' c_j} \quad \text{for every} \quad u, v \in \mathbf{N}.$$

Thus the map $F \to F; x \mapsto x^{c_1' c_j}$ is an automorphism. Since $c_1 U$ and $c_j U$ are different cosets, this is a contradiction. Therefore $a_i = 0$ for all $i \in \{s_1 + 1, \ldots, s\}$, and the result follows. ∎

Remarks. (1) This theorem gives a generalization of a result of Clay and van der Walt [2, (7.44) p. 137].

(2) The development of [2, §7.4] also fits into this frame, and can be simplified by applying the theorem to the diagonalized form of the occuring matrices.

(3) Using the theorem, one can derive an alternative proof of [2, (7.13)] for cyclic Φ. This employs [2, (7.11)].[3]

3. The cyclic case

We now consider the case where N is cyclic. Then N is represented by $\mathbf{Z}_n$, where $n = |N|$. We continue to write $\mathbf{Z}_n^\times$ for the automorphism group of $\mathbf{Z}_n$.

The following lemma is a direct consequence of an example in [2, p. 49].

(3.1) Let p be a prime, ν a natural number and Φ a subgroup of $\mathbf{Z}_{p^\nu}^\times$. Then $(\mathbf{Z}_{p^\nu}, \Phi)$ is a Ferrero pair if and only if Φ is contained in the unique cyclic subgroup of order $p-1$ in $\mathbf{Z}_{p^\nu}^\times$. ∎

(3.2) Theorem. Let $n = \prod_{i=1}^s p_i^{\nu_i} \geq 2$ be the canonical prime factor decomposition of an integer n, such that p_1 is the smallest prime divisor of n. So

$$\mathbf{Z}_n = \mathbf{Z}_{p_1^{\nu_1}} \times \ldots \times \mathbf{Z}_{p_s^{\nu_s}}.$$

(1) If $(\mathbf{Z}_n, \Phi)$ is a Ferrero pair, then Φ is cyclic and the order k divides $p_i - 1$ for every $i \in \{1, \ldots, s\}$. There is an a in $\mathbf{Z}_n$ such that $\Phi_0 a$ is a subgroup of $\mathbf{Z}_n$ if and only if $k = p_1 - 1$. In this case $a \in \mathbf{Z}_{p_1^{\nu_1}}$, and this is the only subgroup of $\mathbf{Z}_n$ of this form.

(2) If k is a positive integer dividing $p_i - 1$ for $i \in \{1, \ldots, s\}$, then there are exactly $\varphi(k)^{s-1}$ subgroups Φ of $\mathbf{Z}_n^\times$ with $|\Phi| = k$ such that $(\mathbf{Z}_n, \Phi)$ is a Ferrero pair. More precisely: Each $\mathbf{Z}_{p_i^{\nu_i}}^\times$ contains exactly one subgroup $\Phi_i = \langle \phi_i \rangle$ of order k. Let $\Phi := \langle (\phi_1, \ldots, \phi_s) \rangle$. As the $d_i, i \in \{2, \ldots, s\}$, run through all integers with $1 \leq d_i \leq k$ and relatively prime to k, then $(\mathbf{Z}_n, \Phi^{(1, d_2, \ldots, d_s)})$ runs through the distinct Ferrero pairs of the above form, and they are pairwise inequivalent.

(3) The Ferrero pairs are circular if and only if each factor is circular.

---

[3] The last statement of this theorem is incorrect: Just use Φ_0 a prime field. However, this does not affect our arguments.

Proof. (1) Since all $\mathbf{Z}_{p_i^{\nu_i}}$ are Φ invariant, Φ operates faithfully on every $\mathbf{Z}_{p_i^{\nu_i}}$, and each $(\mathbf{Z}_{p_i^{\nu_i}}, \Phi)$ is a Ferrero pair by [6, (1.3)]. Now (3.1) shows the first part.

If $\Phi_0 a$ is a subgroup of $\mathbf{Z}_n$, then $k+1$ divides n. Since any prime divisor of $k+1$ is one of the p_i and $k+1 \leq p_1$, we conclude that $k = p_1 - 1$. Thus, the additive order of a is p_1, so $a \in \mathbf{Z}_{p_1^{\nu_1}}$.

If $k = p_1 - 1$, choose any $a \in \mathbf{Z}_n$ of order p_1 and consider $\Phi_0 a$. All elements in Φa have order p_1, and so Φa is the set of all elements in $\mathbf{Z}_n$ of order p_1. This makes $\Phi_0 a$ the unique subgroup of $\mathbf{Z}_n$ of order p_1.

(2) By assumption and (3.1) there are subgroups Φ_i of automorphisms on every $\mathbf{Z}_{p_i^{\nu_i}}$ having order k such that $(\mathbf{Z}_{p_i^{\nu_i}}, \Phi_i)$ is a Ferrero pair. As a direct consequence of (1.2.2), the $\left(\mathbf{Z}_n, \Phi^{(1, d_2, \dots, d_s)}\right)$, $1 \leq d_i \leq k$, and d_i, k relatively prime, are pairwise inequivalent Ferrero pairs, so there are at least $\varphi(k)^{s-1}$ different choices.

Given a Ferrero pair $(\mathbf{Z}_n, \Phi)$, we know that Φ is in $\mathbf{Z}_n^\times$. Furthermore, Φ acts faithfully on $\mathbf{Z}_{p_i^{\nu_i}}$ for all $i \in \{1, \dots, s\}$. Hence, it has to be one of the $\left(\mathbf{Z}_n, \Phi^{(1, d_2, \dots, d_s)}\right)$.

(3) See [2, (5.43)].   ∎

References

[1] M. ASCHBACHER, *Finite group theory.* Cambridge Univ. Press, Cambridge, 1988.

[2] J. R. CLAY, *Nearrings: Geneses and Applications.* Oxford Univ. Press, Oxford, 1992.

[3] J. R. CLAY & H. KARZEL, *Tactical Configurations Derived from Groups Having a Group of Fixed-Point-Free Automorphism.* J. Geom. **27** (1986), 473–502.

[4] N. JACOBSON, *Basic Algebra I*, 2nd ed. Freeman & Co., San Francisco, 1985.

[5] W. F. KE, *On nonisomorphic BIBD with identical parameters.* Annals of Discrete Mathematics (1992), 337–346.

[6] W.-F. KE & H. KIECHLE, *Automorphisms of Certain Design Groups.* J. Algebra **167** (1994), 488–500.

[7] S. LANG, *Algebra*, 2nd ed. Addison-Wesley, Reading, Massachusetts, 1984.

[8] H. WÄHLING, *Theorie der Fastkörper.* Thales Verlag, Essen, 1987.

ON PLANAR LOCAL NEARRINGS AND BACON SPREADS

Emanuel Kolb

Fachbereich Mathematik, Technische Hochschule Darmstadt

64289 Darmstadt, Germany

ABSTRACT

The notion of a planar local nearring is introduced and it is shown, that a planar local nearring N is abelian and gives rise to a certain partial group cover on $N \times N$, called a Bacon spread. Conversely, fixing two normal cover groups $\{X, Y\}$ with $T = X + Y$, one can define on a Bacon spread T a certain centralizer nearring $N_T(X, Y)$, called the nucleus, which is a local nearring. Further $N_T(X, Y)$ is planar and coordinatizes T, if it acts transitively on some group of the cover.

1. Introduction

The object of the present paper is to study the connection between Bacon spreads and planar local nearrings. A Bacon spread is a certain partial cover of a group $(T, +)$ with subgroups. Furthermore, there is a normal subgroup I of T such that T/I together with the factorized cover groups is a spread. Note that Bacon spreads are very useful to describe translation affine Klingenberg planes [1, 24.], in the same way as spreads are used for translation planes. Since planar local nearrings provide good examples of Bacon spreads, it is natural to ask, when a given Bacon spread can be built from a planar local nearring. We will answer this question in Theorem (3.5). The results presented here are a generalization of a method to coordinatize translation planes with nearfields [3].

2. Planar local nearrings

The definition of a local nearring is due to Maxson [4]. In the paper we use the term local nearring in the strong sense, i.e. by a local nearring we mean a zero-symmetric (left-)nearring $(N, +, \cdot)$ with identity such that the set $J = \{ u \in N \mid uN \neq N \}$ forms an ideal. In the definition of Maxson, these nearrings are the local nearrings with $J(N) = J$ ($J(N)$ the intersection of all strictly maximal left ideals of N) [4].

(2.1) Definition (Planar local nearring)
Let $N = (N, +, \cdot)$ be a local nearring with maximal ideal J.
N is a *planar local nearring*, if the planar condition

Y. Fong et al. (eds.), Near-Rings and Near-Fields, 161–167.
© *1995 Kluwer Academic Publishers*

For any $a, b \in N$ with $a - 1 \notin J$ the equation $ax - x = b$ has a unique solution, is satisfied.

(2.2) Remarks

1) The planar condition is equivalent to the following:

   For any $a, a', b \in N$ with $a - a' \notin J$, the equation $ax - a'x = b$ is uniquely solvable.

2) If N is a local nearring with maximal ideal J, then N/J is a nearfield. Furthermore, if N satisfies the planar condition, then N/J is planar.

(2.3) Example For F a field, set $(N, +, \cdot) = (F[x]/(x^3 F[x]), +, \cdot)$ the factorized polynomial ring. Then $N = F + F\epsilon + F\epsilon^2$ with $\epsilon = x + x^3 F[x]$. Denote by p the projection $p : N \to N$, $a_0 + a_1\epsilon + a_2\epsilon^2 \mapsto a_0$ and define a new multiplication $\circ$ by $a \circ b = a \cdot b$, if $a_0 \neq 0$, and $a \cdot p(b)$ else. Then $(N, +, \circ)$ is a planar local nearring, but not a ring and the maximal ideal is given by $J = \{\, a \in N \mid a_0 = 0 \,\}$.

As for nearfields $[6, \S2\ 2.2]$ one can show:

(2.4) Theorem *Let N be a local nearring with maximal ideal J and $1 + 1 = 2 \notin J$. Suppose that N satisfies the condition*
(WP) $ax - x = 0 \to x = 0$ for all $a \in N$ with $a - 1 \notin J$.
Then the following hold.
(i) $x^2 = 1 \Leftrightarrow x = \pm 1$.
(ii) $-1 \in Z(N, \cdot)$.
(iii) N is abelian.

Proof. (i) Let $x^2 = 1$ and set $y = 1 + (-1 + x)2^{-1}$. Then $xy = x + x(-1 + x)2^{-1} = x + (-x + 1)2^{-1} = 1 + (-x + 1)(-1) + (-x + 1)2^{-1} = 1 + (-x + 1)(-1 + 2^{-1}) = 1 + (-x + 1)2^{-1}(-2 + 1) = 1 + (-x + 1)(-1)2^{-1} = 1 + (-(-x + 1))2^{-1} = y$. Hence $xy - y = 0$. We distinct two cases:
(1) $x - 1 \in J$. Then $-1 + x \in J \Rightarrow y \notin J$ and $xy = 1 \cdot y$ implies $x = 1$.
(2) $x - 1 \notin J$. Then $y = 0$ by (WP), which implies $x = -1$.
(ii). Suppose first that $x \notin J$. Then $(x(-1)x^{-1})^2 = x(-1)^2 x^{-1} = 1 \Rightarrow x(-1)x^{-1} = \pm 1$. But $x(-1)x^{-1} = 1$ implies $x(-1) = -x = x \Rightarrow x = 0$. Therefore $x(-1)x^{-1} = -1$. If $x \in J$ observe that $1 + x \notin J$ and hence $(-1)x - 1 = (-1)(1 + x) = (1 + x)(-1) = -x - 1 \Rightarrow (-1)x = x(-1)$.
(iii). By (ii), $-y - x = (x + y)(-1) = (-1)(x + y) = -x - y$ for all $x, y \in N$. $\square$

Note that Theorem (2.4) remains true also for local nearrings with $J(N) \neq J$.

(2.5) Corollary *Every planar local nearring is abelian.*

3. Bacon spreads

A *spread* is defined as a pair $(T, \mathcal{U})$, where $(T, +)$ is a group and $\mathcal{U}$ is a set of T-subgroups such that $U \cap V = \{0\}$, $T = U + V$ for any distinct $U, V \in \mathcal{U}$ and to any $u \in T \backslash \{0\}$ there is exactly one $U \in \mathcal{U}$ with $u \in U$.

(3.1) Definition (Bacon spread)
Let $(T, +)$ be a group, $\mathcal{U}$ a set of T-subgroups and I a normal subgroup of T. $\mathcal{T} = (T, \mathcal{U}, I)$ is a *Bacon spread*, if the following hold:
(i) For all $U \in \mathcal{U}$, $I \subset U + I \subset T$.
(ii) For all $U, V \in \mathcal{U}$, $U + I \neq V + I$ implies $U \cap V = \{0\}$ and $(U + I) \cap V \subseteq I$.
(iii) For all $u \in T \backslash I$, there is exactly one $U \in \mathcal{U}$ with $u \in U$.
(iv) For all $U, V \in \mathcal{U}$, $U + I \neq V + I$ implies $T = U + V$.
A set $\{X, Y\}$ of two normal subgroups with $X + I \neq Y + I$ is a *base* of $\mathcal{T}$.

(3.2) Remarks

1) The term Bacon spread is just another name of Bacon's definition of a group component triple $[1, 24.1.4]$.

2) The Bacon spreads $(T, \mathcal{U}, \{0\})$ are exactly the spreads $(T, \mathcal{U})$.

3) Suppose that $\mathcal{T} = (T, \mathcal{U}, I)$ is a Bacon spread and define $\mathcal{U}/I = \{U + I \mid U \in \mathcal{U}\}$. Then it is easy to see, that $\mathcal{T}/I = (T/I, \mathcal{U}/I)$ is a spread.

4) Let $\{X, Y\}$ be a base of the Bacon spread $\mathcal{T}$. Then $[X, Y] = \{0\}$ $([,]$ the commutator) and thus $T \simeq X \times Y$.

5) From any Bacon spread $\mathcal{T} = (T, \mathcal{U}, I)$ one can obtain an affine Klingenberg plane (see for definition $[2]$) as follows: Choose T as point set, the left cosets $t + U$ with $U \in \mathcal{U}$, $t \in T$ as line set and define a parallelism $\|$ by $t + U \| s + V \leftrightarrow U = V$ and a neigbor relation $\circ$ for points by $t \circ s \leftrightarrow t - s \in I$ (resp. for lines by $t + U \circ s + V \leftrightarrow t + U + I = s + V + I$).

(3.3) Lemma *Let $\mathcal{T} = (T, \mathcal{U}, I)$ be a Bacon spread and $U, V \in \mathcal{U}$ with $U + I \neq V + I$. Then $|t + U \cap s + V| = 1$ for all $t, s \in T$.*

Proof. Find $u \in U$, $v \in V$ such that $-t + s = u - v$. Hence $t + U \cap s + V \neq \emptyset$. Now let $t + u' = s + v'$. Then $-(t + u') + (t + u) = -(s + v') + (s + v) \in U \cap V = \{0\}$ $\Rightarrow u = u'$ and $v = v'$. $\square$

(3.4) Proposition *Let N be a local nearring with maximal ideal J. Define $\mathcal{U}_N = \{(1, a)N, (b, 1)N \mid a \in N, b \in J\}$ and assume that $(2.4)(WP)$ holds for N. Then the triple $(N \times N, \mathcal{U}_N, J \times J)$ satisfies the conditions $(3.1)(i)$-(iii).*

Proof. Throughout the following, set $I = J \times J$.
$(3.1)(i)$: Obviously $(1, a) \in (1, a)N \backslash I$, $(b, 1) \in (b, 1)N \backslash I$, $(0, 1) \notin (1, a)N$ and $(1, 0) \notin (b, 1)N$.

(3.1) (ii): As for the first implication, consider, for example, $(1, a)N + I \neq (1, a')N + I$ with $(1, a)r = (1, a')r'$ for suitable $r, r' \in N$. Then $r = r'$ and $ar - a'r' = 0$. Since a or $a' \notin J$, we obtain, if w.l.o.g. $a \notin J$, $a(a^{-1}a'r - r) = 0 \Rightarrow a - a' \in J$ or $r = 0$ by (WP). But $a - a' \in J$ implies $(1, a)N + I = (1, a')N + I$. Thus $r = 0$. The case $(1, a)N + I \neq (b, 1)N + I$ can be shown similarly. For the second implication observe that $\mathcal{U}/I = \{U + I \mid U \in \mathcal{U}_N\}$ conicides with $\mathcal{U}_{N/J}$ and there the first implication is valid.

(3.1) (iii): Let $u = (x, y) \notin I$. Case 1: $x \notin J$. Then $u \in (1, yx^{-1})N$. Case 2: $x \in J$. Then $y \notin J$ and $u \in (xy^{-1}, 1)N$. It is clear that these subgroups are unique. $\square$

Note that $(N \times N, +)$ endowed with the right-multiplication of N is also an N-group.

(3.5) Theorem *Let N be a local nearring with maximal ideal J, $\mathcal{U} = \mathcal{U}_N$ as in (3.4) and $I = J \times J$. Then $\mathcal{N} = (N \times N, \mathcal{U}, I)$ is a Bacon spread, iff N fulfils the planar condition. Furthermore, in this case, $\{(1, 0)N, (0, 1)N\}$ is a base of $\mathcal{N}$.*

Proof. Assume first that N is planar. Then N satisfies (2.4) (WP) and by (3.4) we only have to prove (3.1) (iv). As for $N \times N = (1, a)N + (1, a')N$ for $a - a' \notin J$, we have to solve the equation $(1, a)r + (1, a')r' = (x, y)$ for any $(x, y) \in N$. Then $r + r' = x$, $ar + a'r' = y \Leftrightarrow r' = -r + x$, $ar - a'r = y - a'x$. Since N is planar a unique solution exists. To prove $N \times N = (1, a)N + (b, 1)N$ for $a \in N$, $b \in J$ we have to solve $r + bs = x$, $ar + s = y$, which is equivalent to $bar - r = by - x$, $s = -ar + y$. Because $ba - 1 \notin J$, a unique solution exists. Conversely, suppose that $\mathcal{N}$ is a Bacon spread and take any $a, b \in N$ with $a - 1 \notin J$. Since $(1, a)N + I \neq (1, 1)N + I$ (for then $(1, a) \in (1, 1)c + I \Rightarrow a - 1 \in c - 1 + J \subseteq J$), it follows from (3.3) that $|(1, a)N \cap ((0, b) + (1, 1)N)| = 1$. Hence $ax = b + x$ is uniquely solvable and N is planar. Finally, since $(1, 0)N, (0, 1)N$ are normal subgroups with $(1, 0)N + I \neq (0, 1)N + I$, the theorem is proved. $\square$

In the Construction of (3.5), $\mathcal{N}$ is called the *Bacon spread over N*.

4. Coordinatization of Bacon spreads with planar local nearrings

Throughout the following let $\mathcal{T} - (T, \mathcal{U}, I)$ be a Bacon spread with base $\{X, Y\}$. By (3.2) (4), every $t \in T$ has a unique decomposition $t = t_X + t_Y$ with $t_X \in X$ and $t_Y \in Y$. Furthermore, $t \in I$, iff $t_X, t_Y \in I$.

(4.1) Definition (Nucleus) The *nucleus* $N_\mathcal{T}(X, Y)$ of $\mathcal{T}$ with respect to $\{X, Y\}$ consists of all maps $\alpha : T \to T$, $t \mapsto t^\alpha$ subject to the following conditions.
(i) $(x + y)^\alpha = x^\alpha + y^\alpha$ for any $x \in X, y \in Y$.
(ii) $U^\alpha \subseteq U$ for any $U \in \mathcal{U}$.
(iii) $(t + I)^\alpha - t^\alpha \subseteq I$ for any $t \in T$.

(4.2) Remarks

1) Any $\alpha \in N_\mathcal{T}(X, Y)$ induces by (4.1) (iii) a nucleus map α/I of the spread

T/I and conversely, any map of T with (4.1)(i),(ii), which can be factorized to a map of T/I, satisfies also (4.1)(iii).

2) We give another description of the nucleus using the concept of the centralizer of a covered group [4, II]. Let $p_X : T \to T$ (resp. $p_Y : T \to T$) be the projections $t \mapsto t_X$ (resp. $t \mapsto t_Y$). Then $S = \{0, 1, p_X, p_Y\}$, where 0 denotes the zero-map and 1 the identity, is a semigroup of endomorphisms of $(T, +)$. Now for a map $\alpha \in T^T$ condition (4.1)(i) is equivalent to $(\forall \sigma \in S) \sigma \alpha = \alpha \sigma$ and therefore $N_T(X, Y)$ is contained in $M_S(T, \mathcal{U} \cup \{I\}) = \{ \alpha \in T^T \mid (\forall U \in \mathcal{U} \cup \{I\}) U^\alpha \subseteq U, (\forall \sigma \in S) \sigma \alpha = \alpha \sigma \}$, the centralizer nearring of $< T, \mathcal{U} \cup \{I\}, S >$.

3) Collecting the results of (1) and (2) we obtain:

$\alpha \in N_T(X, Y)$, iff $\alpha \in M_S(T, \mathcal{U} \cup \{I\})$ and α can be factorized to a map of T/I.

(4.3) Proposition *Let $t \in T \backslash I$ and $\alpha \in N = N_T(X, Y)$. Then the following hold.*

a) *If $t^\alpha = 0$ then $\alpha = 0$.*

b) *If $t^\alpha \in I$ then $T^\alpha \subseteq I$.*

c) *If $t^\alpha \notin I$ then*

    (i) *α is bijective.*

    (ii) *$\alpha|_U : U \to U$ is surjective for any $U \in \mathcal{U}$ and $\alpha^{-1} \in N$.*

Proof. (a) Let $t^\alpha = (t_X)^\alpha + (t_Y)^\alpha = 0 \Rightarrow t_X^\alpha = t_Y^\alpha = 0$. Assume, for example, $x = t_X \notin I$. Then for any fixed $y \in Y$ there is a unique $U \in \mathcal{U}$ with $x + y \in U$ and $U + I \neq Y + I$, for $x + y \notin I$. But then $(x + y)^\alpha = y^\alpha \in U \cap Y = \{0\}$. Therefore $\alpha|_Y = 0$ and similarly $\alpha|_X = 0$, whence $\alpha = 0$.

(b) is clear, since (a) holds also for α/I.

(c.i) Let $(t_X)^\alpha + (t_Y)^\alpha = t^\alpha = s^\alpha = (s_X)^\alpha + (s_Y)^\alpha$. Then $(t_X)^\alpha = (s_X)^\alpha$ and $(t_Y)^\alpha = (s_Y)^\alpha$. Assume, for instance, that $x = t_X \neq s_X = u$. Then for a fixed $y \in Y \backslash I$ there are $U, V \in \mathcal{U}$ with $U + I, V + I \neq X + I$ and $x + y \in U$, $u + y \in V$. It follows that $x^\alpha + y^\alpha = u^\alpha + y^\alpha \in U \cap V$, whence $U = V$, since $y^\alpha \notin I$ by (b). Hence $x - u \in U \cap X = \{0\}$, a contradiction, and α is injective. For the surjectivity consider an $x \in X$ and fix any $y \in Y \backslash I$. Then $y^\alpha \notin I$ and we can find an $U \in \mathcal{U}$ with $U + I \neq X + I$ and $x + y^\alpha \in U$. For $z = x' + y \in (X + y) \cap U$ we conclude $z^\alpha = x'^\alpha + y^\alpha \in (X + y^\alpha) \cap U \Rightarrow z^\alpha = x + y^\alpha \Rightarrow x'^\alpha = x$. Hence $\alpha|_X : X \to X$, and by symmetry $\alpha|_Y : Y \to Y$, is surjective. From (4.1)(i) it follows that α is surjective.

(c.ii) Assume, for example, that $U \in \mathcal{U}$ with $U + I \neq Y + I$. For any $t = t_X + t_Y \in U$ set $s_X = t_X^{\alpha^{-1}}$ and $s \in (s_X + Y) \cap U$. Then $s^\alpha \in (t_X + Y) \cap U$ and thus $s^\alpha = t$.

As for $\alpha^{-1} \in N$ note that conditions (4.1) (i),(ii) are clear. For (4.1)(iii) observe that also α/I is invertible. Now (4.1)(iii) follows from (4.2)(1).
□

On $N_T(X,Y)$ we define now two compositions: An addition $+$ by the pointwise addition of two maps, and a multiplication $\cdot$ by the composition of two maps in $N_T(X,Y)$.

(4.4) Theorem *Let $N = N_T(X,Y)$ be the nucleus of a Bacon spread $T = (T,\mathcal{U},I)$ with base $\{X,Y\}$. Then*
a) $(N,+,\cdot)$ is a local nearring with maximal ideal $J = \{\alpha \in N \mid T^\alpha \subseteq I\}$ and $(T,+)$ with the natural action of N on T is an N-group.
b) $N \backslash J$ is semiregular on $U \backslash I$ for any $U \in \mathcal{U}$.

Proof. (a) First we show that N is closed under $+$ and $\cdot$. Since $M_S(T,\mathcal{U} \cup \{I\})$ with these operations is a zero-symmetric nearring, (4.1)(i),(ii) holds for $\alpha-\beta$, $\alpha\cdot\beta$ and (4.1)(iii) follows from the factorization of α, β to nucleus maps $\alpha/I, \beta/I$ of T/I. Hence $(N,+,\cdot)$ is a zero-symmetric nearring and $\mathbf{1}$ is an identity. Since I is a normal subgroup, we conclude that J is an N-subgroup and a left-ideal. But J is also a right-ideal, because for $\alpha, \beta \in N$, $\gamma \in J$, $t^{(\alpha+\gamma)\beta-\alpha\beta} = (t^\alpha + t^\gamma)^\beta - t^{\alpha\beta} \in I$ by (4.1)(iii). Now consider any $\alpha \in N \backslash J$. By definition there is a $t \in T \backslash I$ with $t^\alpha \notin I$ and, by (4.3)(c.ii), α is bijective with $\alpha^{-1} \in N$. Thus N is local and the remaining assertion is obvious.
(b) From (4.3) it follows that $N \backslash J$ maps $U \backslash I$ bijectively on itself. Suppose that $t^\alpha = t^\beta$ for $\alpha, \beta \in N \backslash J$, $t \in U \backslash I$. Then $t^{\alpha-\beta} = 0$, whence $\alpha = \beta$ by (4.3)(a). □

(4.5) Theorem *Let $T = (T,\mathcal{U},I)$ be a Bacon spread and $N = N_T(X,Y)$ the nucleus of T with respect to the base $\{X,Y\}$. Suppose that the following axiom holds:*
(NT) $N \backslash J$ is transitive on $U \backslash I$ for some $U \in \mathcal{U}$.
Then

  a) $N \backslash J$ is regular on $V \backslash I$ for any $V \in \mathcal{U}$.

  b) For any $V \in \mathcal{U}$ and $v \in V \backslash I$ the map $\Sigma_v : N \to V$, $\alpha \mapsto v^\alpha$ is an N-group isomorphism.

  c) There is an N-group isomorphism Φ from T to N, which maps $\mathcal{U}$ bijectively to the partial cover $\mathcal{U}_N$ defined in (3.4).

  d) N is a planar local nearring and T is isomorphic to the Bacon spread $\mathcal{N} = (N \times N, \mathcal{U}_N, J \times J)$ over N.

Proof. (a) By (4.4)(b) it is only to show that $N \backslash J$ is transitive on any $V \backslash I$ with $V \in \mathcal{U}$. Assume w.l.o.g. that $U + I \neq Y + I$ and take any $x, x' \in X \backslash I$, $u \in (x+Y) \cap U$, $u' \in (x'+Y) \cap U$. Then obviously $u, u' \notin I$ and by (NT) there is an $\alpha \in N \backslash J$ with $u^\alpha = u'$. Since $x = (u+Y) \cap X$, $x' = (u'+Y) \cap X$, it follows that

$x^\alpha = x'$. Thus $N \backslash J$ is transitive on $X \backslash I$. Apply the same argumentation (with X for U and V for X) to any $V \in \mathcal{U}$ with $V + I \neq Y + I$. Then $N \backslash J$ is transitive on $V \backslash I$ and, by symmetry, (a) follows at once.

(b) Clearly for $\Sigma = \Sigma_v$, $\Sigma(\alpha + \beta) = \Sigma\alpha + \Sigma\beta$ and $\Sigma(\alpha\beta) = (\Sigma\alpha) \cdot \beta$ $(\alpha, \beta \in N)$. Thus Σ is an N-group homomorphism and by (4.4)(b) Σ is injective. Suppose now that $v' \in V$. If $v' \in V \backslash I$, $v' \in \text{Im}(\Sigma)$ by (a). If $v' \in I \Rightarrow v' + v \in V \backslash I$ and there is an $\alpha \in N$ with $v^\alpha = v' + v$. But now $v^{\alpha - 1} = v^\alpha - v = v'$.

(c) Fix any $x \in X \backslash I$, $y \in Y \backslash I$ and consider the map $\Phi : N \times N \to T$, $(\alpha, \beta) \mapsto (\Sigma_x \alpha, \Sigma_y \beta) = x^\alpha + y^\beta$. It is obvious, that Φ is an N-group isomorphism and Φ^{-1} is given by $t = t_X + t_Y \to (\Sigma_x^{-1} t_X, \Sigma_y^{-1} t_Y)$. Furthermore, $\alpha, \beta \in J \Leftrightarrow x^\alpha + y^\beta \in I$. Now by (b), $\Phi((\mathbf{1}, \alpha)N) = (x + y^\alpha)^N \in \mathcal{U}$ and $\Phi((\beta, \mathbf{1})N) = (x^\beta + y)^N \in \mathcal{U}$ for any $\alpha \in N$, $\beta \in J$. Conversely, any $U \in \mathcal{U}$ has the form $(x + u_Y)^N$ or $(u_X + y)^N$ for suitable $u_Y \in N$ (resp. $u_X \in J$). Hence $\Phi^{-1}(U) = \Phi^{-1}(x + u_Y)N = (\mathbf{1}, \Sigma_y^{-1} u_Y)N \in \mathcal{U}_N$ (resp. $\Phi^{-1}(U) = \Phi^{-1}(u_X + y)N = (\Sigma_x^{-1} u_X, \mathbf{1})N \in \mathcal{U}_N$). Hence the assertion is proved.

(d) This follows directly from (c) and (3.5).

$\square$

From (2.5) and (4.5)(d) follows at once:

(4.6) Corollary *If the nucleus $N_T(X,Y)$ of a Bacon spread $\mathcal{T} = (T, \mathcal{U}, I)$ with base $\{X, Y\}$ fulfils axiom (4.5)(NT), then $(T, +)$ is abelian.*

5. References

[1] Bacon, P. Y.: An introduction to Klingenberg planes, Volume 3. Published by P. Y. Bacon, 3101 NW 2nd Av., Gainesville, FL 32607 (1979).

[2] Baker, C. A.,; Lane, N. D.; Lorimer, J. W.: An affine characterization of Moufang projective Klingenberg planes. Res. Math. **17** (1990), 27–36.

[3] Kolb, E.: The Schwan/Artin coordinatization for nearfield planes. t. a. in Geo. Ded.

[4] Maxson, C. J.: On local near-rings. Math. Z. **106** (1968), 197–205.

[5] Maxson, C. J.: Near-rings associated with generalized translation structures. J. Geo. **24** (1985), 175–193.

[6] Wähling, H.: Theorie der Fastkörper. Thales Verlag, Essen (1987).

CONSTRUCTION OF FINITE LOOPS OF EVEN ORDER

Alexander Kreuzer

Mathematisches Institut, Technische Universität München
80290 München, Germany

ABSTRACT

Using a construction method of [9], we give examples of non-associative loops with additional properties. These are power associative, left alternative loops which satisfy the automorphic inverse property and the left inverse property but not the Bol identity. It will be shown that, for $n, k \in \mathbb{N}$, non-isomorphic K-loops $(L, \oplus)$ of order $8kn$ exist which are also Bruck loops, having commutative subgroups $(G, \oplus)$ and $(H, \oplus)$ of order $4n$ and $2k$, respectively with $L = G \oplus H$.

1. Introduction

In order to describe sharply 2-transitive groups, H. Karzel introduced in [5] the notion of a neardomain $(F, \oplus, \cdot)$, which is a generalization of a nearfield where $\oplus$ need not be associative (cf. [15]). But until now no example of a neardomain which is not a nearfield is known. To obtain partial results, W. Kerby and H. Wefelscheid investigated separately the additive structure and called $(F, \oplus)$ a K-loop (see definition below), but they could not find a proper example. The interest in K-loops has grown since a famous physical example was found: A. A. Ungar showed that the velocities $\mathbb{R}_c^3 := \{v \in \mathbb{R}^3 : |v| < c\}$ (c denotes the speed of light) with the relativistic velocity composition $\oplus$ form a non-commutative and non-associative structure. Wefelscheid recognized that $(\mathbb{R}_c^3, \oplus)$ is a K-loop (cf. [13, 14]).

It is easy to show that a K-loop is always a Bruck loop. On the other hand, if in a Bruck loop there is no element of the order 2, by [8, Theorem 1] it is also a K-loop (cf. [4, 7]). Several examples of non-associative Bruck loops are known. For odd primes p and q with q dividing $p^2 - 1$, examples of Bol loops of order pq are given by H. Niederreiter and K. H. Robinson [10]. Some of these Bol loops are K-loops. Also T. Kepka constructs in [6] Bruck loops of order p^3. In [8] examples of Bruck and K-loops of order n^4, $n \in \mathbb{N}$, are given. Whereas the Bol loops of order $2^n k$ from [12] are not Bruck loops.

In this note we give examples of loops $(L, \oplus)$ of even order with an automorphism $\delta_{a,b}$ satisfying $a \oplus (b \oplus c) = (a \oplus b) \oplus \delta_{a,b}(c)$ and some additional properties. We see that the axioms of a K-loop are independent, except for the known property that in a loop (K6) implies (K4). In particular, we get examples of Bol loops, Bruck loops and K-loops. Furthermore we show that there are two non-isomorphic examples of order $8nk$.

Y. Fong et al. (eds.), Near-Rings and Near-Fields, 169–179.

Definition. Let L be a set with a binary operation $\oplus$. We call $(L, \oplus)$ a loop, if **(K1)** and **(K2)** are valid. A loop $(L, \oplus)$ is a **weak K-loop** or **WK-loop**, if **(K3)** is fulfilled, and a **K-loop** if in addition **(K4)**, **(K5)** and **(K6)** are valid.

(K1)   Given $a, b \in L$, the equations $a \oplus x = b$ and $y \oplus a = b$ have unique solutions $x, y \in L$.

(K2)   A two-sided neutral element $0 \in L$ exists, i.e. $a \oplus 0 = a = 0 \oplus a$ for every $a \in L$.

Let $a, b \in L$. By **(K1)**, for $x \in L$ there exists a unique element $x' \in L$ with $a \oplus (b \oplus x) = (a \oplus b) \oplus x'$. Hence for any $a, b \in L$, the map $\delta_{a,b} : L \to L, x \mapsto x'$ is a bijection with the property

(α) $$a \oplus (b \oplus x) = (a \oplus b) \oplus \delta_{a,b}(x)$$

(K3) $\delta_{a,b}$ is an automorphism of $(L, \oplus)$.

(K4)   If $a \oplus b = 0$, then $\delta_{a,b} = id$.

From **(K1)** to **(K4)** it follows

(I)   If $a \oplus b = 0$, then $b \oplus a = 0$,

since for $a, x \in L$ with $a \oplus x = 0$, $a \oplus (x \oplus a) = (a \oplus x) \oplus \delta_{a,x}(a) \overset{\textbf{(K4)}}{=} 0 \oplus a = a$, i.e., $x \oplus a = 0$ by **(K1)**. Hence x is the right and left inverse of a and we write $\ominus a = x$.

(K5)   The **automorphic inverse property** is satisfied, i.e., $(\ominus a) \oplus (\ominus b) = \ominus(a \oplus b)$.

(K6) $\delta_{a,b} = \delta_{a,b \oplus a}$.

In a loop, **(K4)** is equivalent to the **left inverse property** $(\ominus a) \oplus (a \oplus b) = b$ (see [9, (2.6)]). We will show in section 2 that **(K4)**, **(K5)** and **(K6)** are independent from the other axioms and we give an example of a WK-loop fulfilling **(K4)** and **(K5)** which is power associative and left alternative, but not fulfilling **(K6)** or **(B)** (see below). On the other hand (see [9, (2.10)]):

(1.1) *In a loop, axiom* **(K6)** *implies* **(K4)**.

Further, we remark that K-loops coincide with the so-called Gyrogroups and WACGs (weakly associative-commutative groupoids), respectively, which Ungar introduced in [13, 14].

A loop $(L, \oplus)$ is called a **Bol loop**, if for all $a, b, c \in L$ the **Bol-identity** is satisfied which is defined below

(B) $\quad a \oplus (b \oplus (a \oplus c)) = (a \oplus (b \oplus a)) \oplus c.$

This property first appeared in the geometric considerations of G. Bol [1] and has also been mentioned by R. H. Bruck [2]. A Bol loop is called a **Bruck loop** (cf. Robinson [11]), if moreover the automorphic inverse property **(K5)** is fulfilled. By [9, (2.2) and 8, (1.2)]:

(1.2) (i) *If in a loop* **(K3)**, **(K4)** *and* **(K5)** *are valid, then* **(K6)** *and* **(B)** *are equivalent.* (ii) *Every K-loop is a Bruck loop.*

A subset G of L is a subloop of a loop $(L, \oplus)$, if G is closed under $\oplus$ and $(G, \oplus)$ forms a loop. We call $(G, \oplus)$ a subgroup of $(L, \oplus)$, if $(G, \oplus)$ is a group.

2. Construction

Let $(G, +)$ and $(H, +)$ be commutative groups with the neutral elements $0 \in G$ and $0 \in H$, and let $U \leq G$ be a subgroup of G. We are interested in maps

$$\mu : G \times G \times H \to U; (a, c, d) \mapsto \mu(a, c, d) = \mu_{a,c,d},$$
$$\circ : H \times \mu(G \times G \times H) \to U; (b, \mu_{a,c,d}) \mapsto b \circ \mu_{a,c,d}$$

with the following properties. For all $a, c, x \in G$ and all $b, d \in H$:

(M1) $\quad$ For every $t \in U$, $b \circ \mu_{a,c+t,d} = b \circ \mu_{a,c,d} = b \circ \mu_{a+t,c,d}.$

(M2) $\quad 0 \circ \mu_{a,c,d} = 0 = b \circ \mu_{a,0,0} = b \circ \mu_{0,0,d}.$

(MI) $\quad b \circ \mu_{a,-a,-b} = 0.$

(M3) $\quad b \circ \mu_{a,c+x,d+y} = b \circ \mu_{a,c,d} + b \circ \mu_{a,x,y}.$

(M4) $\quad -(b \circ \mu_{a,c,d}) = (-b) \circ \mu_{-a,c,d}.$

(M5) $\quad -(b \circ \mu_{a,c,d}) = (-b) \circ \mu_{-a,-c,-d}.$

(M6) $\quad (b+d+b) \circ \mu_{a+c+a,x,y} = (b+d) \circ \mu_{a+c,x,y} - d \circ \mu_{c,x,y} + (b+d) \circ \mu_{a+c,x,y}.$

(MB) $\quad (b+d+b) \circ \mu_{a+c+a,x,y} = b \circ \mu_{a,x,y} + d \circ \mu_{c,x,y} + b \circ \mu_{a,x,y}.$

(M) $\quad$ There are $u, v \in G$ and $h \in H$ with $h \circ \mu_{u,v,-h} \neq 0.$

We set $L := G \times H$ and define for $(a, b), (c, d) \in L$:

(β) $\quad\quad \oplus : L \times L \to L; (a, b) \oplus (c, d) := (a + c + b \circ \mu_{a,c,d}, b + d).$

In [9, section 4], the following propositions (2.1), (2.2), and (2.3) are shown:

(2.1) Theorem. *Let* **(M1)** *and* **(M2)** *be fulfilled. Then:* i) $(L, \oplus)$ *is a loop.*

ii) For $G' = G \times \{0\}$ and $H' = \{0\} \times H$, $(G', \oplus)$ and $(H', \oplus)$ are commutative subgroups of $(L, \oplus)$ with $L = G' \oplus H'$. $(G', \oplus)$ is isomorphic to $(G, +)$ and $(H', \oplus)$ is isomorphic to $(H, +)$.

iii) If **(MI)** is valid, $(L, \oplus)$ satisfies **(I)**, and $\ominus(a, b) = (-a, -b)$ is the inverse of $(a, b) \in L$.

iv) If **(MI)** and **(M)** are valid, $(L, \oplus)$ is not associative.

Now we assume the properties **(M1)**, **(M2)** and **(M3)**. For $\mathfrak{a} = (a, b), \mathfrak{b} = (c, d) \in L$ let

(γ) $\qquad \delta_{\mathfrak{a}, \mathfrak{b}} : L \to L; (x, y) \mapsto (x + b \circ \mu_{a, x, y} + d \circ \mu_{c, x, y} - (b + d) \circ \mu_{a+c, x, y}, y)$

(2.2) $\delta_{\mathfrak{a}, \mathfrak{b}}$ *is an automorphism of* $(L, \oplus)$ *with* $\mathfrak{a} \oplus (\mathfrak{b} \oplus \mathfrak{x}) = (\mathfrak{a} \oplus \mathfrak{b}) \oplus \delta_{\mathfrak{a}, \mathfrak{b}}(\mathfrak{x})$ *for* $\mathfrak{a}, \mathfrak{b}, \mathfrak{x} \in L$.

(2.3) Theorem. *Let* **(M1)**, **(M2)** *and* **(M3)** *be fulfilled. Then:*

i) *Then* $(L, \oplus)$ *is a WK-loop.*

ii) $(L, \oplus)$ *fulfills* **(B)** *if and only if* **(MB)** *is fulfilled.*

iii) *Let* **(MI)** *be valid. Then* $(L, \oplus)$ *fulfills* **(K4)** *if and only if* **(M4)** *is fulfilled.*

iv) *Let* **(MI)** *be valid. Then* $(L, \oplus)$ *fulfills* **(K5)** *if and only if* **(M5)** *is fulfilled.*

v) *Let* **(MI)** *be valid.* $(L, \oplus)$ *is a Bruck loop if and only if* **(M5)** *and* **(MB)** *are fulfilled.*

(2.4) Theorem. *Let* **(M1)**, **(M2)**, *and* **(M3)** *be fulfilled. Then:*

i) **(M6)** *implies* **(M4)**.

ii) $(L, \oplus)$ *fulfills* **(K6)** *if and only if* **(M6)** *is fulfilled.*

iii) *Let* **(M4)** *and* **(M5)** *be valid. Then* **(M6)** *and* **(MB)** *are equivalent.*

Proof. **i)** Set $d = -b$ and $c = -a$ in **(M6)**, hence by **(M2)**, $b \circ \mu_{a, x, y} = -(-b) \circ \mu_{-a, x, y}$. **ii)** By equation (γ), for $\mathfrak{a} = (a, b)$, $\mathfrak{b} = (c, d)$, $\delta_{\mathfrak{a}, \mathfrak{b} \oplus \mathfrak{a}}(x, y) = (x + b \circ \mu_{a, x, y} + (d + b) \cup \mu_{a+c, x, y} - (b + d + b) \cup \mu_{a+c+a, x, y}, y) - \delta_{\mathfrak{a}, \mathfrak{b}}(x, y)$ iff $(d + b) \cup \mu_{a+c, x, y} - (b + d + b) \circ \mu_{a+c+a, x, y} = d \circ \mu_{c, x, y} - (b + d) \circ \mu_{a+c, x, y}$, i.e., iff **(M6)** is valid. **iii)** By [9, (4.1)], **(M2)**, **(M3)**, **(M4)** and **(M5)** imply $\operatorname{ord} b \circ \mu_{a, x, y} = 2$ for all $a, x \in G$ and $b, y \in H$. Hence both **(M6)** and **(MB)** are equivalent to $(b + d + b) \circ \mu_{a+c+a, x, y} = d \circ \mu_{c, x, y}$.

3. Examples of WK-loops.

Using the Theorems **(2.3)** and **(2.4)**, we give examples of WK-loops with additional properties. Let $(G, +)$ and $(H, +)$ be commutative groups and let U

be a subgroup of G, as in section 2. In this section we consider only maps $\mu : G \times G \times H \to U$, $(a, c, d) \mapsto \mu_{a,c,d}$ which depend only on $a, c \in G$ and not on $d \in H$, i.e., $\mu_{a,c,d} = \mu_{a,c,y}$ for every $y \in H$. We denote these maps by $\lambda : G \times G \to U$; $(a, c) \mapsto \lambda_{a,c}$. Therefore we can write the properties of section 2 as follows. Let $a, c \in G$ and $b, d \in H$:

(M1)   For every $t \in U$, $b \circ \lambda_{a,c+t} = b \circ \lambda_{a,c} = b \circ \lambda_{a+t,c}$.

(M2) $0 \circ \lambda_{a,c} = 0 = b \circ \lambda_{a,0}$.

(MI) $b \circ \lambda_{a,-a} = 0$.

(M3) $b \circ \lambda_{a,c+x} = b \circ \lambda_{a,c} + b \circ \lambda_{a,x}$.

(M4) $-(b \circ \lambda_{a,c}) = (-b) \circ \lambda_{-a,c}$.

(M5) $-(b \circ \lambda_{a,c}) = (-b) \circ \lambda_{-a,-c}$.

(M6) $(b + d + b) \circ \lambda_{a+c+a,x} = (b + d) \circ \lambda_{a+c,x} - d \circ \lambda_{c,x} + (b + d) \circ \lambda_{a+c,x}$.

(MB) $(b + d + b) \circ \lambda_{a+c+a,x} = b \circ \lambda_{a,x} + d \circ \lambda_{c,x} + b \circ \lambda_{a,x}$.

(M)   There are $u, v \in G$ and $h \in H$ with $h \circ \lambda_{u,v} \neq 0$.

Let $L := G \times H$ and for $(a, b), (c, d) \in L$: $(a, b) \oplus (c, d) := (a + c + b \circ \lambda_{a,c}, b + d)$.

(3.1) Example of a WK-loop, not satisfying (K4), (K5), and (K6).

For $n \in \mathbb{N} \setminus \{1\}$, let $(G, +) = (\mathbb{Z}_{n^2}, +)$ with the subgroup $U = n\mathbb{Z}_{n^2} = \{0, n, 2n, \dots, (n-1)n\}$ and

$$\lambda : G \times G \to U; (a, c) \mapsto \lambda_{a,c} := \begin{cases} cn, & \text{for } a \in U; \\ 0, & \text{for } a \notin U. \end{cases}$$

For a group $(H, +)$, let $M \subset H \setminus \{0\}$ be an arbitrary subset with $M \neq \emptyset$ and $-y \notin M$ for at least one $y \in M$. Let

$$\circ : H \times U \to U; (b, \lambda_{a,c}) \mapsto b \circ \lambda_{a,c} := \begin{cases} \lambda_{a,c} & \text{for } b \in M; \\ 0 & \text{for } b \notin M. \end{cases}$$

Then **(M1)**, **(M2)**, **(M3)**, **(M)** and **(MI)** are fulfilled, but not **(M4)**, **(M5)**, **(M6)** and **(MB)**.

Proof. **(M1)**: For $t \in U$, $t = sn$ for some $s \in G$. Then $a \in U$ iff $a + sn \in U$, hence $\lambda_{a,c} = \lambda_{a+ns,c}$. On the other hand $(c + sn)n = cn + sn^2 = cn$, i.e., $\lambda_{a,c} = \lambda_{a,c+sn}$. Thus $b \circ \lambda_{a,c} = b \circ \lambda_{a,c+sn} = b \circ \lambda_{a+sn,c}$. **(M2)**: Since $0 \notin M$ and $c \cdot 0 = 0 \in U$, obviously $0 \circ \lambda_{a,c} = 0$ and $\lambda_{a,0} = 0$, hence $b \circ \lambda_{a,0} = 0$. **(M3)**: Let $b \in M$, $a \in U$, then $b \circ \lambda_{a,c+x} = \lambda_{a,c+x} = (c + x)n = cn + xn = b \circ \lambda_{a,c} + b \circ \lambda_{a,x}$. If $b \notin M$ or $a \notin U$, then $b \circ \lambda_{a,c+x} = 0 = b \circ \lambda_{a,c} + b \circ \lambda_{a,x}$. **(MI)**: Let $a \notin U$, then $\lambda_{a,-a} = 0$. If $a = sn \in U$ for $s \in G$, then $\lambda_{a,-a} = -an = -sn^2 = 0$, i.e., $b \circ \lambda_{a,-a} = 0$. **(M)**: Let $b \in M$, then $b \circ \lambda_{0,1} = \lambda_{0,1} = n \neq 0$.

Now let $y \in M$ with $-y \notin M$, then $0 = (-y) \circ \lambda_{0,1} \neq -(y \circ \lambda_{0,1}) = -n \neq (-y) \circ \lambda_{0,-1} = 0$ i.e., (M4) and (M5) are not fulfilled. By [9,(4.1)] and (2.4), (M4) would be a consequence of (M6) or (MB), respectively, also (M6) and (MB) are not fulfilled.

(3.2) Example of a WK-loop with (K5), not satisfying (K4) and (K6).

For $n \in \mathbb{N} \setminus \{1\}$, let $G = \mathbb{Z}_{n^2}$, $U = n\mathbb{Z}_{n^2} = \{0, n, 2n, \ldots, (n-1)n\}$ and $\lambda : G \times G \to U$ as in (3.1). Let $(H, +)$ be a commutative group with a subset $\emptyset \neq M \subset H \setminus \{0\}$ with $-M = M$. $\circ : H \times U \to U$ is defined as in (3.1). Then (M1), (M2), (M3), (M5), (M), and (MI) are fulfilled, but not (M4), (M6) and (MB).

Proof. Since $0 \notin M$, we prove (M1), (M2), (M3), (M) and (MI) in the same way as in (3.1). Since $-M = M$ and U is a group, $b \notin M$ iff $-b \notin M$ and $a \in U$ iff $-a \in U$. Hence $-(b \circ \lambda_{a,c}) = -cn = (-c)n = (-b) \circ \lambda_{-a,-c} \neq (-b) \circ \lambda_{-a,c} = cn$ for $c \neq 0$. Thus (M5) is valid, but not (M4). Hence also (M6) and (MB) are not valid (cf. [9, (4.1)] and (2.4)).

(3.3) Example of a WK-loop with (K4), not satisfying (K5) and (K6).

For $n \in \mathbb{N} \setminus \{1\}$, let $G = \mathbb{Z}_{n^2}$, $U = n\mathbb{Z}_{n^2} = \{0, n, 2n, ..., (n-1)n\}$ and $\lambda : G \times G \to U$ as in (3.1). Let $H := G = \mathbb{Z}_{n^2}$. We define

$$\circ : H \times U \to U; (b, \lambda_{a,c}) \mapsto b \circ \lambda_{a,c} := b \cdot \lambda_{a,c} (\text{multiplication in } (\mathbb{Z}_{n^2}, +, \cdot))$$

Then (M1), (M2), (M3), (M4), (M) and (MI) are fulfilled, but for $n > 2$ not (M5), (M6) and (MB).

Proof. Since $0 \circ \lambda_{a,c} = 0 \cdot \lambda_{a,c} = 0 = b \cdot 0 = b \circ 0 = b \circ \lambda_{a,0}$, (M2) is valid and we prove (M1), (M3), (MI) and (M) as in (3.1).

To prove (M4), the only case we have to consider is $b \neq 0$, $c \neq 0$ with $bcn \neq 0$. Let $a \in U$, then $-(b \circ \lambda_{a,c}) = -bcn = (-b) \circ \lambda_{-a,c} \neq (-b) \circ \lambda_{-a,-c} = (-b)(-c)n = bnc$, i.e., (M4) is valid. But (M5) is not satisfied, if $-bcn \neq bcn$, for example for $b = c = 1$ and $n > 2$. For $n > 2$ we set $b = 0$, $d = 1$, $a = 1$, $c = -2$ and $x < n$, then $1 \circ \lambda_{-1,x} - 1 \circ \lambda_{-2,x} + 1 \circ \lambda_{-1,x} = 0 \neq (b + d + b) \circ \lambda_{a+c+a,x} = 1 \circ \lambda_{0,x} = xn \neq 0 \circ \lambda_{1,x} + 1 \circ \lambda_{-2,x} + 0 \circ \lambda_{1,x} = 0$ since $-2 \notin U$, i.e., $\lambda_{-2,0} = 0$, i.e., (M6) and (MB) are not fulfilled.

For a commutative group $(H, +)$ and $b \in H$ we set $0 \cdot b := 0$ and $n \cdot b := (n-1)b + b$ for $n \in \mathbb{N}$. We consider a non-empty subset $M \subset H$ with the properties:

(i) $0 \notin M$ and $M = -M$.

(ii) If $n \cdot b \in M$ then n is odd and $b \in M$.

(iii) There are $d \in M$ and $b \in H$ with $2 \cdot b + d \notin M$.

(3.4) Examples. The properties (i), (ii) and (iii) are fulfilled for

a) $(\mathbb{Z}_4 \times \mathbb{Z}_4, +)$ and $M = \{(1,2),(3,2)\}$.

b) $(\mathbb{R}^*, \cdot)$ and $M = \{-1\}$.

c) $(\mathbb{Q}^*, \cdot)$ and $M = \{2^m : m \in 2\mathbb{Z} + 1\}$.

Proof. **(i)** is obvious for all examples.

a) Now for $n \in \mathbb{N}$ and $(b_1, b_2) \in \mathbb{Z}_4 \times \mathbb{Z}_4$ let $n \cdot (b_1, b_2) = (nb_1, nb_2) \in \{(1,2),(3,2)\}$. Then $nb_1 \equiv 1 \bmod 4$ or $nb_1 \equiv 3 \bmod 4$ and $nb_2 \equiv 2 \bmod 4$ imply n odd, $b_1 \in \{1,3\}$ and $b_2 = 2$, i.e., $(b_1, b_2) \in M$. For $d = (1,2) \in M$ and $b = (0,1) \in H$, $2 \cdot b + d = (1,0) \neq M$.

b) For $x \in \mathbb{R}^*$, $x^n = -1$ implies $x = -1$ and n odd. $2^2 \cdot (-1) = -4 \notin M$.

c). For $x \in \mathbb{Q}^*$, $x^n = 2^m \in M$ implies $x = 2^k$ with $k \in \mathbb{Z}$ and $k \cdot n = m$. Since m is odd, also k and n are odd, i.e., $x \in M$. Furthermore $3^2 \cdot 2 = 18 \notin M$.

For a loop $(L, \oplus)$, $\mathfrak{a} \in L$ and $n \in \mathbb{N}$ we define $n * \mathfrak{a}$ recursively by $(n+1) * \mathfrak{a} := (n * \mathfrak{a}) \oplus \mathfrak{a}$. $(L, \oplus)$ is called **power associative** if $(n+k) * \mathfrak{a} = n * \mathfrak{a} \oplus k * \mathfrak{a}$ and **left power alternative** if $n * \mathfrak{a} \oplus (k * \mathfrak{a} \oplus \mathfrak{b}) = (n+k) * \mathfrak{a} \oplus \mathfrak{b}$ for any $n, k \in \mathbb{N}$ and $\mathfrak{a}, \mathfrak{b} \in L$. If a loop $(L, \oplus)$ is left power alternative, in particular it is power associative.

(3.5) Example of a left power alternative WK-loop with (K4), (K5) and $\delta_{n*\mathfrak{a}, k*\mathfrak{a}} = id$, not satisfying (B) and (K6).

Let $(G, +)$ be a commutative group with a subgroup T with index 2, i.e., for all $a \in G$ it holds $a + a \in T$ and let $U := \{0, t\}$ be a subgroup of order 2 for some $t \in T$ with $\text{ord}\, t = 2$.

For example let $G := \mathbb{Z}_{4n}$, $T := 2\mathbb{Z}_{4n} = \{0, 2, \ldots, 2n, \ldots, 4n - 2\}$ and $U := \{0, 2n\}$ for $n \in \mathbb{N}$.

We consider the following map

$$\lambda : G \times G \to \{0, t\}; (a, c) \mapsto \lambda_{a,c} := \begin{cases} t & \text{for } a \in T \text{ and } c \notin T; \\ 0 & \text{otherwise.} \end{cases}$$

Let $(H, +)$ be a commutative group and let $M \subset H \setminus \{0\}$ be a subset with the properties **(i)**, **(ii)** and **(iii)**. We define

$$\circ : H \times \{0, t\} \to \{0, t\}; (b, \lambda_{a,c}) \mapsto b \circ \lambda_{a,c} := \begin{cases} \lambda_{a,c} & \text{for } b \in M; \\ 0 & \text{for } b \notin M. \end{cases}$$

Then **(M)**, **(MI)** and **(M1)** to **(M5)** are fulfilled, but not **(M6)** and **(MB)**. Furthermore for $n, k \in \mathbb{N}$ and $\mathfrak{a} \in L$, $\delta_{n*\mathfrak{a}, k*\mathfrak{a}} = id$ (cf. equation (γ)).

Proof. **(M):** Let $b \in M$ and $c \notin T$, then $b \circ \lambda_{0,c} = t \neq 0$. **(MI):** Clearly $a \in T$ iff $-a \in T$, hence $\lambda_{a,-a} = 0$ and $b \circ \lambda_{a,-a} = 0$. **(M1):** For $t \in U \subset T$ we get

$a + t \in T$ iff $a \in T$ and $c \notin T$ iff $c + t \notin T$, hence $\lambda_{a+t,c} = \lambda_{a,c} = \lambda_{a,c+t}$. **(M2)**: Since $0 \in T$ and $0 \notin M$, $b \circ \lambda_{a,0} = b \circ 0 = 0 = 0 \circ \lambda_{a,c}$. **(M3)**: We have to show $\lambda_{a,c+x} = \lambda_{a,c} + \lambda_{a,x}$. Let $a \in T$. If $c, x \in T$, then $a + c \in T$, hence $0 = \lambda_{a,c+x} = \lambda_{a,c} + \lambda_{a,x} = 0 + 0$. If $c \in T$ and $x \notin T$ or $c \notin T$ and $x \in T$, respectively, then $c + x \notin T$, hence $t = \lambda_{a,c+x} = \lambda_{a,c} + \lambda_{a,x} = t$. If $c, x \notin T$, then $c + x \in T$, since T has index 2. Thus $0 = \lambda_{a,c+x} = \lambda_{a,c} + \lambda_{a,x} = t + t = 0$. **(M4)** and **(M5)**: Since T is a group, obviously $\lambda_{a,c} = \lambda_{-a,-c} = \lambda_{-a,c}$. Because $b \in M$ iff $-b \in M$ (cf. **(i)**) and $-t = t$, it follows $-(b \circ \lambda_{a,c}) = b \circ \lambda_{a,c} = (-b) \circ \lambda_{a,c} = (-b) \circ \lambda_{-a,c} = (-b) \circ \lambda_{-a,-c}$.

For $\mathfrak{a} = (a, b) \in L = G \times H$ with $\oplus$ defined as in equation (β), and for $n, k \in \mathbb{N}$, $(n \cdot b) \circ \lambda_{n \cdot a, k \cdot a} = 0$, since $n \cdot b \in M$ implies by **(ii)** n odd, hence $n \cdot a \in T$ if and only if $a \in T$. But then $k \cdot a \in T$, hence $\lambda_{n \cdot a, k \cdot a} = 0$. Therefore $n * \mathfrak{a} = (n \cdot a, n \cdot b)$ and $n * \mathfrak{a} \oplus k * \mathfrak{a} = (n + k) * \mathfrak{a}$, since recursively $n * \mathfrak{a} \oplus k * \mathfrak{a} = (n \cdot a, n \cdot b) \oplus (k \cdot a, k \cdot b) = (n \cdot a + k \cdot a + (n \cdot b) \circ \lambda_{n \cdot a, k \cdot a}, n \cdot b + k \cdot b) = ((n + k) \cdot a, (n + k) \cdot b)$. Hence for $(x, y) \in L$ and $n, k \in \mathbb{N}$, $\delta_{n * \mathfrak{a}, k * \mathfrak{a}}(x, y) = (x + (n \cdot b) \circ \lambda_{n \cdot a, x} + (k \cdot b) \circ \lambda_{k \cdot a, x} - ((n + k) \cdot b) \circ \lambda_{(n+k) \cdot a, x}, y)$. For $l \in \mathbb{N}$, by **(ii)** $(l \cdot b) \circ \lambda_{l \cdot a, x} \neq 0$ iff $l \cdot b \in M$, $l \cdot a \in T$ and $x \notin T$, i.e., by **(iii)** iff l is odd, $b \in M$, $a \in T$ and $x \notin T$. Now let $a \in T$, $b \in M$ and $x \notin T$, then $(l \cdot b) \circ \lambda_{l \cdot a, x} = t$ iff l is odd. Since only two or no one of the three numbers n, k, $n + k$ are odd and $\mathrm{ord}\, t = 2$, $\delta_{n * \mathfrak{a}, k * \mathfrak{a}}(x, y) = (x, y)$, i.e., $\delta_{n * \mathfrak{a}, k * \mathfrak{a}} = id$. It follows that $n * \mathfrak{a} \oplus (k * \mathfrak{a} \oplus \mathfrak{b}) = (n * \mathfrak{a} \oplus k * \mathfrak{a}) \oplus \delta_{n * \mathfrak{a}, k * \mathfrak{a}}(\mathfrak{b}) = (n + k) * \mathfrak{a} \oplus \mathfrak{b}$.

Since $2 \cdot a \in T$, $a + c + a \in T$ iff $c \in T$. Hence by $\mathrm{ord}\, t = 2$, **(MB)** is equivalent to $(2b + d) \circ \lambda_{c,x} = d \circ \lambda_{c,x}$. By **(iii)** $b \in H$ and $d \in M$ exists with $2 \cdot b + d \notin M$, hence for $x \in G \setminus T$, $0 = (2b + d) \circ \lambda_{0,x} \neq d \circ \lambda_{0,x} = \lambda_{0,x} = t$, i.e., **(MB)** and **(M6)** are not valid (cf. (2.4)).

4. Examples of K-loops and Bruck loops

In [9, section 5] two examples of K-loops are given. Here we determine the automorphisms $\delta_{\mathfrak{a}, \mathfrak{b}}$ of these loops and prove that they are not isomorphic. Therefore we determine the left, middle and right nucleus.

For a loop $(L, \oplus)$ denotes $C(L) := \{\mathfrak{a} \in L : \mathfrak{x} \oplus \mathfrak{a} = \mathfrak{a} \oplus \mathfrak{x} \text{ for all } \mathfrak{x} \in L\}$ the **center** of L. Furthermore is

$N_l := \{\mathfrak{a} \in L : \mathfrak{a} \oplus (\mathfrak{b} \oplus \mathfrak{c}) = (\mathfrak{a} \oplus \mathfrak{b}) \oplus \mathfrak{c} \text{ for all } \mathfrak{b}, \mathfrak{c} \in L\}$ the **left nucleus,**
$N_m := \{\mathfrak{b} \in L : \mathfrak{a} \oplus (\mathfrak{b} \oplus \mathfrak{c}) = (\mathfrak{a} \oplus \mathfrak{b}) \oplus \mathfrak{c} \text{ for all } \mathfrak{a}, \mathfrak{c} \in L\}$ the **middle nucleus,**
$N_r := \{\mathfrak{c} \in L : \mathfrak{a} \oplus (\mathfrak{b} \oplus \mathfrak{c}) = (\mathfrak{a} \oplus \mathfrak{b}) \oplus \mathfrak{c} \text{ for all } \mathfrak{a}, \mathfrak{b} \in L\}$ the **right nucleus.**

In this section let $(G, +)$ denote always a commutative group and T a subgroup of $(G, +)$ with index 2. For $m \in T$ with $\mathrm{ord}\, m = 2$, let $U := \{0, m\}$ be a subgroup of order 2. Let $(H, +)$ be a commutative group and let $V \subset H$ be a subgroup with index 2.

For example let $G := \mathbb{Z}_{4n}$, $T := 2\mathbb{Z}_{4n} = \{0, 2, \ldots, 2n, \ldots, 4n - 2\}$ and $U := \{0, 2n\}$ for $n \in \mathbb{N}$. And for $k \in \mathbb{N}$ let $H := \mathbb{Z}_{2k}$, $V := 2\mathbb{Z}_{2k} = \{0, 2, \ldots, 2k - 2\}$ or let $H := \mathbb{Z}$ and $V := 2\mathbb{Z}$.

(4.1) First example of a K-loop and a Bruck loop.

We define

$$\lambda : G \times G \to \{0, m\}; (a, c) \mapsto \lambda_{a,c} := \begin{cases} m & \text{for } a \in T \text{ and } c \notin T; \\ 0 & \text{otherwise}, \end{cases}$$

$$\circ : H \times \{0, m\} \to \{0, m\}; (b, \lambda_{a,c}) \to b \circ \lambda_{a,c} := \begin{cases} \lambda_{a,c} & \text{for } b \notin V; \\ 0 & \text{for } b \in V. \end{cases}$$

Then for $L := G \times H$ and $(a, b) \oplus (c, d) = (a + c + b \circ \lambda_{a,c}, b + d)$, $(L, \oplus)$ is a Bruck loop and a K-loop by **(2.3)** and **(2.4)**.

By definition, $T \times H$ and $G \times V$ are commutative subgroups of $(L, \oplus)$.

For the **proof** see [9, **(5.2.ii)**]. By **(2.3)**, $\delta_{\mathfrak{a},\mathfrak{b}}((x, y)) = (x + b \circ \lambda_{a,x} + d \circ \lambda_{c,x} - (b + d) \circ \lambda_{a+c,x}, y)$ for $\mathfrak{a} = (a, b)$, $\mathfrak{b} = (c, d) \in L$. For example **(4.1)** we can state the automorphisms $\delta_{\mathfrak{a},\mathfrak{b}}$ exactly. Let $\delta : L \to L; (x, y) \mapsto \begin{cases} (x, y) & \text{for } x \in T \\ (x + m, y) & \text{for } x \notin T \end{cases}$. Since $\operatorname{ord} m = 2$, $\delta^2 = id$, i.e., $\delta^{-1} = \delta$.

(4.2) *For all* $\mathfrak{a}, \mathfrak{b} \in L$, $\delta_{\mathfrak{a},\mathfrak{b}} \in \{id, \delta\}$. *Exactly* $\delta_{\mathfrak{a},\mathfrak{b}} = \delta$ *if and only if either* $b \in V$, $d \notin V$, $a \notin T$, *or* $b \notin V$, $d \in V$, $c \notin T$, *or* $b, d \notin V$, $a + c \notin T$.

Proof. For $x \in T$, $\lambda_{a,x} = \lambda_{c,x} = \lambda_{a+c,x} = 0$, hence $\delta_{\mathfrak{a},\mathfrak{b}}((x, y)) = (x, y)$. Now let $x \notin T$. If $b, d \in V$, clearly $b \circ \lambda_{a,x} = d \circ \lambda_{c,x} = (b + d) \circ \lambda_{a+c,x} = 0$. For $b \in V$ and $d \notin V$, it holds $d \circ \lambda_{c,x} = m$ iff $c \in T$ and $(b + d) \circ \lambda_{a+c,x} = m$ iff $a + c \in T$. We obtain $d \circ \lambda_{c,x} + (b + d) \circ \lambda_{a+c,x} = m$, iff $a \notin T$. If $b \notin V$ and $d \in V$ we get in the same way $b \circ \lambda_{c,x} + (b+d) \circ \lambda_{a+c,x} = m$, iff $c \notin T$. For $b, d \notin V$, $b \circ \lambda_{a,x} + d \circ \lambda_{c,x} = m$ iff $a \in T$, $c \notin T$ or $a \notin T$, $c \in T$, i.e. $a + c \notin T$.

(4.3) *Let* $(L, \oplus)$ *be the example of* **(4.1)**, *then* $C(L) = T \times V$, $N_l = N_m = T \times V$ *and* $N_r = T \times H$.

Proof. Let $\mathfrak{a} = (a, b)$, $\mathfrak{x} = (x, y) \in L$. Then $\mathfrak{a} \oplus \mathfrak{x} = \mathfrak{x} \oplus \mathfrak{a}$, iff $b \circ \lambda_{a,x} = y \circ \lambda_{x,a}$. This implies $a \in T$, since for $y = 1$, $x = 0$ and $a \notin T$, $b \circ \lambda_{a,x} = 0 \neq m = y \circ \lambda_{x,a}$. Hence for $a \in T$ and for every $x \in G$, $b \circ \lambda_{a,x} = 0$ iff $b \in V$, i.e., $C(L) = T \times V$.

To determine N_l (N_m, respectively), we must discuss for what $\mathfrak{a} \in L$ ($\mathfrak{b} \in L$, respectively) $\delta_{\mathfrak{a},\mathfrak{b}} = id$ for every $\mathfrak{b} \in L$ ($\mathfrak{a} \in L$, respectively). By **(4.2)** we get "$a \in T$ or $b \notin V$" and "$b \in V$" and "$b \in V$ or $a + c \in T$", hence $\mathfrak{a} = (a, b) \in T \times V$, i.e., $N_l = T \times V$ and analogous $N_m = T \times V$.

N_r is the set of all fixed points of all $\delta_{\mathfrak{a},\mathfrak{b}}$ hence from **(4.2)** $N_r = T \times H$.

Now we consider the situation that the map $\mu : G \times G \times H \to U; (a, c, d) \mapsto \mu_{a,c,d}$ of section 2 depends on $d \in H$.

(4.4) Second example of a K-loop and a Bruck loop.

We define the following maps

$$\mu : G \times G \times H \to \{0, m\}; (a, c, d) \mapsto \mu_{a,c,d} := \begin{cases} m & \text{for } d \in V \text{ and } c \notin T; \\ m & \text{for } d \notin V \text{ and } a + c \notin T; \\ 0 & \text{otherwise,} \end{cases}$$

$$\circ : H \times \{0, m\} \to \{0, m\}; (b, \mu_{a,c,d}) \mapsto b \circ \mu_{a,c,d} := \begin{cases} \mu_{a,c,d} & \text{for } b \notin V; \\ 0 & \text{for } b \in V. \end{cases}$$

Then by (2.3), (2.4), for $L := G \times H$ and $(a, b) \otimes (c, d) = (a + c + b \circ \mu_{a,c,d}, b + d)$, $(L, \otimes)$ is a Bruck loop and a K-loop.

By definition, $T \times H$ and $G \times V$ are commutative subgroups of $(L, \otimes)$.

For the proof see [9, (5.5)]. For $\mathfrak{a} = (a, b)$, $\mathfrak{b} = (c, d) \in L$, the automorphism $\delta_{\mathfrak{a},\mathfrak{b}}$ (cf. (2.3)) of this example $(L, \otimes)$ may have the following form:

$$\varphi : L \to L; (x, y) \mapsto \begin{cases} (x, y) & \text{for } y \in V; \\ (x + m, y) & \text{for } y \notin V. \end{cases}$$

(4.5) *For all $\mathfrak{a}, \mathfrak{b} \in L$, it holds $\delta_{\mathfrak{a},\mathfrak{b}} \in \{id, \varphi\}$ with $\varphi^2 = id$. Exactly $\delta_{\mathfrak{a},\mathfrak{b}} = \varphi$ if and only if $b \in V$, $d \notin V$, $a \notin T$, or $b \notin V$, $d \in V$, $c \notin T$, or $b, d \notin V$, $a + c \notin T$.*

Proof. For $b, d \in V$, $b \circ \mu_{a,x,y} = d \circ \mu_{c,x,y} = (b + d) \circ \mu_{a+c,x,y} = 0$, i.e., $\delta_{\mathfrak{a},\mathfrak{b}} = id$. Now let $b \in V$ and $d \notin V$. If $y \in V$, then $d \circ \mu_{c,x,y} = (b + d) \circ \mu_{a+c,x,y}$, hence $d \circ \mu_{c,x,y} + (b + d) \circ \mu_{a+c,x,y} = 0$. For $y \notin V$, $d \circ \mu_{c,x,y} + (b + d) \circ \mu_{a+c,x,y} = m$, iff either $c + x \notin T$ or $a + c + x \notin T$, thus iff $c + x + a + c + x \notin T$, i.e., iff $a \notin T$. Hence $\varphi_{\mathfrak{a},\mathfrak{b}}((x, y)) = (x + m, y)$, iff $y \notin V$ and $a \notin T$. In the same way we get for $b \notin V$ and $d \in V$, $\varphi_{\mathfrak{a},\mathfrak{b}}((x, y)) = (x + m, y)$, iff $y \notin V$ and $c \notin T$. Now let $b, d \notin V$. Then $b \circ \mu_{a,x,y} + d \circ \mu_{c,x,y} = m$, iff $y \notin V$ and $a + x + c + x \notin T$, i.e., iff $y \notin V$ and $a + c \notin T$.

(4.6) *Let $(L, \otimes)$ be the example of (4.4), then $C(L) = T \times V$, $N_l = N_m = T \times V$ and $N_r = G \times V$.*

Proof. Let $\mathfrak{a} = (a, b)$, $\mathfrak{x} = (x, y) \in L$. Then $\mathfrak{a} \otimes \mathfrak{x} = \mathfrak{x} \otimes \mathfrak{a}$, iff $b \circ \mu_{a,x,y} = y \circ \mu_{x,a,b}$. This implies $a \in T$ and $b \in V$. As in (4.3), $N_l = N_m = T \times V$. N_r is the set of all fixed points of all $\delta_{\mathfrak{a},\mathfrak{b}}$, hence by (4.5), $N_r = G \times V$.

(4.7) *Let $(G, +)$, $(H, +)$ be commutative groups with subgroups $T \subset G$ and $V \subset H$ with index 2, such that $T \times H$ and $G \times V$ are non-isomorphic. Then for $L := G \times H$, the K-loops $(L, \oplus)$ from (4.1) and $(L, \otimes)$ from (4.4) are non-isomorphic.*

Proof. If an isomorphism ψ from $(L, \oplus)$ onto $(L, \otimes)$ exists, then ψ maps the right nucleus $T \times H$ of $(L, \oplus)$ (cf. (4.3)) on the right nucleus $G \times V$ of $(L, \otimes)$ (cf. (4.6)). Since $T \times H$ and $G \times V$, respectively, are subgroups of $(L, \oplus)$ and

$(L, \otimes)$, respectively, ψ would be an isomorphism from the group $T \times H$ onto $G \times V$, a contradiction to the assumption.

(4.7) Corollary. *For $n, k \in \mathbb{N}$ let $n = 2^i \cdot n'$ and $k = 2^j \cdot k'$ with odd numbers $n', k' \in \mathbb{N}$ and $i, j \in \mathbb{N} \cup \{0\}$. If $j \neq i + 1$, in particular if $n = k$, then for $L := \mathbb{Z}_{4n} \times \mathbb{Z}_{2k}$ there exist two non-isomorphic K-loops of order $8kn$ which are also Bruck loops.*

Proof. The K-loops $(L, \oplus)$ and $(L, \otimes)$ exists by **(4.1)** and **(4.4)**. We have $T = 2\mathbb{Z}_{4n} \cong \mathbb{Z}_{2n}$ and $V = 2\mathbb{Z}_{2k} \cong \mathbb{Z}_k$. Then the maximal even order of an element $x \in T \times H = \mathbb{Z}_{2n} \times \mathbb{Z}_{2k}$ is 2^{i+1} or 2^{j+1} and of an element $y \in G \times V = \mathbb{Z}_{4n} \times \mathbb{Z}_k$ is 2^{i+2} or 2^j. Clearly for $j \neq i + 1$, $\operatorname{ord} x \neq \operatorname{ord} y$, hence $T \times H$ and $G \times V$ are non-isomorphic and by **(4.6)**, $(L, \oplus)$ and $(L, \otimes)$ are non-isomorphic.

References

[1] BOL, G., *Gewebe und Gruppen*. Math. Ann. **114** (1937), 414–431

[2] BRUCK, R. H., *A survey of binary systems*. Springer-Verlag, Berlin 1958.

[3] CHEIN, O., PFLUGFELDER, H. O., SMITH, J. D. H., *Quasigroups and Loops, Theory and Applications*. Heldermann Verlag, Berlin 1990.

[4] GLAUBERMAN, G., *On loops of odd order*. J. Algebra **1** (1966), 374–396

[5] KARZEL, H., *Zusammenhänge zwischen Fastbereichen, scharf zweifach transitiven Permutationsgruppen und 2-Strukturen mit Rechtecksaxiom*. Abh. Math. Sem. Univ. Hamburg **32** (1968), 191–206

[6] KEPKA, T., *A construction of Bruck loops*. Commentationes Math. Univ. Carolinae **25** (1984), 591–595.

[7] KIST, G., *Theorie der verallgemeinerten kinematischen Räume*. Beiträge zur Geometrie und Algebra **14**, TUM-Bericht M 8611, München 1986.

[8] KREUZER,A., *Beispiele endlicher und unendlicher K-Loops*. Res. Math. **23** (1993), 355–362.

[9] KREUZER, A. and WEFELSCHEID, H., *On K-loops of finite order*. Res. Math. **25** (1994).

[10] NIEDERREITER, H. and ROBINSON, K. H., *Bol loops of order pq*. Math. Proc. Cambridge Philos. Soc. **89** (1981), 241–256.

[11] ROBINSON, D. A., *Bol-loops*. Trans. Amer. Math. Soc. **123** (1966), 341–354.

[12] ROBINSON, K. H., *A note on Bol loops of order $2^n k$*. Aequationes Math. **22** (1981) 302–306.

[13] UNGAR, A. A., *Thomas rotation and the parametrization of the Lorentz transformation group*. Found. Phys. Lett. **1** (1988), 57–89.

[14] UNGAR, A. A., *Weakly associative groups*. Res. Math. **17** (1990), 149–168.

[15] WÄHLING, H., *Theorie der Fastkörper*. Thales Verlag, Essen 1987.

N-HOMOMORPHISMS OF TOPOLOGICAL N-GROUPS

K. D. Magill, Jr.

106 Diefendorf Hall, SUNY at Buffalo,Buffalo, NY 14214-3093, USA

1. INTRODUCTION AND PRELIMINARY RESULTS

Let N be a right topological nearring. A topological N-group is a pair (G, μ) where G is a topological group and μ is a continuous function from $N \times G$ into G such that $\mu(a + b, x) = \mu(a, x) + \mu(b, x)$ and $\mu(ab, x) = \mu(a, \mu(b, x))$. This is, of course, just the topological analogue of the definition of N-group as given by G. Pilz on page 13 of [4] and a nearring module or N-module as defined by J.R. Clay on page 261 of [2]. We note that Clay deals primarily with left nearrings while Pilz deals with right nearrings as we do here. For any locally compact Hausdorff topological group G, we denote by $N(G)$ the nearring of all continuous selfmaps of G under pointwise addition and composition. Now, fix a continuous selfmap α of G and denote by $N(G)_\alpha$ the nearring whose additive group coincides with that of $N(G)$ but where multiplication is defined by $fg = f \circ \alpha \circ g$ for all $f, g \in N(G)_\alpha$. The nearring $N(G)_\alpha$ is referred to as a *laminated nearring* with *laminating function* or *laminator* α. If α is taken to be the identity map, then $N(G)_\alpha$ coincides with $N(G)$. It was observed in [3] that $N(G)_\alpha$, with the compact-open topology (hereafter referred to as the c-topology) is a topological nearring when G is locally compact and Hausdorff. Throughout this paper, function spaces will have the c-topology. As usual, homomorphisms from one topological algebraic system to another will be assumed to be continuous and isomorphisms will be assumed to be homeomorphisms as well.

In [3] we determined, for a number of topological groups G, precisely what form certain maps μ must take in order for (G, μ) to be a topological $N(G)_\alpha$-group. In this paper, we apply the results from [3] to investigate the $N(G)_\alpha$-homomorphisms from one topological $N(G)_\alpha$-group into another. In a number of cases, we are able to completely describe these maps and we will see that there are quite a few instances in which there is exactly one $N(G)_\alpha$-homomorphism. In particular, there are a number of topological $N(G)_\alpha$-groups where the only $N(G)_\alpha$-endomorphism is the identity map. And now, we need to recall some results and definitions from [3]. Actually, our first result is a special case of Theorem (2.8) of [3]. It tells us that the problem of determining all μ such that (G, μ) is a topological N-group is equivalent to the problem of determining all the homomorphisms from the topological nearring N into the topological nearring $N(G)$.

Theorem (1.1). *Let N be a topological nearring and let G be a locally compact Hausdorff topological group. Let φ be a homomorphism from N into $N(G)$ and define $\mu(a, x) = (\varphi(a))(x)$ for all $a \in N$ and $x \in G$. Then (G, μ) is a topological N-group. Conversely, suppose (G, μ) is a topological N-group and define a map φ from N into $N(G)$ by $(\varphi(a))(x) = \mu(a, x)$ for all $a \in N$ and $x \in G$. Then φ is a homomorphism from N into $N(G)$.*

Y. Fong et al. (eds.), Near-Rings and Near-Fields, 181–191.
© *1995 Kluwer Academic Publishers.*

As we mentioned before, the previous result was proven in more generality in [3] but what we have stated is sufficient for our needs here. Next, we recall some definitions from [3].

Definition (1.2). Let G be a topological group. A selfmap α of G is said to be *G-admissible* if it is continuous, surjective and if K is the kernel of any noninjective endomorphism of G, then α is nonconstant on some coset of K.

Definition (1.3). A topological space is *reversible* if each continuous bijection of the space is a homeomorphism.

The latter spaces were introduced in [5]. Our definition differs from that in [5] but the two definitions are equivalent in view of their Lemma 1. It is immediate that all compact Hausdorff spaces are reversible and it was shown in Theorem 1 of [5, p. 131] that reversible spaces also include all locally Euclidean spaces.

Definition (1.4). A topological group G is *exclusive* if it is a locally compact Hausdorff reversible topological space which is either 0-dimensional or arcwise connected and, in addition, there do not exist any injective endomorphisms onto proper subgroups of G.

Examples of exclusive groups include all of the Euclidean groups as well as the multiplicative group of complex numbers of modulus one.

This next result was verified in [3]. The symbol $\langle x \rangle$ denotes the constant function which maps everything into the point x. Its domain will be clear from context.

Theorem (1.5). *Let G be an exclusive topological group. Let α be a continuous surjection of G and let φ be a homomorphism from the topological nearring $N(G)_\alpha$ into the topological nearring $N(G)_\beta$ such that $\varphi\langle x \rangle \neq \langle 0 \rangle$ for some $x \in G$. Then there exists a continuous surjection h of G and an automorphism t of G such that the following diagram commutes.*

(1.5.1)
$$
\begin{array}{ccccc}
G & \xrightarrow{f} & G & \xrightarrow{\alpha} & G \\
\uparrow{\scriptstyle h} & & \downarrow{\scriptstyle t} & & \uparrow{\scriptstyle h} \\
G & \xrightarrow{\varphi(f)} & G & \xrightarrow{\beta} & G
\end{array}
$$

Theorems (1.1) and (1.5) were then combined to produce the following theorem from [3] which shows exactly what form certain functions μ must take in order for (G, μ) to be a topological $N(G)_\alpha$-group for certain topological groups G.

Theorem (1.6). *Let G be an exclusive topological group. Let α be a G-admissible selfmap of G and let t be an automorphism of G. Define a map μ from $N(G)_\alpha \times G$ into G by*

(1.6.1) $\mu(f, x) = t \circ f \circ \alpha \circ t^{-1}(x).$

Then (G, μ) is a topological $N(G)_\alpha$-group with the property that $\mu(\langle x \rangle, y) \neq 0$ for at least one pair of points $x, y \in G$ and, conversely, every topological $N(G)_\alpha$-group, (G, μ), with this property is obtained in this manner.

These are all the preliminary results we need. We close this section with a few further remarks. It was shown in [3] that if G is exclusive, $N(G)$ is simple and (G, μ) is a topological $N(G)$-group, then either $\mu(f, x) = 0$ for all f and x or there exists an automorphism t of G such that $\mu(f, x) = t \circ f \circ t^{-1}(x)$ for all f and x. So for these groups, we know precisely what all the topological $N(G)$-groups are. It was also shown that there are groups G such that not all topological $N(G)$-groups are of this form. Specifically, the multiplicative group of complex numbers of modulus one is such a group.

2. N-HOMOMORPHISMS

Let N be a topological near-ring, let G and H be two topological groups and let (G, μ) and (H, η) be two topological N-groups.

Definition (2.1). An *N-homomorphism* from (G, μ) to (H, η) is a homomorphism φ from the topological group G into the topological group H which satisfies the following additional property

$$(2.1.1) \qquad \varphi(\mu(a, x)) = \eta(a, \varphi(x))$$

for all $a \in N$ and $x \in G$.

This is, of course, just the topological analogue of Definition 1.25(b) given in [4, p. 15] and Definition 13.8 given in [2, p. 263].

Definition (2.2) Let G be an exclusive topological group. Let α be a G-admissible map and let (G, μ) be a topological $N(G)_\alpha$-group with the property that $\mu(\langle x \rangle, y) \neq 0$ for some pair of points $x, y \in G$. We will then refer to (G, α, μ) as an *admissible triple*.

Theorem (2.3). *Let (G, α, μ) and (G, α, η) be admissible triples. Then there is exactly one $N(G)_\alpha$-homomorphism φ from the topological $N(G)_\alpha$-group (G, μ) into the topological $N(G)_\alpha$-group (G, η). In fact, $\varphi = k \circ t^{-1}$ where t and k are the automorphisms of the topological group G which induce μ and η respectively, so that φ is also an automorphism of the topological group G.*

Proof. From Theorem (1.6) we see that there exist automorphisms t and k of the topological group G such that

$$(2.3.1) \qquad \mu(f, x) = t \circ f \circ \alpha \circ t^{-1}(x) \text{ and } \eta(f, x) = k \circ f \circ \alpha \circ k^{-1}(x)$$

for each $f \in N(G)_\alpha$ and $x \in G$. It is easy to verify that $k \circ t^{-1}$ is an $N(G)_\alpha$-homomorphism from (G, μ) to (G, η). Conversely, let φ be any $N(G)_\alpha$-homomorphism from (G, μ) to (G, η). From (2.1.1) and (2.3.1) we see that

$$(2.3.2) \qquad \varphi \circ t \circ f \circ \alpha \circ t^{-1}(x) = k \circ f \circ \alpha \circ k^{-1} \circ \varphi(x)$$

for all $f \in N(G)_\alpha$ and $x \in G$. For any $x \in G$, we take $f = \langle x \rangle$ in (2.3.2) and we get $\varphi \circ t(x) = k(x)$ and since t and k are automorphisms of the topological group G, we conclude that $\varphi = k \circ t^{-1}$ and, consequently, φ is an automorphism of G.

Corollary (2.4). *Let (G, α, μ) be an admissible triple. Then the only N-endomorphism of the topological $N(G)_\alpha$-group, (G, μ), is the identity map.*

3. TOPOLOGICAL N-GROUPS OF CONTINUOUS FUNCTIONS

Let $AN(G)$ denote the additive group of the nearring $N(G)_\alpha$. We can form from it a topological $N(G)_\alpha$-group, $(AN(G), \rho)$ by defining $\rho(f, g) = fg$ where, we recall, $fg = f \circ \alpha \circ g$ in $N(G)_\alpha$. This is precisely what was done, in general, in Example 1.18a of [4, p. 13] to turn $(N, +)$ into an N-group. We will investigate these $N(G)_\alpha$-groups in further detail but first, we look at other possibilities for defining continuous maps λ so that $(AN(G), \lambda)$ is a topological $N(G)_\alpha$-group. For this portion of the discussion, we will restrict our attention to the additive group R^n of n-tuples. In view of Theorem (1.1), the task of finding these maps λ is equivalent to finding homomorphisms from the laminated nearring $N(R^n)_\alpha$ into the nearring, $N(AN(R^n))$, of all continuous selfmaps of the space of continuous selfmaps of R^n. Now suppose there exists a homomorphism t from the topological group R^n into the topological group $AN(R^n)$ and a continuous function h from $AN(R^n)$ into R^n such that $\alpha = h \circ t$. Then one easily verifies that the map φ defined by $\varphi(f) = t \circ f \circ h$ is a homomorphism from $N(R^n)_\alpha$ into $N(AN(R^n))$. Not all the homomorphisms are of this form but we can get a lot of them this way. The following result shows that the function t must have a fairly simple form.

Lemma (3.1). *A map t from the topological group R^n into the topological group $AN(R^n)$ is a homomorphism if and only if there exist n continuous selfmaps $\{p_i\}_{i=1}^n$ of R^n such that*

$$(3.1.1) \qquad t(r_1, r_2, \ldots, r_n) = \sum_{i=1}^n r_i p_i \text{ for all } (r_1, r_2, \ldots, r_n) \in R^n.$$

Proof. It is a straightforward matter to show that if t is defined as in (3.1.1), then t is a homomorphism from R^n into $AN(R^n)$. Suppose, conversely, that t is a homomorphism from R^n into $AN(R^n)$. Let $\{e_i\}_{i=1}^n$ be the standard basis for the real vector space R^n and let $t(e_i) = p_i$. Then $p_i = t(e_i) = t(m(1/m)e_i) = mt((1/m)e_i)$ for all positive integers m which means $t((1/m)e_i) = (1/m)p_i$ for all positive integers m. From this, we get $t((m/s)e_i) = mt((1/s)e_i) = (m/s)p_i$. That is, $t(re_i) = rp_i$ for all positive rational numbers r. Then $-t(re_i) = t(-re_i) = -rp_i$ and we conclude that $t(re_i) = rp_i$ for all rational numbers r. This holds for $1 \leq i \leq n$ and since the left and right hand sides of (3.1.1) are equal on a dense subset of R^n, it follows that (3.1.1) is valid.

The previous result tells precisely how we must define the homomorphisms t. There is much greater latitude for the functions h from $AN(R^n)$ into R^n and we will describe some of these. For any compact subset K of R^n and any open subset G of R^n, we let $[K,G] = \{f \in AN(R^n) : f[K] \subseteq G\}$. We recall that $\{[K,G] : K$ is compact and G is open$\}$ is a subbasis for the c-topology on $AN(R^n)$. We will denote the set $[\{x\}, G]$ more simply by $[x, G]$.

Lemma (3.2). *Choose any point* $a \in R^n$ *and define* $h(f) = f(a)$. *Then* h *is a continuous function from* $AN(R^n)$ *onto* R^n.

Proof. Suppose $f \in AN(R^n)$ and G is any open subset of R^n containing $h(f) = f(a)$. Then $[a, G]$ is a neighborhood of f and $h[a, G] \subseteq G$. Thus, h is continuous and it is evident that it is surjective.

In our next result we will use the notation and terminology of [1]. In particular, the norm of a point $x \in R^n$ will be denoted by $|x|$.

Lemma (3.3). *Let* D *be any n-dimensional cube in* R^n *and define* $h(f) = \int_D f$. *Then* h *is a continuous function from* $AN(R^n)$ *onto* R^n.

Proof. Let $f \in AN(R^n)$ and $\epsilon > 0$ be given. For each $a \in D$, we let

$$G_a = \left\{ x \in R^n : |x - f(a)| < \tfrac{\epsilon}{2A(D)} \right\},$$

$$H_a = \left\{ x \in D : |f(x) - f(a)| < \tfrac{\epsilon}{2A(D)} \right\},$$

$$K_a = \left\{ x \in D : |f(x) - f(a)| \leq \tfrac{\epsilon}{2A(D)} \right\}$$

where $A(D)$ is the content of D. Now $\{H_a\}_{a \in D}$ is an open cover of D and hence, some finite subcollection $\{H_{a_i}\}_{i=1}^m$ also covers D. Since each K_a is compact, $\cap_{i=1}^m [K_{a_i}, G_{a_i}]$ is an open subset of $AN(R^n)$ and it contains f. Now let $g \in \cap_{i=1}^m [K_{a_i}, G_{a_i}]$ and let any $x \in D$ be given. Then $x \in K_{a_i}$ for some i and hence $f(x), g(x) \in G_{a_i}$ which implies

$$|f(x) - f(a_i)| < \tfrac{\epsilon}{2A(D)} \quad \text{and} \quad |g(x) - f(a_i)| < \tfrac{\epsilon}{2A(D)}$$

Therefore, we have

$$|f(x) - g(x)| \leq |f(x) - f(a_i)| + |g(x) - f(a_i)| < \tfrac{\epsilon}{A(D)}$$

and, in view of Theorem 24.16, page 327 of [1], this implies that

$$|h(f) - h(g)| = \left| \int_D f - \int_D g \right| = \left| \int_D (f - g) \right| \le \epsilon.$$

Consequently, h is continuous and it is easily verified that it is surjective.

Let h_i ($1 \le i \le n$) be a continuous function from $AN(R^n)$ into R. Then the map H defined by $H(g) = (h_1(g), h_2(g), \ldots, h_n(g))$ is a continuous map from $AN(R^n)$ into R^n. This next result shows us how to get various continuous functions h from $AN(R^n)$ into R.

Lemma (3.4). *Choose any compact subset D of R^n, and define*

$$h(f) = max\{|f(x)| : x \in D\}.$$

Then h is a continuous function from $AN(R^n)$ into R.

Proof. Let $f \in AN(R^n)$ and $\epsilon > 0$ be given. For each $a \in D$, let

$$G_a = \{x \in R^n : |x - f(a)| < \tfrac{\epsilon}{2}\},$$
$$H_a = \{x \in D : |f(x) - f(a)| < \tfrac{\epsilon}{2}\},$$
$$K_a = \{x \in D : |f(x) - f(a)| \le \tfrac{\epsilon}{2}\}.$$

The family $\{H_a\}_{a \in D}$ is an open cover of D and hence, some finite subcollection $\{H_{a_i}\}_{i=1}^n$ also covers D. Then $W = \cap_{i=1}^n [K_{a_i}, G_{a_i}]$ is a neighborhood of f and it readily follows that if $g \in W$, then $|f(x) - g(x)| < \epsilon$ for all $x \in D$. Suppose f assumes its maximum value on D at c and g assumes its maximum value on D at d. Then

$$h(f) - \epsilon = f(c) - \epsilon < g(c) \le g(d) < f(d) + \epsilon \le f(c) + \epsilon = h(f) + \epsilon.$$

Since $h(g) = g(d)$, we have $|h(g) - h(f)| < \epsilon$ and the proof is complete.

Now we use some of these results and Theorem (1.1) to define continuous functions λ so that $(AN(R^n), \lambda)$ is a topological $N(R^n)_\alpha$-group.

Theorem (3.5). *Choose n continuous selfmaps $\{p_i\}_{i=1}^n$ of R^n, a continuous map h from $AN(R^n)$ into R^n and define a continuous selfmap α of R^n by*

$$\alpha(r_1, r_2, \ldots, r_n) = h(\sum_{i=1}^n r_i p_i) \quad \text{for all } (r_1, r_2, \ldots, r_n) \in R^n.$$

Define a function λ from $N(R^n)_\alpha \times AN(R^n)$ into $AN(R^n)$ by $\lambda(f, g) = \sum_{i=1}^n s_i p_i$ where $f \circ h(g) = (s_1, s_2, \ldots, s_n)$. Then $(AN(R^n), \lambda)$ is a topological $N(R^n)_\alpha$-group.

Proof. Define $t(r_1, r_2, \ldots, r_n) = \sum_{i=1}^{n} r_i p_i$ and $\varphi(f) = t \circ f \circ h$. Then $\alpha = h \circ t$ and one easily verifies that φ is an algebraic homomorphism from $N(R^n)_\alpha$ into $N(AN(R^n))$. Moreover, h is continuous by hypotheses and t is continuous in view of Lemma (3.1). Thus φ is continuous and Theorem (1.1) tells us that $AN(R^n), \lambda)$ is a topological $N(R^n)_\alpha$-group where $\lambda(f, g) = t \circ f \circ h(g) = \sum_{i=1}^{n} s_i p_i$ and $f \circ h(g) = (s_1, s_2, \ldots, s_n)$.

Corollary (3.6). *Choose n continuous selfmaps $\{p_i\}_{i=1}^{n}$ of R^n, a point $a \in R^n$ and define a continuous selfmap α of R^n by $\alpha(r_1, r_2, \ldots, r_n) = \sum_{i=1}^{n} r_i p_i(a)$ for all $(r_1, r_2, \ldots, r_n) \in R^n$. Define a function λ from $N(R^n)_\alpha \times AN(R^n)$ into $AN(R^n)$ by $\lambda(f, g) = \sum_{i=1}^{n} s_i p_i$ where $f \circ g(a) = (s_1, s_2, \ldots, s_n)$. Then $(AN(R^n), \lambda)$ is a topological $N(R^n)_\alpha$-group.*

Proof. Take the function h in Theorem (3.5) to be defined by $h(g) = g(a)$. Then h is continuous by Lemma (3.2) and the conclusion now follows from Theorem (3.5).

Our next two results concern $N(R^n)_\alpha$-homomorphisms between topological $N(R^n)_\alpha$-groups of the form described in Theorem (3.5) and those of the form (R^n, μ).

Theorem (3.7). *Let $(AN(R^n), \lambda)$ be a topological $N(R^n)_\alpha$-group where α is defined as in Theorem (3.5), let (R^n, μ) be a topological $N(R^n)_\alpha$-group such that (R^n, α, μ) is an admissible triple. Then there is exactly one $N(R^n)_\alpha$-homomorphism φ from (R^n, μ) into $(AN(R^n), \lambda)$ and it is defined by*

$$(3.7.1) \quad \varphi(r_1, r_2, \ldots, r_n) = \sum_{i=1}^{n} s_i p_i \text{ where } (s_1, s_2, \ldots, s_n) = t^{-1}(r_1, r_2, \ldots, r_n)$$

and t is the linear automorphism of the vector space R^n which induces μ.

Proof. Suppose φ is defined as in (3.7.1). One easily verifies that φ is a homomorphism from the topological group R^n into the topological group $AN(R^n)$. We want to show that it satisfies condition (2.1.1) as well. That is, we wish to verify that $\varphi(\mu(f, r)) = \lambda(f, \varphi(r))$ for all $f \in N(R^n)_\alpha$ and $r \in R^n$. By Theorem (1.6), we have $\mu(f, r) = t \circ f \circ \alpha \circ t^{-1}(r)$. Thus, from (3.7.1) we get

$$(3.7.2) \quad \varphi(\mu(f, r)) = \varphi(t \circ f \circ \alpha \circ t^{-1}(r)) = \sum_{i=1}^{n} a_i p_i$$

where $t^{-1}(t \circ f \circ \alpha \circ t^{-1}(r)) = (a_1, a_2, \ldots, a_n)$ which means

$$(3.7.3) \quad f \circ \alpha \circ t^{-1}(r) = (a_1, a_2, \ldots, a_n).$$

From Theorem (3.5), we get

$$(3.7.4) \quad \lambda(f, \varphi(r)) = \lambda(f, \sum_{i=1}^{n} s_i p_i) = \sum_{i=1}^{n} b_i p_i$$

where

(3.7.5)
$$t^{-1}(r_1, r_2, \ldots, r_n) = (s_1, s_2, \ldots, s_n) \text{ and } f \circ h(\textstyle\sum_{i=1}^n s_i p_i) = (b_1, b_2, \ldots, b_n).$$

We must show that $a_i = b_i$ for all i. From Theorem (3.5), we see that

$$h(\sum_{i=1}^n s_i p_i) = \alpha(s_1, s_2, \ldots, s_n).$$

From this and both (3.7.5) and (3.7.3), we get

$$(b_1, b_2, \ldots, b_n) = f \circ h(\textstyle\sum_{i=1}^n s_i p_i) = f \circ \alpha(s_1, s_2, \ldots, s_n) =$$

$$f \circ \alpha \circ t^{-1}(r_1, r_2, \ldots, r_n) = (a_1, a_2, \ldots, a_n).$$

Hence, condition (2.1.1) is satisfied and we conclude that φ is an $N(R^n)_\alpha$-homomorphism from (R^n, μ) into $(AN(R^n), \lambda)$.

On the other hand let φ be any $N(R^n)_\alpha$-homomorphism whatsoever from (R^n, μ) into $AN(R^n), \lambda)$. Then φ must satisfy condition (2.1.1) so that we have $\varphi(\mu(f, x)) = \lambda(f, \varphi(x))$ for all $f \in N(R^n)_\alpha$ and $x \in R^n$. Because of Theorem (1.6), this translates into

$$\varphi(t \circ f \circ \alpha \circ t^{-1}(x)) = \sum_{i=1}^n y_i p_i \text{ where } f \circ h(\varphi(x)) = (y_1, y_2, \ldots, y_n).$$

This must hold for any f (and any x) and we choose f to be the constant function $\langle s \rangle$ where $s = (s_1, s_2, \ldots, s_n)$. Then $f \circ h(\varphi(x)) = s$ and we get $\varphi(t(s)) = \sum_{i=1}^n s_i p_i$. Let $t(s) = r$ and get $\varphi(r_1, r_2, \ldots, r_n) = \sum_{i=1}^n s_i p_i$ for all $(r_1, r_2, \ldots, r_n)$ where $(s_1, s_2, \ldots, s_n) = t^{-1}(r_1, r_2, \ldots, r_n)$.

Theorem (3.7) tells us that there is exactly one $N(R^n)_\alpha$-homomorphism from (R^n, μ) into $(AN(R^n), \lambda)$. This next result provides us with an example of an $N(R^n)_\alpha$-homomorphism from a topological $N(R^n)_\alpha$-group of the form $(AN(R^n), \lambda)$ into one of the form (R^n, μ). The problem remains to determine all such $N(R^n)_\alpha$-homomorphisms.

Theorem (3.8). *Choose n continuous selfmaps $\{p_i\}_{i=1}^n$ of R^n and choose a point $a \in R^n$ such that the linear map t from R^n into R^n defined by $t(r_1, r_2, \ldots, r_n) = \sum_{i=1}^n r_i p_i(a)$ is nonsingular. Then define $\lambda(f, g) = \sum_{i=1}^n r_i p_i$ where $f \circ g(a) = (r_1, r_2, \ldots, r_n)$ for each pair of continuous functions $f, g \in AN(R^n)$, let $\alpha = t$ and define $\mu(f, x) = t \circ f \circ \alpha \circ t^{-1}(x)$ for every continuous selfmap f of R^n and all $x \in R^n$. Then $(AN(R^n), \lambda)$ and (R^n, μ) are topological $N(R^n)_\alpha$-groups and the mapping φ defined by $\varphi(f) = f(a)$ is an $N(R^n)_\alpha$-homomorphism from $(AN(R^n), \lambda)$ into (R^n, μ).*

Proof. We have seen from our previous considerations that both $AN(R^n), \lambda)$ and (R^n, μ) are topological $N(R^n)_\alpha$-groups. In order for φ to be an $N(R^n)_\alpha$-homomorphism, we must have $\varphi(\lambda(f,g)) = \mu(f, \varphi(g))$ for all $f, g \in AN(R^n)$ according to condition (2.1.1). But this translates into

$$(3.8.1) \qquad \sum_{i=1}^{n} r_i p_i(a) = t \circ f \circ \alpha \circ t^{-1}(g(a)) \text{ for all } f, g \in R^n$$

where $f \circ g(a) = (r_1, r_2, \ldots, r_n)$. Since $\alpha = t$, we have

$$t \circ f \circ \alpha \circ t^{-1}(g(a)) = t \circ f(g(a)) = \sum_{i=1}^{n} r_i p_i(a).$$

Thus, (3.8.1) is satisfied and we have verified the fact that φ is an $N(R^n)_\alpha$-homomorphism.

Now we return to the topological $N(G)_\alpha$-group $(AN(G), \rho)$ where $\rho(f,g) = fg = f \circ \alpha \circ g$. Let (G, μ) be another topological $N(G)_\alpha$-group where μ is induced by an automorphism t of G. We first look at $N(G)_\alpha$-homomorphisms from (G, μ) into $(AN(G), \rho)$. One easily verifies that the map φ defined by $\varphi(x) = \langle t^{-1}(x) \rangle$ for all $x \in G$ is an $N(G)_\alpha$-homomorphism from the topological $N(G)_\alpha$-group (G, μ) into the topological $N(G)_\alpha$-group $(AN(G), \rho)$. Our next result shows that this is the only $N(G)_\alpha$-homomorphism.

Theorem (3.9). *Let (G, α, μ) be an admissible triple and let φ be an $N(G)_\alpha$-homomorphism from (G, μ) into $(AN(G), \rho)$. Then $\varphi(x) = \langle t^{-1}(x) \rangle$ for all $x \in G$ where t is the automorphism which induces μ.*

Proof. By condition (2.1.1), we have $\varphi(t \circ f \circ \alpha \circ t^{-1}(x)) = f \circ \alpha \circ \varphi(x)$ for all $f \in N(G)_\alpha$ and $x \in G$. Take $f = \langle x \rangle$ and get $\varphi(t(x)) = \langle x \rangle$ for all $x \in G$. This implies that $\varphi(x) = \langle t^{-1}(x) \rangle$ for all x.

Thus, we see that there is precisely one $N(G)_\alpha$-homomorphism from (G, μ) into $(AN(G), \rho)$. However, the case for $N(G)_\alpha$-homomorphisms from $(AN(G), \rho)$ into (G, μ) is far different. One can verify that if we choose any $a \in G$ and define $\varphi(f) = t \circ f \circ \alpha \circ t^{-1}(a)$ where t is the automorphism of G which induces μ, then φ is an $N(G)_\alpha$-homomorphism from $(AN(G), \rho)$ into (G, μ). We will refer to such a homomorphism as *natural*. Our next result shows that every $N(G)_\alpha$-homomorphism coincides with a natural $N(G)_\alpha$-homomorphism on what can be a rather large $N(G)_\alpha$-subgroup of $(AN(G), \rho)$. In what follows, we let $H(G)_\alpha = \{f \circ \alpha : f \in N(G)_\alpha\}$. $H(G)_\alpha$ is, indeed, an $N(G)_\alpha$-subgroup of $(AN(G), \rho)$ since $\rho(f,g) \in H(G)_\alpha$ for each $f \in N(G)_\alpha$ and $g \in H(G)_\alpha$.

Theorem (3.10). *Let (G, α, μ) be an admissible triple and let φ be an $N(G)_\alpha$-homomorphism from $(AN(G), \rho)$ into (G, μ). Then there exists a natural $N(G)_\alpha$-homomorphism from $(AN(G), \rho)$ into (G, μ) which coincides with φ on the $N(G)_\alpha$-subgroup, $H(G)_\alpha$.*

Proof.   Let t be the automorphism of the topological group G which, according to Theorem (1.6), induces μ. In view of (2.1.1), we have $\varphi(\rho(f,g)) = \mu(f, \varphi(g))$ and this means

$$(3.10.1) \qquad \varphi(f \circ \alpha \circ g) = t \circ f \circ \alpha \circ t^{-1}(\varphi(g)) \text{ for all } f, g \in N(G)_\alpha.$$

Let δ be the map defined by $\delta(x) = x$ for all x and let $\varphi(\delta) = a$. Let $g = \delta$ in (3.10.1) and get

$$(3.10.2) \qquad \varphi(f \circ \alpha) = t \circ f \circ \alpha \circ t^{-1}(a) \text{ for all } f, g \in N(G)_\alpha.$$

Since α is surjective, there exists an element $b \in G$ such that $\alpha(t^{-1}(b)) = t^{-1}(a)$ and we define a natural $N(G)_\alpha$-homomorphism from $(AN(G), \rho)$ into (G, μ) by $\psi(f) = t \circ f \circ \alpha \circ t^{-1}(b)$ for all $f \in N(G)_\alpha$. For any f, we now have

$$\psi(f \circ \alpha) = t \circ f \circ \alpha \circ \alpha \circ t^{-1}(b) = t \circ f \circ \alpha \circ t^{-1}(a) = \varphi(f \circ \alpha).$$

Consequently, φ agrees with the natural $N(G)_\alpha$-homomorphism ψ on the $N(G)_\alpha$-subgroup, $H(G)_\alpha$, and the proof is complete.

Corollary (3.11).   *Let (G, δ, μ) be an admissible triple. Then every $N(G)$-homomorphism from $(AN(G), \rho)$ into (G, μ) is natural. Specifically, for each $N(G)$-homomorphism φ from $(AN(G), \rho)$ into (G, μ) there exists an element $a \in G$ such that $\varphi(f) = t \circ f \circ t^{-1}(a)$ for each $f \in N(G)$, where t is the automorphism of G which induces μ.*

Proof.   Take α in Theorem (3.10) to be the identity map δ. Then $\rho(f, g) = f \circ g$ and $H(G)_\delta$ coincides with $AN(G)$ so this corollary is an immediate consequence of that theorem.

Some Remarks: One might be tempted to show that the $N(G)_\alpha$-subgroup, $H(G)_\alpha$, is dense in $AN(G)$ for then φ and ψ would have to coincide on all of $AN(G)$ and each $N(G)_\alpha$-homomorphism from $(AN(G), \rho)$ into (G, μ) would have to be natural. But little would be gained for it is not difficult to verify that α must be injective in order for $H(G)_\alpha$ to be dense. For compact groups and locally Euclidean groups, this means that α is a homeomorphism, in which case $H(G)_\alpha$ coincides with $AN(G)$.

Corollary (2.4) provided us with a class of $N(G)_\alpha$-groups with the property that the only $N(G)_\alpha$-endomorphism is the identity map. In contrast to this, our concluding result provides us with a topological $N(G)_\alpha$-group with many $N(G)_\alpha$-endomorphisms.

Theorem (3.12).   *Choose any $a \in G$ and define $\varphi(f) = \langle f(\alpha(a)) \rangle$. Then φ is an $N(G)_\alpha$-endomorphism of the topological $N(G)_\alpha$-group $(AN(G), \rho)$.*

Proof.   This is verified in a completely straightforward manner with no additional conditions on the group G other than the requirement that it be locally compact and Hausdorff and no conditions on the sandwich function α other than continuity. These conditions are needed in order to insure that $(AN(G), \rho)$ is a topological $N(G)_\alpha$-group.

Acknowledgement: The author is indebted to the referee for a very careful reading of a previous version of this manuscript which resulted in the detection of a number of errors.

References

1. Bartle, R.G., "The elements of real analysis," John Wiley & Sons, Inc., New York, (1964).
2. Clay, James R., "Nearrings, Geneses and Applications", Oxford University Press, New York (1992).
3. Magill, K.D., Jr., *Topological N-groups*, Geometriae Dedicata, 46 (1993) 181–196.
4. Pilz, G., "Near-rings," North Holland Math. Studies, 23, Revised ed., Amsterdam, (1983).
5. Rajagopolan, M. and A. Wilansky, *Reversible topological spaces*, J. Austral. Math. Soc. 6 (1966) 129–138.

THE BICENTRALIZER NEARRINGS OF R

K. D. Magill, Jr.

106 Diefendorf Hall, SUNY at Buffalo, Buffalo, NY 14214-3093 USA

P. R. Misra

College of Staten Island, CUNY, Staten Island, NY 10301, USA

Let Γ be an additive (not necessarily abelian) topological group. As is customary, an automorphism of Γ will be assumed to be a homeomorphism as well and endomorphisms will be assumed to be continuous. For any group G, of automorphisms of Γ, we will denote by $N_G(\Gamma)$ the bicentralizer nearring of Γ, with respect to G, which consists of all continuous selfmaps of Γ which commute with all elements of G. In this paper, we take $\Gamma = R$, the additive group of real numbers and we determine all the bicentralizer nearrings of R. It is well known that any automorphism t of R is of the form $t(x) = ax$ for some $a \neq 0$ and we will denote this map by t_a. We will denote by $Aut(R)$ and $End(R)$, the automorphism group and the endomorphism ring, respectively, of R. Finally, it is known that the only endomorphism of R which is not an automorphism is the map which sends everything into zero so that $End(R) = \{t_0\} \cup Aut(R)$. Moreover, the mapping which sends a into t_a is an isomorphism from R_M, the multiplicative group of nonzero real numbers, onto $Aut(R)$. We will adhere to the convention that if G is any subgroup of R_M, then $[G]$ is the subgroup of $Aut(R)$ which corresponds to G. That is $[G] = \{t_a : a \in G\}$. For our first result, we choose the smallest (with respect to cardinality) nontrivial subgroup of R_M. Its proof consists of one line and even that is probably not necessary.

Theorem 1. *Let* $G = \{-1, 1\}$. *Then* $N_{[G]}(R)$ *consists of all the odd functions.*

Proof. $f \in N_{[G]}(R)$ if and only if $f \circ t_{-1} = t_{-1} \circ f$, which is to say, $f(-x) = -f(x)$ for all $x \in R$.

Now let any $a > 1$ be given. Let $X = [-a, -1]$ and $Y = [1, a]$. Then let $C(X)$ denote the collection of all continuous maps f from X into R such that $f(-a) = af(-1)$ and let $C(Y)$ denote the collection of all continuous maps g from Y into R such that $g(a) = ag(1)$. For each pair $(f, g) \in C(X) \times C(Y)$, we associate a function, which we denote by $\langle\!\langle f, g \rangle\!\rangle$, as follows. Suppose $x < 0$. Then there exists an integer n (positive or negative) such that $-(a^{n+1}) < x \leq -(a^n)$ and we define $\langle\!\langle f, g \rangle\!\rangle(x) = a^n f(x/a^n)$. If $x > 0$, there exists an integer n such that $a^n \leq x < a^{n+1}$ and we define $\langle\!\langle f, g \rangle\!\rangle(x) = a^n g(x/a^n)$. Finally, define $\langle\!\langle f, g \rangle\!\rangle(0) = 0$. One can verify that the function $\langle\!\langle f, g \rangle\!\rangle$ is continuous. The fact that $f(-a) = af(-1)$ and $g(a) = ag(1)$ guarantees that $\langle\!\langle f, g \rangle\!\rangle$ is continuous at the points $-a^n$ and the points a^n, respectively. As for the point zero, note that if x is

Y. Fong et al. (eds.), Near-Rings and Near-Fields, 193–198
© 1995 Kluwer Academic Publishers.

close to zero, then the exponent n in both the expressions $a^n f(x/a^n)$ (for $x < 0$) and $a^n g(x/a^n)$ (for $x > 0$) must be negative but large in absolute value so that $\langle\!\langle f, g\rangle\!\rangle(x)$ will be close to zero. Now we let

$$N_1(a) = \{\langle\!\langle f, g\rangle\!\rangle : (f, g) \in C(X) \times C(Y)\}.$$

We chose the notation $N_1(a)$ because this particular family of functions turns out to be a subnearring of $N(R)$ but we don't need to prove this directly as it falls out of our considerations in the next result. The subscript 1 indicates that still other nearrings are yet to be defined. For any subset A of R_M, we denote by $\langle A\rangle$ the subgroup of R_M which is generated by A.

Theorem 2. *Let $a > 1$ and let $G = \langle a\rangle$. Then $N_{[G]}(R) = N_1(a)$.*

Proof. Suppose $\langle\!\langle f, g\rangle\!\rangle \in N_1(a)$. Now, a function $h \in N(R)$ belongs to $N_{[G]}(R)$ if and only if $h(ax) = ah(x)$ for all x. Suppose $x > 0$. Then $a^n \leq x < a^{n+1}$ for some integer n and we have $\langle\!\langle f, g\rangle\!\rangle(ax) = a^{n+1}g(ax/a^{n+1}) = a^{n+1}g(x/a^n)$ which implies that $\langle\!\langle f, g\rangle\!\rangle(ax) = a\langle\!\langle f, g\rangle\!\rangle(x)$. Next suppose $x < 0$. Then $-(a^{n+1}) < x \leq -(a^n)$ for some integer n and therefore $\langle\!\langle f, g\rangle\!\rangle(ax) = a^{n+1}f(ax/a^{n+1}) = a^{n+1}f(x/a^n) = a\langle\!\langle f, g\rangle\!\rangle(x)$. Noting that $\langle\!\langle f, g\rangle\!\rangle(0) = 0$, we get $\langle\!\langle f, g\rangle\!\rangle \in N_{[G]}(R)$.

Now suppose $h \in N_{[G]}(R)$. We must show that $h \in N_1(a)$. Let $f = h|[-a, -1]$ and $g = h|[1, a]$. We assert that $h = \langle\!\langle f, g\rangle\!\rangle$. Suppose $x > 0$. Then $a^n \leq x < a^{n+1}$ for some integer n and we have $1 \leq x/a^n < a$. Now $g(x/a^n) = h(x/a^n) = (h(x))/a^n$ since $h \in N_{[G]}(R)$ and therefore, $h(a^n x) = a^n h(x)$ for all $x > 0$ and all $n \in Z$ where Z denotes the integers. Consequently, $h(x) = a^n g(x/a^n) = \langle\!\langle f, g\rangle\!\rangle(x)$ for all $x > 0$. One shows, in a similar manner, that $h(x) = \langle\!\langle f, g\rangle\!\rangle(x)$ for negative x as well and the proof is complete.

Note that so long as $a > 0$, we can, with no loss of generality, take $a > 1$ as well since, if $a < 1$, we can take $1/a$ as our generator. Now let $a < -1$. Let $C([1, a^2])$ denote the family of all continuous functions f from $[1, a^2]$ into R such that $f(a^2) = a^2 f(1)$. For each $f \in C([1, a^2])$ we associate a function f_* as follows. Let $x > 0$. Then there exists an integer n such that $a^{2n} \leq x < a^{2n+2}$ and we define $f_*(x) = a^{2n} f(x/a^{2n})$. For $x < 0$, we have $a^{2n+1} < x \leq a^{2n-1}$ for some integer n and we define $f_*(x) = a^{2n-1} f(x/a^{2n-1})$. Finally, we define $f_*(0) = 0$. It can be verified that f_* is continuous and we let

$$N_2(a) = \{f_* : f \in C([1, a^2])\}.$$

Theorem 3. *Let $a < -1$ and let $G = \langle a\rangle$. Then $N_{[G]}(R) = N_2(a)$.*

Proof. Here also, $f \in N_{[G]}(R)$ if and only if $f(ax) = af(x)$ for all x, from which it follows that $f(a^n x) = a^n f(x)$ for all $x \in R$ and $n \in Z$ whenever $f \in N_{[G]}(R)$. Suppose $f_* \in N_2(a)$ and let $x > 0$. Then $a^{2n} \leq x < a^{2n+2}$ for some integer n and we have $f_*(x) = a^{2n} f(x/a^{2n})$ where $f \in C([1, a^2])$. But ax is negative and we also have $a^{2n+3} < ax \le a^{2n+1}$ and hence,

$$f_*(ax) = a^{2n+1} f(ax/a^{2n+1}) = a^{2n+1} f(x/a^{2n}) = af_*(x).$$

One shows that $f_*(ax) = af_*(x)$ for negative x as well and we conclude that $f_* \in N_{[G]}(R)$.

Now suppose $g \in N_{[G]}(R)$. We must show that $g = f_*$ for some $f \in C([1, a^2])$. Let $f = g|[1, a^2]$. Evidently $f \in C([1, a^2])$. Suppose $x > 0$. Then $a^{2n} \le x < a^{2n+2}$ for some integer n and we have $1 \le x/a^{2n} < a^2$. Since $g \in N_{[G]}(R)$, we have $g(a^n x) = a^n g(x)$ for all x and each integer n. Because of this, we have $f(x/a^{2n}) = g(x/a^{2n}) = (g(x))/a^{2n}$. This implies $g(x) = a^{2n} f(x/a^{2n}) = f_*(x)$. One can also show that $g(x) = f_*(x)$ for negative x and, consequently, $g = f_* \in N_2(a)$.

R_M^+ will denote the subgroup of R_M consisting of all the positive real numbers and R^- will denote the collection of negative real numbers. For each ordered pair, (a, b), of real numbers, we associate a function $\langle\langle a, b \rangle\rangle$ as follows.

$$\langle\langle a, b \rangle\rangle(x) = \begin{cases} ax & \text{for } x \le 0 \\ bx & \text{for } x \ge 0. \end{cases}$$

We then let $N_3 = \{\langle\langle a, b \rangle\rangle : (a, b) \in R \times R\}$.

Theorem 4. *Let G be a subgroup of R_M^+ which is not cyclic. Then $N_{[G]}(R) = N_3$.*

Proof. The *log* function is an isomorphism from R_M^+ onto R_A, the additive group of reals and hence, $log[G]$ is a subgroup of R_A which is not cyclic. Proposition 1 of [1], page 7 tells us that the closed subgroups of R_A are R_A, $\{0\}$ and groups of the form aZ, $a > 0$ where Z denotes the group of integers. Since $log[G]$ is not cyclic, it follows that $cl\, log[G] = R_A$ and this implies that $cl\, G = R_M^+$. Now suppose $f \in N_{[G]}(R)$. Then $f(vx) = vf(x)$ for every $v \in G$ and all $x \in R$. Take $x = 1$ and $b = f(1)$ and get $f(v) = bv$ for all $v \in G$. Since G is dense in R_M^+, it follows that $f(x) = bx$ for all $x \ge 0$.

We also have $f((-v)x) = f(v(-x)) = vf(-x)$ for all $v \in G$ and $x \in R$. Take $x = 1$ and $a = -f(-1)$ and get $f(-v) = -av$ for all $v \in G$. Let $x = -v$ and note that we have $f(x) = ax$ for all $x \in -G$. Since G is dense in R_M^+, $-G$ is dense in R^- and we see that $f(x) = ax$ for all $x \le 0$. Consequently, $f = \langle\langle a, b \rangle\rangle \in N_3$. One easily verifies that if $f \in N_3$, then $f(vx) = vf(x)$ for each $v \in G$ (in fact, for each $v > 0$) which means $f \in N_{[G]}(R)$. This completes the proof.

The following result is an immediate consequence of the preceding theorem.

Corollary 5. *Let $G = \langle 2, 3 \rangle$. Then $N_{[G]}(R) = N_3$.*

Recall that t_a is the map which is defined by $t_a(x) = ax$.

Theorem 6. *Let G be a subgroup of R_M such that $G \cap R^- \neq \emptyset$ and $G \cap R_M^+$ is not cyclic. Then $N_{[G]}(R) = End(R)$.*

Proof. Suppose $f \in N_{[G]}(R)$. Then $f(vx) = vf(x)$ for all $v \in G$ and $x \in R$. Let $x = 1$ and $f(1) = a$ and get $f(v) = av$ for all $v \in G$. Just as in the proof of Theorem 4, $G \cap R_M^+$ is dense in R_M and it readily follows that $f(x) = ax$ for all $x \geq 0$. Choose any element $b \in G \cap R^-$. Then the function t_b maps R_M^+ homeomorphically onto R^- and, consequently, $t_b[G \cap R_M^+]$ is a dense subset of R^-. Since it is also a subset of G, we must have $f(vx) = vf(x)$ for all x and all $v \in t_b[G \cap R_M^+]$. Again we take $x = 1$ and since $a = f(1)$, we have $f(v) = av$ for all $v \in t_b[G \cap R_M^+]$ and it follows that $f(x) = ax$ for all $x \leq 0$. Thus, $f = t_a \in End(R)$. Since $End(R) \subseteq N_{[G]}(R)$, the proof is complete.

Corollary 7. *Let $G = \langle -2, 3 \rangle$. Then $N_{[G]}(R) = End(R)$.*

Now let $a > 1$ and let $C([1, a])$ denote the collection of all continuous functions from $[1, a]$ into R such that $f(a) = af(1)$. For each function $f \in C([1, a])$, we associate a continuous selfmap $f_\diamond$ of R. Suppose $x > 0$. Then $a^n \leq x < a^{n+1}$ for some integer n. Define $f_\diamond(x) = a^n f(x/a^n)$. For $x < 0$, define $f_\diamond(x) = -f_\diamond(-x)$ and, finally, let $f_\diamond(0) = 0$. We then let

$$N_4(a) = \{f_\diamond : f \in C([1, a])\}.$$

Theorem 8. *Suppose $-1 \in G$ and $G \cap R_M^+$ is cyclic with generator $a > 1$. Then $N_{[G]}(R) = N_4(a)$.*

Proof. Since $-1 \in G$, it follows that $G = \langle a \rangle \cup (-\langle a \rangle)$. Suppose $f_\diamond \in N_4(a)$. We want to show that $f_\diamond(ax) = af_\diamond(x)$ for all $x \in R$. First, suppose $x > 0$. Then $a^n \leq x < a^{n+1}$ for some n and $f_\diamond(x) = a^n f(x/a^n)$ where f is a continuous function from $[1, a]$ into R. Now $a^{n+1} \leq ax < a^{n+2}$ and $f_\diamond(ax) = a^{n+1} f(ax/a^{n+1}) = a^{n+1} f(x/a^n)$. Evidently, $f_\diamond(ax) = af_\diamond(x)$. For negative x, we have $f_\diamond(ax) = -f_\diamond(-ax) = -f_\diamond(a(-x)) = -af_\diamond(-x) = a(-f_\diamond(-x)) = af_\diamond(x)$. Thus, $f_\diamond(ax) = af_\diamond(x)$ for all $x \in R$. Since $f_\diamond$ is an odd function, it readily follows that $f_\diamond(bx) = bf_\diamond(x)$ for all $x \in R$ and all $b \in G$. Consequently, $f_\diamond \in N_{[G]}(R)$.

Now suppose $f \in N_{[G]}(R)$. Then $f(a^n x) = a^n f(x)$ for all x. We must show that $f = g_\diamond$ for some $g \in C([1, a])$. Let $g = f|[1, a]$ and suppose $x > 0$. Then $a^n \leq x < a^{n+1}$ for some n and $g_\diamond(x) = a^n g(x/a^n) = a^n f(x/a^n) = f(x)$. Thus, f and $g_\diamond$ agree on the positive real numbers. But f is an odd function since $-1 \in G$ and $g_\diamond$ is an odd function by definition. Thus, f and $g_\diamond$ must be identical, which means $f \in N_4(a)$.

Corollary 9. *Let $G = \langle -1, 2 \rangle$. Then $N_{[G]}(R) = N_4(a)$.*

The only remaining case to consider is that where $G \cap R_M^+$ is cyclic, $G \cap R^- \neq \emptyset$ but $-1 \notin G$ and this is the substance of our next result.

Theorem 10. *Suppose $G \cap R_M^+$ is cyclic with generator $a > 1$, $G \cap R^- \neq \emptyset$ but $-1 \notin G$. Then $G = \langle -\sqrt{a} \rangle$ and $N_{[G]}(R) = N_2(-\sqrt{a})$.*

Proof. Let $b \in G \cap R^-$. Then $b^2 = a^m$ for some integer m. Moreover, m must be odd, otherwise we would have $b = -a^{m/2}$ which would imply $-1 = (-a^{m/2})/a^{m/2} \in G$ and this is a contradiction. Therefore, we have $b = -(\sqrt{a})^m$ where m is an odd integer. Then $a^n b \in G$ for all integers n which means $-(\sqrt{a})^{m+2n} \in G$ for all n. Consequently, $-(\sqrt{a})^n \in G$ for all odd integers n and it follows from this that $G = \langle -\sqrt{a} \rangle$. It now follows from Theorem 3 that $N_{[G]}(R) = N_2(-\sqrt{a})$.

We denote by $NO(R)$ the nearring of all continuous odd functions from R to R.

Theorem 11. *The nearrings $NO(R)$, $N_1(a)$, $N_2(a)$, N_3, $N_4(a)$ and $End(R)$ are all mutually nonisomorphic.*

Proof. First of all, $End(R)$ is isomorphic to the field of real numbers and none of the other nearrings are even rings, much less fields. Thus, $End(R)$ is not isomorphic to any of the previous five types of nearrings and we now turn our attention exclusively to those nearrings. We show first that $N_1(a)$, $N_2(a)$, N_3 and $N_4(a)$ are mutually nonisomorphic. Now t_0 is the additive identity of each of the nearrings. Define $\gamma(x) = -x$ and let δ denote the identity map. The map δ is, of course, the multiplicative identity of each of the nearrings. Note that if γ belongs to one of the nearrings, then it is the unique map for which $\gamma + \delta = t_0$. Note also that a function f is odd if and only if $f \circ \gamma = \gamma \circ f$. Now we list some facts about the four nearrings. The verifications are routine so we omit them.

(11.1) $N_1(a)$ contains odd functions and also functions which are not odd.

(11.2) $N_2(a)$ contains no odd functions.

(11.3) N_3 contains odd functions and also functions which are not odd.

(11.4) $SO(N_3)$, the multiplicative semigroup of all odd functions of N_3, coincides with $\{t_a : a \in R\}$.

(11.5) $N_4(a)$ consists of odd functions.

Now the map γ has been algebraically characterized and so have the odd functions. It follows from this and properties (11.1) to (11.5) inclusive that $N_1(a)$ and N_3 cannot be isomorphic to either $N_2(a)$ or $N_4(a)$. It also follows that $N_2(a)$ cannot be isomorphic to $N_4(a)$. We next show that $N_1(a)$ cannot be isomorphic to N_3. In view of (11.4), $SO(N_3)$ is a group with zero, that is, every nonzero element has an inverse. It is not difficult to show that $N_1(a)$ contains many odd functions which do not have inverses. For an example, let $g(x) = sin(bx - b)$ for $x \in [1, a]$ where $b = \pi/(a-1)$ and then define $f(x) = -g(-x)$ for $x \in [-a, -1]$. Then $\langle\langle f, g \rangle\rangle$ is an odd function in $N_1(a)$ which has no inverse since it is not injective. This completes the verification that $N_1(a)$, $N_2(a)$, N_3 and $N_4(a)$ are mutually nonisomorphic.

It remains to show that $NO(R)$ is not isomorphic to any one of $N_1(a)$, $N_2(a)$, N_3 or $N_4(a)$. $NO(R)$ is not isomorphic to either $N_1(a)$, $N_2(a)$ or N_3 since $NO(R)$ contains only odd functions. It is not isomorphic to $N_4(a)$ since $N_4(a)$ contains a unit, different from the identity (namely t_a), which commutes with each element of $N_4(a)$ while $NO(R)$ contains no such unit. More specifically, given any $a \neq 0$, there exist many functions $f \in NO(R)$ such that $f(ax) \neq af(x)$ for some x. This concludes the proof.

Some closing remarks are in order. First of all, we note that the particular element a played no role in the previous proof. For instance, if $a, b > 1$, then $N_1(a)$ is not isomorphic to, say, $N_4(b)$. Next it follows from our previous considerations that two groups of automorphisms of R may be isomorphic and yet produce nonisomorphic bicentralizer nearrings and, on the other hand, two nonisomorphic groups may produce isomorphic bicentralizer nearrings. For example, let $G = \langle -2 \rangle$ and $H = \langle 2 \rangle$. Then $[G]$ and $[H]$ are both infinite cyclic groups of automorphisms but it follows from Theorems 2, 3, and 11 that $N_{[G]}(R)$ and $N_{[H]}(R)$ are not isomorphic. Now let $G = \langle 2, 3 \rangle$ and $H = \langle 5, 7, 11 \rangle$. Then $[G]$ and $[H]$ are certainly not isomorphic and yet $N_{[G]}(R)$ and $N_{[H]}(R)$ are identical (they both coincide with N_3) according to Theorem 4.

References

1. Bourbaki, N., *Elements of Mathematics, General Topology, Part 2*, Addison-Wesley, Mass. (1966).

WHEN IS $M_A(G)$ A RING?

C.J. Maxson

Department of Mathematics, Texas A & M University
College Station, TX 77843, U.S.A.

ABSTRACT
Necessary and sufficient condition in terms of the group G and the automorphism group $\mathcal{A}$ are presented for $M_A(G)$ to be a ring.

I. Introduction

In this note we let $(G, +)$ be a group with identity 0, but not necessarily abelian and let $\mathcal{A}$ be a subgroup of automorphisms of G. As is well-known, the set $M_A(G) := \{f : G \to G \mid f(0) = 0, f\alpha = \alpha f, \forall \alpha \in \mathcal{A}\}$ is a zero-symmetric near-ring with identity under the operations of function addition and composition. We call this near-ring the *centralizer near-ring determined by the pair* $(\mathcal{A}, G)$. We recall a few well-known facts.

Lemma A. $\forall f \in M_0(G)$, $f \in M_A(G) \iff \forall x \in G$, $st(x) \subseteq st(f(x))$ *where as usual,* $st(x) = \{\alpha \in \mathcal{A} \mid \alpha(x) = x\}$.

Lemma B. (Betsch's Lemma). $\forall x, y \in G$, $\exists f \in M_A(G)$ *with* $f(x) = y \iff st(x) \subseteq st(y)$. *[See [3], Prop. 9.199.]*

We recall also that the pair $(\mathcal{A}, G)$ is said to be *regular* if $\forall x, y \in G^* = G \backslash \{0\}$, $st(x) \subseteq st(y) \implies st(x) = st(y)$ and the pair $(\mathcal{A}, G)$ is said to satisfy the *finiteness condition* (F.C.) if $\forall x \in G$, $\forall \alpha \in \mathcal{A}$, $st(x) \subseteq st(\alpha(x)) \implies st(x) = st(\alpha(x))$. Since $st(\alpha(x)) = \alpha st(x)\alpha^{-1}$, if G is finite, $(\mathcal{A}, G)$ always satisfies (F.C.).

Lemma C. ([4], Lemma (3.3)). *Let* $(\mathcal{A}, G)$ *satisfy (F.C.). If* $M_A(G)$ *is a ring then* $(\mathcal{A}, G)$ *is regular.*

Proof. Let $x, y \in G^*$ with $st(x) \subsetneq st(y)$. From (F.C.) we see that $y \notin \mathcal{A}x$. Define $f : G \to G$ by

$$f(v) = \begin{cases} \alpha(y), & \text{if } v = \alpha(x) \text{ for some } \alpha \in \mathcal{A}, \\ 0, & \text{if } v \notin \mathcal{A}x. \end{cases}$$

It is straightforward to verify that $f \in M_A(G)$. Let e_x denote the orbit idempotent on $\mathcal{A}x$, i.e., $e_x(v) = v$ if $v \in \mathcal{A}x$ and $e_x(v) = 0$ if $v \notin \mathcal{A}x$.

If $x + y \in \mathcal{A}x$ then $e_x(e_x + f)(x) = e_x(x + y) = x + y$ and $(e_x e_x + e_x f)(x) = x + e_x(y) = x$. Since $M_A(G)$ is a ring we get $y = 0$, a contradiction. Also, if $x + y \notin \mathcal{A}x$, $e_x(e_x + f)(x) = 0$ while $(e_x e_x + e_x f)(x) = x$, again a contradiction. Hence the result follows. ∎

Henceforth in this paper we assume that the pair $(\mathcal{A}, G)$ satisfies the finiteness condition, (F.C.). It is the purpose of this note to develop necessary and sufficient conditions in terms of the pair $(\mathcal{A}, G)$ for $M_A(G)$ to be a ring.

Y. Fong et al. (eds.), Near-Rings and Near-Fields, 199–202.

II. Main results

As above, for $x \in G^*$, let $\mathcal{A}x := \{\alpha(x) \mid \alpha \in \mathcal{A}\}$ be the orbit of x in G determined by the action of $\mathcal{A}$ on G. Here we work only with non-zero orbits. For notational purposes, we let $\mathcal{A}^\circ$ denote the group of autormorphisms $\mathcal{A}$ with the zero function, θ, adjoined. Hence $\mathcal{A}^\circ x = \mathcal{A}x \cup \{0\}$. We say two orbits $\mathcal{A}a$ and $\mathcal{A}b$ are *compatible* if for some $x \in \mathcal{A}a$, $y \in \mathcal{A}b$ we have $st(x) \subseteq st(y)$ or $st(y) \subseteq st(x)$. Combining methods from [2] and [4] we obtain the next result.

Lemma 1. *If $M_\mathcal{A}(G)$ is a ring, then $\forall x, y \in G^*$, $st(x) \subseteq st(y) \Longrightarrow \mathcal{A}x = \mathcal{A}y$, i.e., there are no compatible orbits.*

Proof. From Lemma B there exists an $f \in M_\mathcal{A}(G)$ with $f(x) = y$. As above, for any $v \in G^*$, let e_v denote the orbit idempotent on $\mathcal{A}v$.

If $x+y = 0$ then $0 = e_x(x+y) = e_x(e_x+f)(x) = e_x(x)+e_xf(x) = x+e_x(y) = x+e_x(-x)$. So $e_x(-x) = -x$ which means $-x \in \mathcal{A}x$, i.e., $y \in \mathcal{A}x$. If $x+y \neq 0$, then $x+y = e_{x+y}(e_x+f)(x) = e_{x+y}(x)+e_{x+y}(y)$. From this we see that $x, y \in \mathcal{A}(x+y)$ so $\mathcal{A}x = \mathcal{A}(x+y) = \mathcal{A}y$. ■

If there are no compatible orbits, we say the pair $(\mathcal{A}, G)$ is *incompatible*, so when $M_\mathcal{A}(G)$ is a ring, $(\mathcal{A}, G)$ is incompatible. Let $\mathcal{R} := \{a_\sigma \in G \mid \sigma \in \Sigma\}$ be an arbitrary but fixed set of representatives for the (non-zero) orbits. For $a \in \mathcal{R}$, let $H_a := C_G(st(a)) = \{v \in G \mid \alpha(v) = v, \forall \alpha \in st(a)\}$, a subgroup of G. The proof of the following characterization is straightforward.

Lemma 2. *If $(\mathcal{A}, G)$ is regular, $H_a = \{u \in G \mid st(u) = st(a)\} \cup \{0\}$.*

Now, for $a \in \mathcal{R}$, let $\mathcal{A}_a := \{\alpha \in \mathcal{A} \mid \alpha(H_a) = H_a\}$, a subgroup of $\mathcal{A}$ where we continue to assume that $(\mathcal{A}, G)$ is regular. Then $\mathcal{A}_a$ is a group of fixed point free automorphisms of H_a. In fact, if for some $x \in H_a$, $\beta \in \mathcal{A}_a$, $\beta(x) = x$ then $\beta(u) = u$, $\forall u \in H_a$ since $u \in H_a$ implies $st(u) = st(a) = st(x)$. Hence β acts as the identity on H_a. Therefore by considering the automorphisms in $\mathcal{A}_a$ restricted to H_a we get that $\mathcal{A}_a$ is indeed fixed point free on H_a.

Lemma 3. *If $M_\mathcal{A}(G)$ is a ring, then $H_a = \mathcal{A}_a a \cup \{0\}$ and $M_{\mathcal{A}_a}(H_a)$ is a near-field, for each $a \in \mathcal{R}$.*

Proof. We first show $H_a = \mathcal{A}_a a \cup \{0\}$. Let $0 \neq x \in H_a$. Then $st(x) = st(a)$ and since $(\mathcal{A}, G)$ is incompatible, x must be in the orbit $\mathcal{A}a$. Consequently, there is some $\beta \in \mathcal{A}$, $x = \beta a$. We show $\beta \in \mathcal{A}_a$. To this end, we take $y \in H_a$ and show $\beta y \in H_a$. In fact, $st(\beta y) = \beta st(y)\beta^{-1} = \beta st(a)\beta^{-1} = st(\beta a) = st(x) = st(a)$ since $x, a \in H_a$. Thus $\beta y \in H_a$ as desired. Since $\mathcal{A}_a a \subseteq H_a$ we have $H_a = \mathcal{A}_a a \cup \{0\}$. Since $\mathcal{A}_a$ is fixed point free on H_a, we have from [1] (Corollary II.2) that $M_{\mathcal{A}_a}(H_a)$ is a near-field. ■

Theorem 4. ([2], [4]). *$M_\mathcal{A}(G)$ is a ring $\iff$ $M_\mathcal{A}(G)$ is a product of division rings.*

Proof. Since the sufficiency of the condition is clear, we need only establish the necessity. To this end, for $a \in \mathcal{R}$, let $\overline{\mathcal{A}a} = G^* \setminus \mathcal{A}a$ and consider the left ideal,

$l(\overline{\mathcal{A}a}) := \{f \in M_{\mathcal{A}}(G) \mid f(x) = 0, \forall x \notin \mathcal{A}a\}$. We show $l(\overline{\mathcal{A}a})$ is a two-sided ideal and $M_{\mathcal{A}}(G) = \prod_{a \in \mathcal{R}} l(\overline{\mathcal{A}a})$. Let $f \in l(\overline{\mathcal{A}a})$ and $g \in M_{\mathcal{A}}(G)$. For $y \in \overline{\mathcal{A}a}$, $st(y) \subseteq$ $st(g(y))$ and since $(\mathcal{A}, G)$ is regular, $st(g(y)) = st(y)$. This means $g(y) \notin \mathcal{A}a$ since $(\mathcal{A}, G)$ is incompatible. Thus $fg(y) = 0$ so $fg \in l(\overline{\mathcal{A}a})$. Clearly $\prod_{a \in \mathcal{R}} l(\overline{\mathcal{A}a}) \subseteq$ $M_{\mathcal{A}}(G)$. Now let $f \in M_{\mathcal{A}}(G)$ and define $f_a : G \to G$ by

$$f_a(x) = \begin{cases} f(x), & x \in \mathcal{A}a, \\ 0, & \text{otherwise.} \end{cases}$$

Then $f_a \in l(\overline{\mathcal{A}a})$ and $f = \prod_{a \in \mathcal{R}} f_a \in \prod_{a \in \mathcal{R}} l(\overline{\mathcal{A}a})$. This gives $M_{\mathcal{A}}(G) = \prod_{a \in \mathcal{R}} l(\overline{\mathcal{A}a})$.

To complete the proof we show that $l(\overline{\mathcal{A}a}) \cong M_{\mathcal{A}_a}(H_a)$. Then, from Lemma 3, $l(\overline{\mathcal{A}a})$ is a near-field contained in the ring $M_{\mathcal{A}}(G)$, and hence is a division ring. Let $f \in M_{\mathcal{A}_a}(H_a)$ and define $\hat{f} : G \to G$ by

$$\hat{f}(x) = \begin{cases} \beta f(a), & x = \beta a \in \mathcal{A}a, \\ 0, & \text{otherwise.} \end{cases}$$

For $\gamma \in \mathcal{A}$, $x \in G$, we have $\gamma x \in \mathcal{A}a$ if and only if $x \in \mathcal{A}a$. Hence if $x \notin \mathcal{A}a$, $\gamma \hat{f}(x) = 0 = \hat{f}(\gamma x)$ while if $x \in \mathcal{A}a$, say $x = \beta a$, then $\gamma \hat{f}(x) = \gamma \beta f(a) = \hat{f}(\gamma \beta a) = \hat{f}(\gamma x)$. Consequently, $\hat{f} \in l(\overline{\mathcal{A}a})$ and so the map $\Phi : M_{\mathcal{A}_a}(H_a) \to l(\overline{\mathcal{A}a})$ is a near-ring morphism. If $\Phi(f) = \hat{f}$ is the zero function in $l(\overline{\mathcal{A}a})$ then $f(a) = 0$ so f is identically zero in $M_{\mathcal{A}_a}(H_a)$, i.e., Φ is injective. Finally, let $g \in l(\overline{\mathcal{A}a})^*$. Then $g(a) \neq 0$, so by Lemma A, $st(a) \subseteq st(g(a))$ and from Lemma C, $st(a) = st(g(a))$. Then from Lemma 2, $g(a) \in H_a^* = \mathcal{A}_a a$. For $x \in H_a^*$, $x = \alpha a$ for some $\alpha \in \mathcal{A}$ and $g(x) = \alpha g(a) \in \mathcal{A}_a a \subseteq H_a$. Hence $g|_{H_a} \in M_{\mathcal{A}_a}(H_a)$ and $\widehat{g|_{H_a}} = g$, i.e., Φ is surjective. ∎

Therefore, to obtain our characterization result, we must find, in terms of $\mathcal{A}$ and G, when $M_{\mathcal{A}_a}(H_a)$ is a division ring. We have the following general result.

Theorem 5. *$M_{\mathcal{A}}(G)$ is a division ring $\Longleftrightarrow$ 1) $(\mathcal{A}, G)$ is regular, 2) $G = \mathcal{A}a \cup \{0\}$, and 3) for all $\alpha, \beta \in \mathcal{A}_a$, $(\alpha a + \beta a = \gamma a, \gamma \in \mathcal{A}_a^0 \Longrightarrow \alpha \delta a + \beta \delta a = \gamma \delta a$, for all $\delta \in \mathcal{A}_a)$.*

Proof. From [1] (Theorem III.1.c), $M_{\mathcal{A}}(G)$ is a near-field if and only if $(\mathcal{A}, G)$ is regular and $G = \mathcal{A}a \cup \{0\}$. Suppose $M_{\mathcal{A}}(G)$ is a division ring and let $\alpha, \beta \in \mathcal{A}_a$ with $\alpha a + \beta a = \gamma a$, $\gamma \in \mathcal{A}_a^0$. Since $\alpha, \beta \in \mathcal{A}_a$, $\alpha a, \beta a \in H_a$, hence $\alpha a + \beta a \in H_a$ and $st(\alpha a) = st(\beta a) = st(a)$ by Lemma 2. Let $\delta \in \mathcal{A}_a$ and let $f, g, h \in M_{\mathcal{A}}(G)$ be defined by $f(a) = \alpha(a)$, $g(a) = \beta(a)$, $h(a) = \delta(a)$ on $\mathcal{A}a$ and 0 elsewhere. Then $\gamma \delta a = \gamma h a = h \gamma a = h(\alpha a + \beta a) = h(f + g)(a) = (hf + hg)(a) = h \alpha a + h \beta a = \alpha h a + \beta h a = \alpha \delta a + \beta \delta a$.

Conversely suppose the conditions are satisfied. Then we know $M_{\mathcal{A}}(G)$ is a near-field. Let $f, g, h \in M_{\mathcal{A}}(G)$. Then $f(a) = \alpha(a)$, $g(a) = \beta(a)$ and $h(a) = \delta(a)$ for some $\alpha, \beta, \delta \in \mathcal{A}_a$. That α, β, δ are in $\mathcal{A}_a$ follows as in the proof of Lemma 3. Then $h(f + g)(a) = h(\alpha a + \beta a)$. Let $\alpha a + \beta a = \gamma a$ and note that $\gamma \in \mathcal{A}_a^0$.

But then $\alpha\delta a + \beta\delta a$ so $h(\alpha a + \beta a) = h(\gamma a) = \gamma h(a) = \gamma\delta a = \alpha\delta a + \beta\delta a = \alpha h(a) + \beta h(a) = h(\alpha a) + h(\beta a) = (hf + hg)(a)$. Since $G = \mathcal{A}a \cup \{0\}$ we have $h(f + g)(x) = (hf + gh)(x)$ for all $x \in G$, i.e., $M_\mathcal{A}(G)$ is a division ring. ■

Combining the above results we obtain our characterization.

Theorem 6. $M_\mathcal{A}(G)$ is a ring $\iff$ 1) $(\mathcal{A}, G)$ is regular and incompatible, and 2) for each orbit $\mathcal{A}a$, for each $\alpha, \beta \in \mathcal{A}_a$, $(\alpha a + \beta a = \gamma a$, $\gamma \in \mathcal{A}_a^0 \implies \alpha\delta a + \beta\delta a = \gamma\delta a$, for all $\delta \in \mathcal{A}_a)$.

We conclude by presenting an alternate way of arriving at property 2) of the above characterization. If $(\mathcal{A}, G)$ is regular and incompatible. then from Lemma 3, $M_{\mathcal{A}_a}(H_a)$ is a near-field. We define a function $\psi : M_{\mathcal{A}_a}(H_a) \to \mathcal{A}_a$ by $\psi(\theta) = \theta$ and for $f \neq \theta$, $\psi(f) = \alpha$ where $f(a) = \alpha(a) \in H_a$, $\alpha \in \mathcal{A}_a$. Since $\mathcal{A}_a$ is fixed point free on H_a, it is straightforward to verify that ψ is a bijection and $\psi(fg) = \psi(g)\psi(f)$ for $f, g \in M_{\mathcal{A}_a}(H_a)$. Also, this bijection induces an addition, $\oplus$, on $\mathcal{A}_a^0$ by $\alpha \oplus \beta = \gamma$, where $\alpha a + \beta a = \gamma a$. In this way $(\mathcal{A}_a^0, \oplus, \circ)$ is a left near-field and hence, $M_{\mathcal{A}_a}(H_a)$ is a division ring if and only if the right distributive law holds in $(\mathcal{A}_a^0, \oplus, \circ)$ i.e., if and only if $(\alpha \oplus \beta)\delta = \alpha\delta \oplus \beta\delta$, $\alpha, \beta, \delta \in \mathcal{A}_a^0$. But this is precisely property 2) of Theorem 6.

References

[1] C. J. Maxson and J. D. P. Meldrum, *Centralizer representations of near-fields*, Comm. in Alg. **89** (1984), 406–415.

[2] C. J. Maxson, M. R. Pettet and K. C. Smith, *On semisimple rings that are centralizer near-rings*, Pac. J. Math. **101** (1982), 451–461.

[3] G. F. Pilz, Near-rings, second, revised edition, North Holland, Amsterdam, 1983.

[4] M. Zeller, *Centralizer near-rings on infinite groups*, Ph. D. dissertation, Texas A & M University, College Station Texas, 1980.

ANSHEL-CLAY NEAR-RINGS AND SEMIAFFINE PARALLELOGRAMSPACES

Hans H. Ney

Pickardrst. 21, D-66346 Püttlingen, Germany

ABSTRACT

One of the best known theorems in commutative geometry is: *From each vectorspace over a skewfield of dimension not less then three arises an affine space, and to each affine space of geometric dimension not less than three, there is a skewfield F and a vectorspace coordinatisating it.*

A natural question arising is: Does a kind of nearring and a generalization of affine space exist, giving a similar situation? First steps in this direction has been made in J. André's works on *nearaffine spaces* and *nearvectorspaces* using *nearfields*. A wider class of nearrings doing the job is the class of abelian leftidealfree *Anshel-Clay-Nearrings ACNs*.

Proofs of the following may be found in [9].

Terminology

We need some terminology from non-commutative geometries and lattice theory.

Definition :
A triple $(X, \sqcup, \|)$ is called a *join-space with parallelity* iff X is a nonvoid set of points, and

$$\sqcup : X \times X \ni (x,y) \mapsto x \sqcup y := \sqcup(x,y) \in \sqcup(X \times X) \subseteq 2^X$$

is a mapping into the powerset of X, called *join*, the set of all $x \sqcup y$ is denoted by **L**, the join of two different points is called a *honest line*, and $\| \subseteq \mathbf{L} \times \mathbf{L}$ is an equivalence relation, and the following properties are true:

(L0) $x \sqcup x = \{x\}$

(L1) $x, y \in x \sqcup y$

(L2) $z \in (x \sqcup y) \setminus \{x\} \Rightarrow x \sqcup y = x \sqcup z$

(R) $x \neq y \Rightarrow |x \sqcup y| \geq 3$

(P0) $x \sqcup x \| y \sqcup y$

(P1) $x \in X$ and $L \in \mathbf{L} \Rightarrow \dot{\exists} L' \in \mathbf{L}$ and $\exists y \in X$: $L' = x \sqcup y$ and $L' \| L$ (We write $(x \| L) := L'$.)

(P2) $x \sqcup y \| x' \sqcup y' \Rightarrow y \sqcup x \| y' \sqcup x'$

In a join-space with parallelity the following conditions may be true or not.

Y. Fong et al. (eds.), Near-Rings and Near-Fields, 203–207.

(Tam) $x, y, z \in X$ and $x', y' \in X$ and $x \sqcup y \| x' \sqcup y'$
$\quad \Rightarrow \exists z' \in X$ such that $x \sqcup z \| x' \sqcup z'$ and $y \sqcup z \| y' \sqcup z'$
(Pgm) $x, y, z \in X \Rightarrow (y \| x \sqcup z) \cap (z \| x \sqcup y) \neq \emptyset$

Subspaces and lattice theory

Definition ([3]):
A subset $U \subseteq X$ is called a *subspace* iff $(x \| y \sqcup z) \subseteq U$ for all $x, y, z \in U$.

Theorem ([3], [9]):
In a join-space with parallelity and (Tam) the set

$$(x \| U) := \bigcup_{L \in \mathbf{L} \, \cap 2^U} (x \| L)$$

is a subspace iff U is a subspace.

We denote by $\mathbf{U}(X)$ or simply by $\mathbf{U}$ the system of all subspaces of X. If $0 \in X$ is a fixed point, $\mathbf{U}_0$ denotes the subsystem of subspaces containing this 0.

Subspaces with more than one point, but containing no other subspaces than the one-elementary subspaces or the void subspace, are called *primitive*.

The collection of all primitive subspaces, is denoted by $\mathbf{P}$. The collection of all primitive subspaces, containing $0 \in X$ is denoted by $\mathbf{P}_0$.

Theorem ([2]):
$(\mathbf{U}, [\,], \sqcap)$ is a lattice if $[\,]$ and $\sqcap$ are definded by

$$[\mathbf{V}] := \bigcap_{\bigcup \mathbf{V} \subseteq U \in \mathbf{U}} U \quad \text{for } \mathbf{V} \subseteq \mathbf{U}$$

and $\sqcap$ is defined by set-theoretic intersection. $(\mathbf{U}_0, [\,], \sqcap)$ is a sublattice.

Theorem ([9] I. §4(4.22)):
If $(X, \sqcup, \|)$ is a join space with parallelity, where (Tam) and (Pgm) are true, a so-called abelian space, then we have: $(\mathbf{U}_0, [\,], \sqcap)$ is a projective lattice iff

$(E_{\mathbf{P}_0})$ $\forall U \in \mathbf{U}_0 \, \exists P \in \mathbf{P}_0$: $U \cap P = \{0\}$, and
(Req) $\forall \{P, Q\} \subset \mathbf{P}_0$ with $P \neq Q \, \exists R \in \mathbf{P}_0$: $[P, Q] = [P, R] = [Q, R]$.

Definition ([9] I. §4 (4.19)):
An abelian space, where $\mathbf{U}_0$ is a projective lattice for each $0 \in X$, is called a *parallelogramspace*. The cardinality of a minimal generating set of atoms in the lattice $\mathbf{U}_0$ is called the *dimension of the parallelogramspace X*.

Theorem ([1] and J. André personal communications):
In a parallelogramspace of dimension not less than three, the *diagonal condition* is true:

$(Diag)$ $(a, b, c, d \in X$, a, b, c, d pairwise different, $a, b \in P$, $b, c \in Q$, $c, d \in R$, $d, a \in S$, $a, c \in T$, $P, Q, R, S, T \in \mathbf{P}$ pairwise different) $\Rightarrow$ $\exists U \in \mathbf{P}$: $b, d \in U$.

Semiaffinity, Flataxioms.

If U is a subspace of X, in [3] V.§2(1) is defined.

$F1(U) :\Leftrightarrow (a \notin U, b, c \in U, a \sqcup b = a \sqcup c) \Rightarrow b = c;$

$F2(U) :\Leftrightarrow (a, b \in U, a \neq b, c, d \in X \setminus U, \quad a \sqcup c = a \sqcup d, b \sqcup c = b \sqcup d) \Rightarrow c = d.$

If $\mathbf{M}$ is a set of subspaces, where $F1(M)$ is true for all M in $\mathbf{M}$, we write $F1(\mathbf{M})$. $F2(\mathbf{M})$ is analogously defined.

Theorem ([9] I.§5):
In a parallelogramspace we have

$$F1(\mathbf{P}) \Rightarrow F1(\mathbf{U}) \quad \text{and} \quad F1(\mathbf{P}) \text{ and } F2(\mathbf{P}) \Rightarrow F2(\mathbf{U}).$$

Definition:
A parallelogramspace where $F1$ and $F2$ are true for all subspaces is called a *semiaffine parallelogramspace*.

Configurational conditions (Schlieungsaussagen).

In a semiaffine parallelogramspace we define the following configurational conditions [9] I.§6.

Little $\mathbf{P}$-Desargues-Condition $dP_{\mathbf{P}}$:
$(x, x' \in P \in \mathbf{P}, x \neq x', y, z \in X \setminus P, y', z' \in X$
$x \sqcup x' \| y \sqcup y' \| z \sqcup z', x \sqcup y \| x' \sqcup y',$
$x \sqcup z \| x' \sqcup z') \Rightarrow y \sqcup z \| y' \sqcup z'$

Great $\mathbf{P}$-Desargues-Condition $DP_{\mathbf{P}}$:
$(u, x \in P \in \mathbf{P}, u \neq x, y, z \in X \setminus P, x', y', z' \in X,$
$u \sqcup x = u \sqcup x', u \sqcup y = u \sqcup y', u \sqcup z = u \sqcup z',$
$x \sqcup y \| x' \sqcup y', x \sqcup z \| x' \sqcup z') \Rightarrow y \sqcup z \| y' \sqcup z'$

Little $\mathbf{P}$-Pappos-Condition $pP_{\mathbf{P}}$:
$(x, y, z \in P \in \mathbf{P}, u, v, w \in (u \| P), P \cap (u \| P) = \emptyset,$
$u \sqcup y \| v \sqcup z, x \sqcup v \| y \sqcup w) \Rightarrow x \sqcup u \| z \sqcup w$

Theorem ([9] I.§6 (6.22), (6.24), (6.28)):
In a semiaffine parallelogramspace of dimension not less than three, these three configurational-conditions are true.

Spaces on abelian leftidealfree ACNs.

Definition ([4], [5], [9] II§1 (1.14)):
An ACN is a zero-symmetric nearring $(N, +, \cdot)$ [with the left distributive law $a(b+c) = ab + ac$] with

(0) $N \setminus \{0\}$ can be split into no void disjoint subsets A_r $(r \in R)$:

$$N = \{0\} \,\dot\cup\, \bigcup_{r \in R} A_r$$

(1) $\forall r \in R \,|A_r| \geq 2$

(2) $\forall r \in$ R $(A_r, \cdot)$ is a group with unit element 1_r

(3) $\forall r, s \in$ R the mapping $a \mapsto a1_s$ is a group isomorphism from $(A_r, \cdot)$ to $(A_s, \cdot)$

(4) $\forall a \in N$ and $r \in$ R $1_r a = a$

N is called *abelian* iff $n + m = m + n$ for all $n, m \in N$. N is called *leftidealfree* iff no other subgroups $(S, +)$ from $(N, +)$ other than $\{0\}$ and N have the property

$$N \cdot S := \{n.s \mid n \in N, s \in S\} \subseteq S.$$

In the following we are looking only at abelian, leftidealfree ACNs N. Nearfields are such ACNs with $|R| = 1$. Further examples may be found in [5].

The operator group $\left(N, N^{(I)}, +, \cdot\right)$

For an ACN N and a set I of no less than two indices, we set

$$N^{(I)} := \{(n_i)_{i \in I} \mid n_i \in N, \text{ all but finitely many } n_i \text{ being } 0\}.$$

Addition in $N^{(I)}$ is defined componentwise

$$\underline{n} := (n_i)_{i \in I}, \quad \underline{m} := (m_i)_{i \in I}, \quad \underline{n} + \underline{m} := (n_i + m_i)_{i \in I}$$

Left operation of N on $N^{(I)}$ is defined by

$$n \cdot (\underline{n}) := (n \cdot n_i)_{i \in I}$$

The space $\left(N^{(I)}, \sqcup, \|\right)$

Take $X := N^{(I)}$ as points, $\underline{x} \sqcup \underline{y} := \underline{x} + N \cdot \left(-\underline{x} + \underline{y}\right)$ as lines, $\underline{x} \sqcup \underline{y} \| \underline{x}' \sqcup \underline{y}' : \Leftrightarrow$ $N \cdot (-\underline{x} + \underline{y}) = N \cdot (-\underline{x}' + \underline{y}')$ as parallelity.

Main Theorem ([9] II.§4 (4.13), §5 (5.10), (5.13)):
$\left(N^{(I)}, \sqcup, \|\right)$ is a semiaffine parallelogramspace with $(Diag)$ where the configurational conditions $dP_{\mathbf{P}}, DP_{\mathbf{P}}, pP_{\mathbf{P}}$ are true.

A connection to Frobeniusgroups. ([9] II.§2 (2.4), (2.5))

The set of mappings

$$G := \left\{[\underline{v}, a] : N^{(I)} \ni \underline{x} \mapsto \underline{v} + a.\underline{x} \in N^{(I)} \mid \underline{v} \in N^{(I)}, a \in A_r\right\}$$

with the law of composition

$$[\underline{v}, a] \circ [\underline{w}, b] := [\underline{v} + a \cdot \underline{w}, a \cdot b]$$

is a generalized Frobeniusgroup. Especially: If N and I are finite, G is a Frobeniusgroup.

Characterization of $\mathbf{P_0}$ in $\left(N^{(I)}, \sqcup, \|\right)$

A *(left)-endomorphism of* N is a mapping $f : N \to N$ with $f(n + m) = f(n) + f(m)$ and $f(l \cdot n) = l \cdot f(n)$ for all $n, m, l \in N$. In a leftidealfree abelian ACN all endomorphisms are automorphisms or the zero-mapping. The automorphisms are denoted by $A^\star$.

To describe the primitive subspaces of $N^{(I)}$ which contain 0, we need another representation of the elements in $N^{(I)}$. If $n\underline{e}_j$ denotes that element in $N^{(I)}$ with $n_j = n$ and $n_i = 0$ for all $i \neq j$, then $\underline{n} = \sum n_i\underline{e}_i$.

Theorem ([9] II.§5 (5.17), (5.18)):
$$P \in \mathbf{P}_0\left(N^{(I)}\right) \Leftrightarrow \exists l, \text{ a natural number, and, } \exists f_{i_1}, \ldots, f_{i_l} \in A^\star:$$
$$P = \left\{ \sum_{k=1}^{l} f_{i_k}(n)\underline{e}_{i_k} \mid n \in N \right\}.$$

Proofs may be made by elementary calculation, making use of the generalized Frobeniusgroup G and the theory of groupspaces developed by J. André or may be found in [9].

Theorem ([9] III.§4):
If $(X, \sqcup, \|)$ is a semiaffine parallelogramspace with $(Diag)$ and the three configurational conditions, then there exists an abelian, leftidealfree ACN and a set of indices with $(X, \sqcup, \|) = \left(N^{(I)}, \sqcup, \|\right)$.

Theorem ([9] III):
If $(X, \sqcup, \|)$ is a parallelogramspace with dimension not less than three, with F1 and F2 true for all primitive subspaces, then the semiaffinity, the condition $(Diag)$ and the three configurational conditions can be proved.

REFERENCES

[1] J. André, *Affine Geometrien Über Fastkörpern.* Math. Z. **136** (1975), 295-313.

[2] J. André, *Über die Schließung von Parallelogrammen und Dimension in nichtkommutativen Räumen.* Mitt. Math. Sem. Gießen **149** (1981), 77-83.

[3] J. André, *Endliche nichtkommutative Geometrie.* Annales Universitatis Saraviensis, Series Mathematicae **2** (1988), 1-136.

[4] J. André, *On finite noncommutative spaces over certain nearrings.* Near-rings and Near-fields, Contributions to general algebra **8** (1992), 5-14.

[5] J. André and H. Ney, *On Anshel-Clay-Nearrings,* Proceedings of the conference on Near-Rings and Near-Fields Oberwolfach, 1989, to appear in 1994.

[6] O. Bachmann, *Über eine Klasse verallgemeinerter affiner Räume.* Monatshefte f. Math. **79** (1979), 285-297.

[7] L. Dubreil-Jacotin, L. Lesieur et R. Croisot, *Leçons sur la théorie des treillis, des structures algébriques ordonnées et des treillis géométriques.* Gauthier-Villars, Paris, 1953.

[8] H. Karzel, K. Sörensen, und D. Windelberg, *Einführung in die Geometrie.* Vandenhoeck u. Ruprecht, Göttingen, 1973,

[9] H. Ney, *Semiaffine Parallelogrammräume,* Diss. Saarbrücken, 1991.

[10] R. Scapellato, *On Geometric Near-Rings,* Bolletino della Unione Matematica Italiana (6) **2-A** (1983), 389-393.

[11] O. Tamaschke, *Projektive Geometrie I,* B I, 1969.

[12] E. Theobald, *Near-rings and noncommutative Geometry,* Proceedings of the conference on Near-Rings and Near-Fields, S. Benedetto del Tronto, 1981.

ON SEMI-ENDOMORPHAL MODULES OVER
ORE DOMAINS

Dorota Niewieczerzał

Institute of Mathematics, Warsaw Uniwersity,
ul. Banacha 2, 02-097 Warszawa, Poland

ABSTRACT

In this paper we present a partial answer to the Question 3 formulated by
J. Hausen in [4]. This question is strictly connected with some local properties
of homogeneous maps of modules.

1. Principal ideal domains

In this note all rings are associative with $1 \neq 0$ and all modules are right. In
[3] the class $\mathcal{R}$ was defined in the following way: a ring R belongs to this class iff
for every R-module M the near-ring $\mathrm{M}_R(M)$ of all R-homogeneous maps is a ring.
The following theorem was proved there:

Theorem 1 (Th. II.1 from [3].) *For every ring R the following conditions are*
equivalent:

1. *R belongs to $\mathcal{R}$;*
2. *$\mathrm{M}_R(M) = \mathrm{End}_R(M)$ for every R-module M.*

Unfortunately the class $\mathcal{R}$ does not contain such "good" rings as domains. For
example from [7] we know that if R is a domain then $\mathrm{M}_R(R^2)$ is never a ring. From
the other side for some module M a near-ring $\mathrm{M}_R(M)$ could be a ring sometimes
being equal $\mathrm{End}_R(M)$, sometimes not. It depends on assumptions on modules.

Jutta Hausen proposed in [4] a different approach to the studying near-rings
of homogeneous maps. At the beginning of her paper she has formulated the
following questions:

Question 1 *Given a ring R, find necessary and/or sufficient conditions for an*
R-module M such that $\mathrm{M}_R(M)$ is a ring.

Question 2 *Given a ring R find necessary and/or sufficient conditions for an*
R-module M such that $\mathrm{M}_R(M)$ is equal to $\mathrm{End}_R(M)$.

Let M be an R-module. As in [4] we will say that M is *semi-endomorphal* if
$\mathrm{M}_R(M)$ is a ring, and *endomorphal* if $\mathrm{M}_R(M)$ is equal to $\mathrm{End}_R(M)$.

J. Huasen fully answered the above questions for **Z**-modules. For the solution
she used a notion of type, very well known in the theory of abelian groups. Using
the concept of type, she defined an absolutely anisotropic group as a group without
independent elements having comparable types.

Y. Fong et al. (eds.), Near-Rings and Near-Fields, 209–212.

Theorem 2 (Th. 3.8 from [4].) *An abelian group G is semi-endomorphal but not endomorphal if and only if G is an absolutely anisotropic torsion-free group of rank at least two.*

This theorem gives nontrivial examples of such groups because the existence of absolutely anisotropic abelian groups was in fact established in [1].

J. Hausen also made some observations for modules over principal ideal domains (PID's). For such modules one can define types of elements and consequently absolutely anisotropic modules. In [4] the following fact is proved:

Proposition 1 (Prop. 4.1 from [4].) *Let R be a commutative principal ideal domain and let M be a nonzero torsion-free absolutely anisotropic R-module. If R has only finitely many prime ideals then M has rank one.*

In the similar way one can prove:

Proposition 2 *Let R be a commutative principal ideal domain and let M be a nonzero torsion-free absolutely anisotropic R-module. If the cardinality of R is greater than the cardinality of the set of all prime ideals of R, then M has Goldie rank one.*

Proof. Let us assume that M has Goldie rank at least two. If, as in [4], we take elements x, y independent in M, then the set $V = \{x + yr : r \in R\}$ has the same cardinality as R and any two elements in R are independent. So by assumption elements of V have pairwise incomparable types.

Further for any $m \in M$ by $\tau(m)$ we will denote the type of m. Using standard calculations on types (comp. [2]) one can prove that for any $r \neq s$, $r,s \in R$ we have:

$$\tau(x + yr) \cap \tau(x + ys) = \tau(x) \cap \tau(y). \tag{1}$$

Now by assumption for every $r \in R$ there exists a prime ideal P_r which belongs to the type of $x + yr$ with infinite exponent but belongs to the type $\tau(x) \cap \tau(y)$ with finite exponent. From the equality 1 we have that if $r \neq s$ then $P_r \neq P_s$. In this way we have chosen a subset of prime ideals of R which has the same cardinality as R, which is impossible. ∎

Example. Let $C[x]$ be a ring of polynomials over the field C of complex numbers, N be a set of nonrational numbers from C and let S be the multiplicative semigroup generated by $\{x - a : a \in N\}$. Next let R be the localization $C[x]S^{-1}$. Of course R is a PID. It is easy to see that the set of prime ideals of R is countable while R is not countable.

2. Ore domains

In [4] the author also formulated the following

Question 3 *Given a ring R and a semi-endomorphal R-module M does there exist a family $\{X_i\}_{i \in I}$ of submodules of M such that*

1. $M = \cup_{i \in I} X_i$;
2. *For each $f \in M_R(M)$ and for each $i \in I$, restriction $f|_{X_i} \in \mathrm{Hom}_R(X_i, X_i)$?*

The source of this problem as well as theory of piecewise endomorphisms is, in fact, the example 2.1 from [6]. Now we would like to present a partial answer for the above question

Let R be a right Ore domain. This means that R has the classical total ring of right quotients. Of course commutative domains are right Ore domains. Let us recall that a module of Goldie rank 1 is called uniform. It is easy to see that a domain R is a right Ore domain if and only if R as a right R-module is uniform.

Further R will always denote a right Ore domain, $Q(R)$ its ring of right quotients, and M a torsion-free R-module. It is easy to prove the following lemmas.

Lemma 1 *A submodule $N \subset M$ is a maximal uniform submodule of M if and only if N is uniform and for every $r \in R$ and every $m \in M$ the following implication is true:*

$$\text{if } mr \neq 0, \text{ and } mr \in N \text{ then } m \in N. \quad \blacksquare$$

Lemma 2 *For every submodule N of M the following conditions are equivalent:*

1. *$N = N' \cap M$ where N' is a subspace of dimension 1 in the right linear space $M \otimes_R Q(R)$ over $Q(R)$;*
2. *N is a maximal uniform submodule of M;*
3. *N is of Goldie rank 1 and M/N is torsion-free.* $\quad \blacksquare$

Let $\mathcal{U}$ be the family of all maximal uniform submodules of M. It is a cover of M. Indeed, by our global assumptions, for every $0 \neq m \in M$ the cyclic submodule mR is uniform. From the above lemmas we have that for such mR there exists (only one) maximal uniform submodule containing it. Let us define the following relation in $\mathcal{U}$.

Definition 1 *Let $U, V \in \mathcal{U}$. We say that $U \leq V$ iff $\mathrm{Hom}_R(U, V) \neq 0$. We say that U, V are incomparable iff $U \not\leq V$ and $V \not\leq U$, which means that $\mathrm{Hom}_R(U, V) = 0$ and $\mathrm{Hom}_R(V, U) = 0$.*

Note that if G is a torsion-free abelian group then all elements of any uniform subgroup have the same type. So one knows the type of such subgroup. It is known that if H and F are maximal uniform subgroups then the type of H is less or equal than the type of F if and only if $\mathrm{Hom}(H, F) \neq 0$. So Definition 1 is a generalization of the abelian group case.

Definition 2 *An R-module M is absolutely anisotropic iff every two elements in $\mathcal{U}$ are incomparable.*

Theorem 3 *Let R be a right Ore domain and M be a torsion-free right R-module. Then the following conditions are equivalent:*

1. *M is semi-endomorphal;*
2. *M is absolutely anisotropic;*
3. *For every $f \in M_R(M)$ and for every $U \in \mathcal{U}$ the restriction $f|_U$ belongs to $\mathrm{End}_R(U)$.*

Proof. $(1) \Longrightarrow (2)$. This is in fact contents of Prop. 2.6 in [4].

$(2) \Longrightarrow (3)$. Let M be an absolutely anisotropic R-module, $f \in M_R(M)$ and $U \in \mathcal{U}$. By Lemma 1.1 in [5] we can go through to the localization $M' = M \otimes_R Q(R)$ with our homogeneous map f. So let $f' : M' \to M'$ be defined in the natural way as: $f'(m \otimes q) = f(m) \otimes q$ for every $q \in Q(R)$ and $m \in M$. Let $U' = U \otimes_R Q(R)$. From the choice of U it follows that U' is a one-dimensional right vector space over $Q(R)$. Because $f'|_{U'} : U' \to M'$ is a homogeneous map it is easy to see that $f'|_{U'}$ is additive, so $f|_U$ is additive too. This means that $f|_U$ is an R-homomorphism from U to M. Now by assumed (2), $f|_U$ is an endomorphism of U.

$(3) \Longrightarrow (1)$. This is in fact contents of Lemma 3.5 in [4]. ∎

References

[1] R. A. Beaumont and R. S. Pierce, *Torsion free groups of rank two*, Mem. AMS **38** (1961).

[2] L. Fuchs, Infinite abelian groups vol. II, Academic Press, 1973.

[3] P. Fuchs, C. J. Maxson and G. Pilz, *On rings for which homogeneous maps are linear*, Proc. AMS **112** (1991), 1–7.

[4] J. Hausen, *Abelian groups whose semi-endomorphisms form a ring*, Proc. of the Curacao Conference, Marcel Dekker, 1993.

[5] J. Krempa and D. Niewieczerzał, *On homogeneous mappings of modules*, Contr. to General Algebra **8** (1992), 123–135.

[6] C. J. Maxson and K. C. Smith, *Centralizer near-rings that are endomorphism rings*, Proc. AMS **80** (1980) 189–195.

[7] C. J. Maxson and A. P. J. Van der Walt, *Centralizer near-rings over free ring modules*, J. Austral. Math. Soc., (to appear).

SUBIDEALS AND NORMALITY OF NEAR-RING MODULES

Gary L. Peterson

Department of Mathematics, James Madison University
Harrisonburg, VA 22807, USA

ABSTRACT

In this note we will introduce the property of normality of near-ring modules and we will see how a number of results from group theory, especially those dealing with subnormal subgroups, extend to normal modules of d. g. near-rings.

1   Introduction

A subgroup H of a group G is called *subnormal* if there is a series of subgroups

$$G = H_0 \supseteq H_1 \supseteq \ldots \supseteq H_n = H \tag{1.1}$$

such that each H_{i+1} is normal in H_i. The series in equation (1.1) is called a *subnormal series*. H. Wielandt is usually credited with initiating the study of subnormal subgroups of groups when, in his 1939 paper [8], he proved that the join of two subnormal subgroups of a finite group is again a subnormal subgroup. Since then, the study of subnormal subgroups has been an active area of study in group theory. The purpose of this paper is to initiate a similar study for near-ring modules.

To begin, we shall say that an R-subgroup H of an R-module G where R is a near-ring is an *R-subideal* of G if there is a series as in equation (1.1) only with each H_{i+1} an R-ideal of H_i. Such a series will be called an *R-subideal series*. Note that subnormal subgroups and series are special cases occurring when a group is viewed as a near-ring module over the integers. Unless a group is being considered as a module over more than one near-ring, we shall delete the prefix R and use ideal for R-ideal, subideal for R-subideal, and so on. If H is a subideal of G, the length of a subideal series from G to H of shortest length will be called the *subideal index* of H in G and will be denoted $s(G : H)$.

We next introduce the concept of normality of near-ring modules which will be crucial in obtaining many of our results. Suppose G is again an R-module where R is a near-ring. The quotient near-ring $R/Ann_R(G)$ will be denoted R_G. We may then view R_G as a subnear-ring of $M(G)$. The inner automorphism group of G, $Inn(G)$, is also a subset of $M(G)$. For $g \in G$, the inner automorphism of G induced by g will be denoted τ_g. We shall say that *R is normal on G* or G is a *normal R-module* if $Inn(G)$ normalizes the action of R on G--that is, if

$$\tau_g^{-1} R_G \tau_g = R_G$$

Y. Fong et al. (eds.), Near-Rings and Near-Fields, 213–225.

for all $g \in G$. Note that if H is an R-subgroup of G, then any conjugate of H by an element g of G, H^g, will again be an R-subgroup of G when R is normal on G. This fact will play an important role in many of our proofs.

Our main reference on group theory and, in particular, on subnormal subgroups will be [6]. A more extensive reference on subnormal subgroups is [1]. Except for that fact that our groups G will be additive instead of multiplicative, we shall use the notation of [6]. In particular, if H and K are subsets of G:

$\langle H, K \rangle$ denotes the subgroup generated by H and K.

$[H, K] = \langle [h, k] \mid h \in H, k \in K \rangle$.

$[H, iK] = [[H, (i-1)K], K]$.

$H^K = \langle h^k \mid h \in H, k \in K \rangle$.

$H^{G,0} = G, H^{G,1} = H^G, \ldots, H^{G,i+1} = H^{H^{G,i}}, \ldots$

In the case when H and K are subgroups of G, we shall often use the facts that

$$H^K = H + [H, K] \text{ and } H^{G,i+1} = H + [G, iH].$$

Finally, the commutator identity

$$[a + b, c] = [a, c]^b + [b, c]. \tag{1.2}$$

also will be useful to us.

For this remainder of this paper, we shall assume that our near-ring R is distributively generated by a multiplicative semigroup S with identity and that G is a unital (R, S)-module. Notice that if H and K are R-subgroups of G, then each of the subgroups in the previous paragraph are also R-subgroups of G. In addition, note that if H is a subideal of G and if

$$G = H_0 \supseteq H_1 \supseteq \ldots \supseteq H_n = H$$

is a subideal series, then $H_i \supseteq H^{G,i}$ for each i and hence the subideal index of H in G is the least positive integer m such that $H^{G,m} = H$.

Many of our proofs will be similar to those in group theory. While this similarity sorely tempted this author to omit these proofs, they have been included for the convenience of the reader.

Since the study of subnormal subgroups began with Wielandt's join theorem, let us start by considering joins of subideals.

2   Joins of Subideals

By the join of two subideals H and K of our R-module G, we will mean $\langle H, K \rangle$. It is not true in general that the join of two subideals is a subideal as this is not even true in the special case of subnormal subgroups as first noted by Zassenhaus [9, exercise 23, p. 235]. There are, however, many theorems about when the join of two subnormal subgroups is subnormal. No attempt will be made here to make a complete examination of whether all such theorems extend to subideals; rather, in this section, we shall simply see a sample of how some of the join results in Section 13.1 of [6] extend to subideals. Following the approach of [6], we begin with:

Lemma 2.1 *If H is a subideal and K is an ideal of G, then $H + K$ is a subideal of G.*

Proof. This follows from the fact that if

$$G \supseteq H_1 \supseteq \ldots \supseteq H_n = H$$

is a subideal series, then

$$G \supseteq H_1 + K \supseteq \ldots \supseteq H_n + K = H + K$$

is a subideal series.

Lemma 2.2 *If H and K are subideals of G and if K normalizes H, then $H + K$ is a subideal of G.*

Proof. Setting $H_i = H^{G,i} = H + [G, (i-1)H]$, note that K normalizes H_i. By Lemma 2.1, we have that $H_{i+1} + K$ is a subideal of $H_i + K$ from which it follows that $H + K$ is a subideal of G.

Lemma 2.3 *If H and K are subideals of G and if $J = \langle H, K \rangle$, then the following are equivalent:*
(i) J is a subideal of G.
(ii) H^K is a subideal of G.
(iii) $[H, K]$ is a subideal of G.

Proof. The facts that $[H, K]$ is a normal subgroup of H^K and H^K is a normal subgroup of J are elementary group theory results. $(i) \Rightarrow (ii) \Rightarrow (iii)$ now follows since $[H, K]$ is then an ideal of H^K which is then in turn an ideal of J. We next obtain $(iii) \Rightarrow (ii)$ by first noting that H normalizes $[H, K]$ by the commutator identity in equation (1.2). Thus it follows that $H^K = H + [H, K]$ is a subideal of G by Lemma 2.2. Finally, we obtain $(ii) \Rightarrow (i)$ by observing that $J = K + H^K$ and then again applying Lemma 2.2.

We shall say that G has the *subideal join property* if the join of any two subideals of G is again a subideal of G. A class of modules satisfying the subideal join property is given by:

Theorem 2.1 *If G' is a nilpotent group, then G has the subideal join property.*

To prove Theorem 2.1 we shall need:

Proposition 2.1 *If G is a nilpotent group, then each R-subgroup of G is a subideal of G.*

Proof. Suppose H is an R-subgroup of G. Since

$$H^{G,i+1} = H + [G, iH] \subseteq H + [G, iG]$$

and since the nilpotency of G forces $[G, iG] = 0$ for some i, we have $H^{G,i+1} = H$ for some i and hence H is a subideal.

Proof of Theorem 2.1. Note that $[H, K]$ is an R-subgroup of G' and hence is a subideal of G' by Proposition 2.1. As G' is an ideal of G, we then have $[H, K]$ is a subideal of G. Thus $\langle H, K \rangle$ is a subideal of G by Lemma 2.3.

Up to now, we have not required normality of our module to obtain our join results. This has been caused by the fact that we have not had to deal with conjugates of subideals. Proofs of many join theorems do, however, require that we deal with conjugates of subideals in which case we will need normality. To illustrate, we shall prove a near-ring module version of a join result of D. J. S. Robinson from which Wielandt's original join theorem for finite groups follows as a corollary.

Theorem 2.2 *If R is normal on G and if G' has maximum condition on subideals, then G has the subideal join property.*

Proof. Suppose H and K are subideals of G and $J = \langle H, K \rangle$. We shall use induction on $n = s(G : H)$. Let $x_1, \ldots, x_m \in K$ and set

$$L = \langle H, H^{x_1}, \ldots, H^{x_m} \rangle.$$

Since $s(H^G : H) = n - 1$, L is a subideal of H^G and hence of G. As $\langle H, H^x \rangle = \langle H, [H, x] \rangle$ for any $x \in G$, $L = \langle H, N \rangle$ where $N = \langle [H, x_1], \ldots, [H, x_m] \rangle$. From the commutator identity in equation (1.2), we have that $[H, x_i]$ is normalized by H and hence N is an R-ideal of L. Thus N is a subideal of G and hence of G'. By the maximum condition on subideals of G', $[H, K] = N$ for some such N. Therefore $[H, K]$ is a subideal of G and hence J is a subideal of G by Lemma 2.3.

3 Persistence

Another notion introduced by Wielandt is that of persistence. Following the presentation in Appendix G of [9], we say that a property P of groups is *normally (subnormally) persistent* if

(1) any group isomorphic to a group having property P has property P and

(2) the subgroup of a group generated by normal (subnormal) subgroups having property P has property P.

Being a p-group, being a periodic group, and being a locally nilpotent group are some examples of normally persistent properties. Clearly subnormal persistence implies persistence. But, the converse is also true [9, Theorem 28]. In this section we shall extend persistence to near-ring modules, although it remains to be seen whether this will be a useful concept in the study of near-ring modules.

We shall say that a property P of normal R-modules is *ideally (subideally) persistent* if (1) and (2) hold when groups are replaced by normal R-modules and normal subgroups (subnormal subgroups) are replaced by ideals (subideals). Let us now show that we obtain a normal module version of Theorem 28 of [9]:

Theorem 3.1 *For normal R-modules, ideal persistence implies subideal persistence.*

Proof. Suppose that P is an ideally persistent property, G is a normal R-module and $\{H_\lambda \mid \lambda \in \Lambda\}$ is a collection of subideals of G having property P. We show that

$$H = \langle H_\lambda \mid \lambda \in \Lambda \rangle$$

has property P by breaking the proof into two cases:

Case 1. Suppose $\{s(G : H_\lambda) \mid \lambda \in \Lambda\}$ is bounded.

We proceed by induction on the least upper bound n of this set of subideal indices. Clearly H has property P when $n = 0$ since $H_\lambda = G$ for all λ, so assume $n > 0$. Fix a $\lambda \in \Lambda$. For $h \in H$, H_λ^h has property P and since $s(H_\lambda^G : H_\lambda) < n$, $\langle H_\lambda^h \mid h \in H \rangle = H_\lambda^H$ has property P. But H_λ^H is an ideal of H and hence P being ideally persistent gives us $\langle H_\lambda^H \mid \lambda \in \Lambda \rangle = H$ has property P.

Case 2. Suppose $\{s(G : H_\lambda) \mid \lambda \in \Lambda\}$ is not bounded.

Fixing a λ, we have that $\langle H_\lambda^h \mid h \in H \rangle = H_\lambda^H$ has property P by case 1. Thus $\langle H_\lambda^H \mid \lambda \in \Lambda \rangle = H$ has property P.

4 Socles

As first defined for near-ring modules in [2], the socle of a near-ring module G, denoted $Soc(G)$, is the sum of the minimal ideals of G. The socle series constructed by setting $Soc_1(G) = Soc(G)$ and $Soc_{i+1}(G)/Soc_i(G) = Soc(G/Soc_i(G))$ has proven to be very useful for studying tame near-ring modules as, for example, can be seen from Chapter 10 of [4]. In [5], modified forms of socle series (called S-socle series) again proved useful. Subnormal socles formed by the join of the minimal subnormal subgroups of a group have received study as well (see Section 13.3 of [6]). In this and the next section we shall consider subideal socles. To begin our development, we first note the following generalization of an elementary result in group theory ([6], 3.3.12).

Lemma 4.1 *Suppose that $G = \oplus \sum_{\lambda \in \Lambda} G_\lambda$ where each G_λ is a nonabelian minimal ideal of G. If H is a subideal of G, then H is a direct sum of some of the $G_\lambda, \lambda \in \Lambda$.*

Proof. If H is an ideal of G, this follows immediately from 3.3.12 of [6] by using the image of S in R_G for the set of endomorphisms Ω in 3.3.12. The result now follows for a subideal H by considering H as a subideal of the ideal H^G of G and using induction on the subideal index of H in G.

Theorem 10.20 of [4] describes the structure of the socle of a module of a tame endomorphism near-ring. With the aid of Lemma 4.1, the reader can verify that this theorem extends to the following result under our d. g. assumptions.

Theorem 4.1 $Soc(G) = A \oplus B$ *where A is the sum of the abelian minimal ideals of G and B is the sum of the nonabelian minimal ideals of G. Further, A is a direct sum of some of the abelian minimal ideals of G, B is the direct sum of all of the nonabelian minimal ideals of G, and B is a perfect group.*

It will be convenient to pattern our development of subideal socles after the result of Theorem 4.1 by considering the parts generated by the abelian and nonabelian minimal subideals. The following generalization of another result of Wielandt (cf. [6, 13.3.1]) is crucial for the nonabelian part:

Lemma 4.2 *Suppose that R is normal on G and H and K are subideals of G with $H \cap K = 0$. If H is a nonabelian minimal subideal of G, then $[H, K] = 0$.*

Proof. Let $J = \langle H, K \rangle$ and $n = s(J : H)$. We use induction on n. If $n \leq 1$, H is an ideal of J. Thus $[H, K] \subseteq H$ and, by the commutator identity in equation (1.2), $[H, K]$ is a then an ideal of H. Suppose $[H, K] \neq 0$. Then $[H, K] = H$ which forces H to be contained in K^J. But then $J = K^J$ and, since K is a subideal of J, $K = J$ giving us $H \cap K = H$, a contradiction. Thus $[H, K] = 0$.

If $n > 1$, note that $H \neq H^k$ for some $k \in K$. Since H and H^k are minimal subideals of G, $H \cap H^k = 0$. As $s(H^J : H) = n - 1$, $[H, H^K] = 0$. For any $h_1, h_2 \in H$, the equation

$$0 = [h_1^k, h_2] = [h_1 + [h_1, k], h_2] = [h_1, h_2]^{[h_1, k]} + [[h_1, k], h_2]$$

$$= [h_1, h_2]^{[h_1, k]} - [h_1, k] + [h_1, k]^{h_2} \tag{4.1}$$

along with the fact that H normalizes $[H, K]$ by equation (1.2) gives us $[h_1, h_2] \in [H, K]$. Now we have $H = H' \subseteq [H, K]$. As in the last paragraph, we again obtain $J = K^J = K$ which leads to the contradiction of the fact that $H \cap K = 0$.

Corollary 4.1 *A nonabelian minimal subideal of G normalizes every subideal of G.*

Proof. If H is a nonabelian minimal subideal and K is a subideal of G, then either $H \cap K = 0$ in which case $[H, K] = 0$ or $H \subseteq K$. In either case H normalizes K.

Now we form the part generated by the nonabelian minimal subideals of what will become our subideal socle. The reader should notice the similarity between the ideal D of the following theorem and the ideal B of Theorem 4.1.

Theorem 4.2 *Suppose R is normal on G. Let*

$$\mathcal{D} = \{H \mid H \text{ is a nonabelian minimal subideal of } G\}$$

and

$$D = \langle H \mid H \in \mathcal{D} \rangle \tag{4.2}$$

where we take $D = 0$ if $\mathcal{D}$ is the empty set. Then D is an ideal of G, $D = \oplus \sum_{H \in \mathcal{D}} H$, and D is a perfect group. Further, the subideal index of any H in $\mathcal{D}$ is at most 2.

Proof. D is an ideal since $H^g \in \mathcal{D}$ for each $g \in G$. The direct sum result is obtained by the following commonly used Zorn's lemma argument. Calling a subset $\mathcal{I}$ of $\mathcal{D}$ independent if

$$\langle H \mid H \in \mathcal{I} \rangle = \oplus \sum_{H \in \mathcal{I}} H,$$

the set of independent subsets of $\mathcal{D}$ is partially ordered under inclusion and has a maximal element $\mathcal{M}$ by Zorn's lemma. We must have

$$D = \oplus \sum_{H \in \mathcal{M}} H \tag{4.3}$$

for if not there exists $K \in \mathcal{D}$ such that $(\oplus \sum_{H \in \mathcal{M}} H) \cap K = 0$. But then

$$\langle \oplus \sum_{H \in \mathcal{M}} H, K \rangle = \oplus \sum_{H \in \mathcal{M}} H \oplus K$$

contradicting the maximality of $\mathcal{M}$. Equation (4.3) and Lemma 4.1 now give us

$$D = \oplus \sum_{H \in \mathcal{D}} H. \tag{4.4}$$

From equation (4.4) we obtain that D is perfect since each $H \in \mathcal{D}$ is perfect and the subideal index of each $H \in \mathcal{D}$ is at most 2 since each such H is an ideal of D and D is an ideal of G.

Let us turn our attention to what the abelian minimal subideals generate by setting

$$C = \langle K \mid K \text{ is an abelian minimal subideal of } G \rangle,$$

where $C = 0$ if G has no abelian minimal subideals. If R is normal on G, K^g is an abelian minimal subideal whenever K is for each $g \in G$ and hence C is an ideal of G. Moreover, because C is generated by abelian subnormal subgroups, C is a Baer group (see Section 12.2 of [6] or Section 2.5 of [1] for a discussion of Baer groups) and hence is a locally nilpotent group. We now form our *subnormal socle* of a normal module G, denoted $Sis(G)$, by setting

$$Sis(G) = C + D. \tag{4.5}$$

Note that it is not apparent that we have a direct sum in equation (4.5) as we do in Theorem 4.1. In the case of subnormality we do have a direct sum [6, 13.3.5], but whether this extends to the subideal setting remains unresolved at this time. We next consider one case in which we do obtain a direct sum. Another will be considered in the next section.

Following the terminology of S. D. Scott [7], we say that G is a *tame* R-module if each R-subgroup of G is an ideal of G. Scott also has introduced stronger versions of tameness. The strongest occurs when $Inn(G) \subseteq R$ in which case G is called a *compatible* R-module. (Actually, this is not Scott's definition of compatiblity, but is equivalent to it in our d. g. setting by Proposition 1.2 of [5].)

In [5], the concepts of tame and compatible are generalized by saying that G is *weakly tame* (*weakly compatible*) if there is a subideal series

$$G = G_0 \supseteq G_1 \supseteq \ldots \supseteq G_n = 0$$

with each G_i/G_{i+1} tame (compatible).

Theorem 4.3 *Suppose R is normal on G. If G is weakly compatible and R satisfies the descending chain condition on right ideals, then $Sis(G) = C \oplus D$. Also, C is a solvable group.*

Proof. We need $C \cap D = 0$, so suppose this is not the case. By Lemma 4.1 and Theorem 4.2, $C \cap D$ is a direct sum of some of the elements of $\mathcal{D}$ and hence contains a minimal subideal M. Since M is a minimal subideal, we then have that M is both a compatible and a minimal module. Hence M is finite by [4, Lemma 10.39]. Since M is contained in the locally nilpotent group C, we then have M is nilpotent which is impossible because M is perfect as well. Thus we have $Sis(G) = C \oplus D$.

To see why C is solvable, first note that C is weakly compatible and hence has a subideal series S

$$C = C_0 \supseteq C_1 \supseteq \cdots \supseteq C_m = 0$$

with each factor C_i/C_{i+1} a compatible R-module. Since R satifies the descending chain condition on right ideals, C has an S-socle series as constructed on page 1168 of [5]. By Lemma 1.3 of [5], each S-socle factor of C is a direct sum of minimal ideals each of which is either abelian or perfect. But just as in the previous paragraph, we cannot have any perfect summands since C is locally nilpotent. Hence each S-socle factor is abelian and consequently C is solvable.

Locally nilpotent groups can be quite nasty groups as is illustrated by McLain's example [6, pp. 347–349]. The solvability of C in Theorem 4.3, however, does give us a much better behaved group and S-socle series of C could well prove to be useful when examining C. This raises the question as to whether other conditions might lead to an amenable group for C. For instance, suitable chain conditions on locally nilpotent groups often force the groups to be nilpotent. This has lead this author to wonder whether some appropriate chain conditions on near-rings acting on G might not force C to be nilpotent. In addition to giving us a better behaved group for C, the nilpotency of C will give us the existence of a socle series for C when R satisfies the descending chain condition on right ideals by Corollary 1.8 of [5] which might prove useful for studying subideal socles. In the next section we shall see a case where C is nilpotent.

5   Compatible Envelopes

Recall that R_G denotes $R/Ann_R(G)$. Let us denote the image in R_G of our generating set S for R by S_G. The subnear-ring of $M(G)$ generated by $Inn(G)$ and R_G will be called the *compatible envelope* of R on G and will be denoted $C(R_G)$. Notice that the compatible envelope is then the smallest near-ring containing R_G for which G is a faithful compatible module. Also notice that a subset H of G is an R-ideal if and only if H is a $C(R_G)$-ideal of G so that we will not have to distinguish between R-ideals and $C(R_G)$-ideals. Finally, when R is normal on G, $C(R_G)$ is additively generated by the elements $\alpha\tau_g$ where $\alpha \in S_G$, $g \in G$, and τ_g again denotes the inner automorphism g induces on G. In this section we shall see how the compatible envelope can be employed to give us results about the subideal socle.

Theorem 5.1 *Suppose that R is normal on G. If $C(R_G)$ satisfies the descending chain condition on right ideals, then $D = B$, C is solvable, and $Sis(G) = C \oplus D$.*

Proof. Notice that if H is a nonabelian minimal subideal of G, then

$$H^G = \sum_{g \in G} H^g$$

is a minimal ideal of G. Hence $D \subseteq B$. Conversely, suppose that M is a nonabelian minimal ideal of G. By Lemma 10.39 of [4], M is finite and hence contains a minimal R-subideal H. But then we have $M = H^G$ so that M and hence B is contained in D. Thus we have $D = B$.

The solvability of C follows in much the same manner as it did in Theorem 4.3. Considering the socle series of C, no summand of a socle factor can be a finite perfect group because C is locally nilpotent. Hence each socle factor of C is abelian. Thus C is solvable since its socle series terminates in C after a finite number of steps [4, Theorem 10.37]. Finally, $D \cap C = 0$ since $D \cap C$ is both perfect and solvable and hence $Sis(G) = C \oplus D$.

We conclude this section by showing that we can get the nilpotency of C if we add $|S| < \infty$ to the hypothesis of Theorem 5.1. We will need the following generalization of a result of Mal'cev and McLain [6, 12.1.6]:

Lemma 5.1 *Suppose that R is normal on G, G is a locally nilpotent group, and $|S| < \infty$. If M is a minimal ideal of G, then M is contained in the center of G.*

Proof. Suppose that $S = \{\alpha_1 = 1, \alpha_2, \ldots, \alpha_n\}$. If M is not contained in the center of G, there exists $m \in M$ and $g \in G$ such that $b = [m, g] \neq 0$. Note that

$$M = (bR)^G = \langle b\alpha_1, \ldots, b\alpha_n \rangle^G.$$

Hence it follows that there exist $g_1, \ldots, g_k \in G$ such that $m \in H$ where

$$H = \langle (b\alpha_i)^{g_j} | 1 \leq i \leq n, 1 \leq j \leq k \rangle.$$

Setting

$$K = \langle m\alpha_i, g\alpha_i, g_1\alpha_i, \ldots, g_k\alpha_i | 1 \leq i \leq n \rangle,$$

we have K is an R-subgroup of G and, being finitely generated, is nilpotent. Consider the R-subgroup

$$A = (mR)^K = \langle m\alpha_1, \ldots, m\alpha_n \rangle^K$$

of K. Note that b lies in the R-subgroup $[A, K]$ from which it follows that H is contained in $[A, K]$. This gives us $m \in [A, K]$ which forces $A = [A, K]$. But then we have $[A, iK] = A \neq 0$ for all i violating the nilpotency of K.

Now we prove:

Theorem 5.2 *Suppose R is normal on G and $C(R_G)$ satisfies the descending chain condition on right ideals. If $|S| < \infty$, then C is nilpotent.*

Proof. Again consider the socle series of C. By the previous lemma, the socle series of C is a central series and hence C is nilpotent.

6   Wielandt Ideals

Wielandt formed a normal subgroup of a group G which has since become known as the Wielandt subgroup of G by intersecting the normalizers of the subnormal subgroups of G. In this section we shall extend this construction to form an ideal of our R-module G. To do so, we will have to consider normalizers of subideals. Normalizers of R-subgroups of modules of d. g. near-rings have been introduced in [3]. If H is an R-subgroup of our R-module G, we will define the *module normalizer* of H to be

$$MN_G(H) = \{x \in G | -xr + H + xr = H \text{ for all } r \in R\}.$$

In [3], the module normalizer is called the R-normalizer and is denoted $N_R(H)$; the terminology and notation is changed here because we shall have to identify the module in which H lies rather than the near-ring in our work. By Proposition 4.5 of [3], we have that $MN_G(H)$ is the largest R-subgroup of G which contains H as an ideal.

Suppose now that R is normal on G. For the *Wielandt ideal* of G, denoted $\omega(G)$, we shall use the intersection of the normalizers of the subideals of G. Certainly $\omega(G)$ is an R-subgroup of G and is, in fact, an ideal by the second part of the following lemma.

Lemma 6.1 *Suppose that R is normal on G.*
(i) For each $g \in G$, the mapping $r \longrightarrow \tau_g^{-1} r \tau_g, r \in R_G$, is a near-ring automorphism of R_G.
(ii) If A is an R-subgroup of G and $g \in G, MN_G(A)^g = MN_G(A^g)$.

Proof. The proof of (i) is routine and left to the reader. To see (ii), note that for any $x, g \in G$ and $r \in R$,

$$(-xr + A + xr)^g = -x^g(\tau_g^{-1} r \tau_g) + A^g + x^g(\tau_g^{-1} r \tau_g)$$

and the result then follows since the map in part (i) is onto.

One might expect that Wielandt ideals are often trivial. In the group theory setting (which is a special case occurring when R is the ring of integers), however, Wielandt has shown that every normal subgroup which satisfies the minimum condition on normal subgroups is contained in the Wielandt subgroup [6, 13.3.7]. In particular, the Wielandt subgroup of any nontrivial finite group is then nontrivial. We can obtain a similar result in the normal module setting:

Theorem 6.1 *Suppose R is normal on G. If M is a minimal ideal of G and if M satisfies the minimum condition on ideals, then $M \subseteq \omega(G)$.*

Proof. Choose a minimal ideal N of M. We then have $M = N^G = \sum_{g \in G} N^g$. By Proposition 5.1 of [7], $M = \oplus \sum_{g \in X} N^g$ for some subset X of G. Also, $|X| < \infty$ because of the minimum condition on ideals of M. Note that N is a minimal subideal of M; otherwise N would properly contain a nontrivial ideal which would be an ideal of M since M is a direct sum of ideals which are conjugates of N. The finiteness of X now forces M to have minimum condition on subideals.

Let H be a subideal of G. We prove $M \subseteq MN_G(H)$ by induction on $n = s(G : H)$. Of course, this is immediate when $n \leq 1$. To do the induction step, note that $H^G \cap M = 0$ or $M \subseteq H^G$. In the former case, $[M, H^G] = 0$ from which it easily follows that $M \subseteq MN_G(H)$, so assume $M \subseteq H^G$. Let K be a minimal ideal of H^G contained in M. Since M has minimum condition on subideals, K has minimum condition on ideals. The induction hypothesis gives us

$$K \subseteq MN_{H^G}(H) \subseteq MN_G(H).$$

The same holds for any conjugate of K and hence

$$M = \sum_{g \in G} K^g \subseteq MN_G(H).$$

Note that we have $Soc(G) \subseteq \omega(G)$ when G satisfies minimum condition on subideals as an immediate corollary to Theorem 6.1.

Much more has been done to study Wielandt subgroups which suggests that more study of Wielandt ideals is in order as well. We leave this for future investigations.

7 Chain conditions

We end this paper with one more instance where normality of near-ring modules proves to be important. Chain conditions on ideals of a near-ring module in general are not inherited by R-subgroups of the module. In the group theory setting (or when R is the ring of integers), a theorem of J. S. Wilson [6, 3.1.8] states that subgroups of finite index do inherit chain conditions on normal subgroups. Let us show:

Theorem 7.1 *Suppose that R is normal on G. If G has minimum (maximum) condition on ideals and if H is an R-subgroup of finite index in G, then H has minimum (maximum) condition on ideals.*

Proof. Suppose G satisfies minimum condition on ideals and H does not. Note that $core(H) = \cap_{g \in G} H^g$ is an ideal of G of finite index. Thus an infinite chain of descending ideals $K_1 \supset K_2 \supset \ldots$ of H will terminate after a finite number of steps modulo $core(H)$. It then follows that the chain $K_1 \cap core(H) \supseteq K_2 \cap core(H) \supseteq \ldots$ will not terminate after a finite number of steps. Hence $core(H)$ does not satisfy minimum condition on ideals, so we might as well assume H is an ideal of G.

Let T be a transversal for H in G and let K be an ideal of G contained in H which is minimal with respect to not having minimum condition on ideals of H. Suppose

$$K = K_1 \supset K_2 \supset \ldots \tag{7.1}$$

is an infinite chain of ideals of H. Note that K_i^T is an ideal of G contained in K. If $K_i^T \subset K$, then the series $K_i \supset K_{i+1} \supset \ldots$ terminates after a finite number of steps by the minimality of K. Consequently, $K_i^T = K$ for each i.

Among all the series of the type in (7.1), choose a subset X of T of smallest order such that $K_i^X = K$ for every i and every such series. Note that if $x \in X$,

$K_i^{X+\{-x\}} = K$ since $K^x = K$ so that we may assume $0 \in X$. Of course, we cannot have $X = \{0\}$ for then $K_i = K$. Hence $Y = X \setminus \{0\}$ is nonempty. By the minimality of X, we have that there is a series as in (7.6) with $K_j^Y \subset K$ for some j. Using this series, set $L_i = K_i \cap K_i^Y$ which is an ideal of H. If $L_i = L_{i+1}$,

$$K_i = K_i \cap K_{i+1}^X = K_i \cap (K_{i+1} + K_{i+1}^Y) = K_{i+1} + (K_i \cap K_{i+1}^Y)$$

$$\subseteq K_{i+1} + L_i = K_{i+1} + L_{i+1} = K_{i+1},$$

a contradiction. Thus $L_{i+1} \subset L_i$ for all i and hence $L_i^X = K$ for all i. But now we have

$$K_i = K_i \cap L_i^X = K_i \cap (L_i + L_i^Y) = L_i + (K_i \cap L_i^Y)$$

$$\subseteq L_i + (K_i \cap K_i^Y) = L_i$$

which gives us $K_i \subseteq K_i^Y$ and hence $K_i^Y = K_i^X = K$ for all i, a contradiction. Consequently we have proven that H has minimal condition on ideals.

For the proof in the maximal condition setting, again suppose H does not have the chain condition. Using an approach similar to the one in the first paragraph of this proof, we again find we may assume that H is an ideal of G.

Let T again be a transversal for H in G. If A is a subset of G, A_T will denote $\cap_{t \in T} A^t$. Choose an ideal K of G contained in H which is maximal with respect to the property that H/K does not have maximum condition on its ideals. Suppose that

$$K = K_1 \subset K_2 \subset \ldots \tag{7.2}$$

is an infinite chain of ideals of H. Note that each $(K_i)_T$ is an ideal of G. If $K \subset (K_i)_T$, $H/(K_i)_T$ has maximum condition on ideals and hence $K_j = K_{j+1}$ for some $j \geq i$. Hence $(K_i)_T = K$ for all i.

Let X be a subset of T of smallest order such that $(K_i)_X = K$ for each term of every series of the type in (7.2). As in the proof in the minimum condition case, we may assume that $0 \in X$ and we must have $Y = X \setminus \{0\}$ is nonempty. Proceeding as in the third paragraph of this proof, choose a series of the type in (7.7) with $K \subset (K_j)_Y$ for some j. Now we interchange intersections and sums and set $L_i = K_i + (K_i)_Y$. If $L_i = L_{i+1}$,

$$K_i = K_i + (K_{i+1})_X = K_i + (K_{i+1} \cap (K_{i+1})_Y) = (K_i + (K_{i+1})_Y) \cap K_{i+1}$$

$$\supseteq L_i \cap K_{i+1} = L_{i+1} \cap K_{i+1} = K_{i+1},$$

a contradiction. Thus $L_i \subset L_{i+1}$ and $(L_i)_X = K$ for all i. Since

$$K_i = K_i + (L_i)_X = K_i + (L_i \cap (L_i)_Y) = L_i \cap (K_i + (L_i)_Y)$$

$$\supseteq L_i \cap (K_i + (K_i)_Y) = L_i,$$

we have $K_i \supseteq (K_i)_Y$ and hence $(K_i)_Y = (K_i)_X = K$ for all i, a contradiction. This completes the proof in the maximum condition case.

References

[1] J. C. Lennox and S. E. Stonehewer, *Subnormal Subgroups of Groups*, Oxford University Press, Oxford, 1987

[2] C. G. Lyons and J. D. P. Meldrum, N-series and tame near-rings, *Proc. Royal Soc. Edinburgh*, **86A** (1980), 153-163

[3] G. Mason, Solvable and nilpotent near-ring modules, *Proc. Amer. Math. Soc.*, **40** (1973), 351-357

[4] J. D. P. Meldrum, *Near-Rings and their Links with Groups*, Pitman, London, 1985

[5] G. L. Peterson, Weakly tame near-rings, *Comm. Alg.*, **19** (1991), 1165- 1181

[6] D. J. S. Robinson, *A Course in the Theory of Groups*, Springer-Verlag, New York, 1982

[7] S. D. Scott, Tame near-rings and N-groups, *Proc. Edinburgh Math. Soc.*, **23** (1980), 275-296

[8] H. Wielandt, Eine Verallgemeinerung der invarianten Untergruppen, *Math Z.*, **45** (1939), 209-244

[9] H. Zassenhaus, *The Theory of Groups*, Chelsea, New York, 1958

ENDOMORPHISM NEARRINGS ON FINITE GROUPS, A REPORT

Gerhard Saad & Momme Johs Thomsen

Universität der Bundeswehr Hamburg 22039 Hamburg, Germany

Sergei A. Syskin

Institute of Mathematics, Novosibirsk 630090, Russia

ABSTRACT

After a historical introduction this report announces in detail seventeen theorems about endomorphism nearrings on finite groups for various groups and classes of groups, mainly without proofs or only sketching some proofs, but indicating instead which theorems with their full proofs we plan to publish together in one paper and what headings these papers will have.

1 Introduction

To set the scene we begin this report with a short overview of some results about endomorphism nearrings on finite groups which can be found in the mathematical literature. This is done also to introduce on the way some notation and mainly of course to describe the platform from which we started our work about endomorphism nearrings on some finite groups.

For a given group $(G, \cdot)$ let $(M(G), +, \circ)$ be the set of all mappings from G to G with pointwise addition and with the composition of mappings as multiplication, i.e. with

$$g(\alpha + \beta) = g\alpha \cdot g\beta \quad \text{and} \quad g(\alpha \circ \beta) = (g\alpha)\beta$$

for $g \in G$ and $\alpha, \beta \in M(G)$. Then $(M(G), +, \circ)$ is a nearring. The largest zero-symmetric subnearring of $(M(G), +, \circ)$ is $(M_0(G), +, \circ)$ consisting of all mappings of $M(G)$ which fix the identity of G.

Let $(\operatorname{Inn} G, \circ)$ and $(\operatorname{Aut} G, \circ)$ be the group of all inner automorphisms and the group of all automorphisms of $(G, \cdot)$, and let $(\operatorname{End} G, \circ)$ be the semigroup of all endomorphisms of $(G, \cdot)$. These three subsets of $M(G)$ generate three subnearrings $I(G) \subseteq A(G) \subseteq E(G)$ of $(M_0(G), +, \circ)$. These last three so-called endomorphism nearrings on G are the prototypes for distributively generated nearrings. Their structures and their possible coincidences have been extensively studied for various groups or classes of groups.

The first important result in this direction was obtained in 1958 by A. Fröhlich in [6] where he showed that if $(G, \cdot)$ is a finite simple non-abelian group than $I(G) = A(G) = E(G) = M_0(G)$.

The next steps in this direction were done by C. G. Lyons and J. J. Malone from 1970 onwards. In [10] they presented results about nearrings of the form $E(G)$ and

Y. Fong et al. (eds.), Near-Rings and Near-Fields, 227–238.

applied these results to give a detailed picture of $E(G)$ for the symmetric group $G = S_3$ of degree 3, i. e. for the smallest group G such that $E(G) \neq \text{End}\, G$. Then in [11], [12] they considered the three endomorphism nearrings on finite dihedral groups D_{2n} of order $2n$. There they gave also formulas for the order of these nearrings. For example in [11] they showed that $I(D_{2n}) = A(D_{2n}) = E(D_{2n})$ and that the order of this nearring is $2 \cdot n^3$, when n is odd. As D_6 is just S_3 this includes $I(S_3) = A(S_3) = E(S_3)$.

In a further step J. J. Malone investigated in [15] such endomorphism nearrings on generalized quaternion groups Q_n of order 2^n with $n \geq 4$. There he proved also that $I(Q_n)$ has order 2^{3n-5} and that $A(Q_n) = E(Q_n)$ has order 2^{3n-4}. One can add that $I(Q_3)$ has order 2^4 whereas $A(Q_3) = E(Q_3)$ has order 2^8, the last not being covered by the preceeding formula. Recently these investigations were generalized by C. G. Lyons and G. Mason in [13].

To stay further in chronological order we next report on the work of Y. Fong and J. D. P. Meldrum which they did from 1978 onwards in [17], [3], [4], [5]. There they studied these endomorphism nearrings for symmetric groups S_n. In [17] is proved that $I(S_n) = A(S_n) = E(S_n)$ if $n \geq 5$, which is indeed also true if $n \leq 3$. In [4] the detailed structure of $E(S_n)$ for $n \geq 5$ is presented. Finally in [5] the detailed structure of $E(S_4)$ is presented and the order is determined as $|E(S_4)| = 2^{35} \cdot 3^3 = 927\ 712\ 935\ 936$. So concerning symmetric groups only the two endomorphism nearrings $I(S_4)$ and $A(S_4)$ were not treated.

Finally we mention the work of J. D. P. Meldrum in [18], where he considered the three nearrings on general linear groups $GL(n, q)$ consisting of all invertible $n \times n$ matrices over $\text{GF}(q)$, the Galois field with q elements. He assumed that $n \geq 2$ to avoid trivial cases. For $n = 2$ he furthermore assumed that $q > 3$. This means he considered the cases when $GL(n, q)$ is not solvable. So two nontrivial cases were excluded in [18] because they did not fall within the scope of methods used there. But the case $GL(2, 2)$ was the first to be studied at all, $GL(2, 2)$ being just S_3 or D_6. So actually only the case $GL(2, 3)$ remained untreated. Also the question of the coincidence of the three endomorphism nearrings on $GL(n, q)$ was not treated.

For more information about nearrings in general, we refer to the three existing books on nearrings by G. Pilz [20], J. D. P. Meldrum [19] and J. R. Clay [2]. For more details about the reported material in particular, we recommend [19], whose notation we follow here closely. However, we write the groups on which our near-rings act multiplicatively. Latin letters will denote these groups and their elements whereas Greek letters are used for mappings defined on these groups. Notation from group theory will be standard, see for example the book of J. J. Rotman [21].

2   Endomorphism nearrings on some exceptional groups

We start with giving our theorems about endomorphism nearrings on some exceptional groups. Essentially these results were announced also already in [22], [23]. The first four theorems are from our paper [24] entitled "*On endomorphism nearrings of symmetric groups*", which appears 1994. There we begin with the following theorem thus closing the gap left open by Y. Fong and J. D. P. Meldrum in [17], [3], [4], [5].

Theorem 1. *The symmetric group S_4 satisfies* $\mathrm{I}(S_4) = \mathrm{A}(S_4) = \mathrm{E}(S_4)$.

Theorem 1 and above-mentioned results can be summarized in the next theorem thus unifying the existing results about endomorphism nearrings on symmetric groups.

Theorem 2. *All symmetric groups S_n satisfy* $\mathrm{I}(S_n) = \mathrm{A}(S_n) = \mathrm{E}(S_n)$.

Note that with Theorem 2 also the order and the detailed structure of all endomorphism nearrings on all symmetric groups is known because, as mentioned before, $\mathrm{E}(S_n)$ is already investigated for all n (see [10], [4], [5]).

Next we turn in [24] our attention to the most important subgroup of S_n, the alternating group A_n of degree n consisting of all even permutations of S_n. Since A_n for $n \geq 5$ is a finite simple non-abelian group, the before-mentioned result of A. Fröhlich implies

$$\mathrm{I}(A_n) = \mathrm{A}(A_n) = \mathrm{E}(A_n) = \mathrm{M}_0(A_n) \ .$$

But for $n \leq 4$ A_n is solvable. For $n = 3$, A_3 being just the cyclic group of order 3, we clearly have

$$\mathrm{I}(A_3) = \mathrm{A}(A_3) = \mathrm{E}(A_3) = \mathrm{End}\,A_3 = \mathrm{GF}(3) \ .$$

In [24] we show that the only remaining case $n = 4$ is somewhat out of turn.

So for our group A_4 let us introduce some notation. We define $R = \mathrm{I}(A_4)$ and $K = \mathrm{E}(A_4)$. Let V_4 be the Klein four-subgroup of A_4 consisting of the three involutions $(1,2)(3,4)$, $(1,3)(2,4)$, $(1,4)(2,3)$ and the identity. Finally we define

$$Q = \{\alpha \in R \,|\, A_4\alpha \subseteq V_4 \,, \ V_4\alpha = 1\}$$

and

$$P = \{\alpha \in K \,|\, A_4\alpha \subseteq V_4 \,, \ V_4\alpha = 1\} \ .$$

Concerning the group A_4 we prove in [24] the following two theorems about the nearrings

$$R = \mathrm{I}(A_4) \subseteq \mathrm{A}(A_4) \subseteq \mathrm{E}(A_4) = K \ :$$

Theorem 3. *Q is an ideal of $R = \mathrm{I}(A_4)$ such that $Q^2 = 0$ and R/Q is isomorphic to $\mathrm{GF}(4) \times \mathrm{GF}(3)$, the direct product of the fields of order 4 and 3. The order of Q is 2^8 and so the order of R is $2^{10} \cdot 3$.*

Theorem 4. *It is true that $\mathrm{A}(A_4) = \mathrm{E}(A_4) = K$. Furthermore P is an ideal of K such that $P^2 = 0$ and K/P is isomorphic to $\mathrm{End}\,V_4 \times \mathrm{GF}(3)$, the direct product of the endomorphism ring of V_4 and the field of order 3. The order of P is 2^{12} and so the order of K is $2^{16} \cdot 3$. Thus K does not coincide with R.*

Note that V_4 is just the 2-dimensional vector space over $\mathrm{GF}(2)$, so $\mathrm{End}\,V_4$ is just the ring of all 2×2 matrices over $\mathrm{GF}(2)$.

The theorems just given about the endomorphism nearrings on S_4 and A_4 are used in our next forthcoming paper [25] "*Some linear groups and their endomorphism nearrings*". There we consider the three endomorphism nearrings on the general linear group $GL(2,3)$ and on the special linear group $SL(2,3)$. Both are solvable groups. In [18] the case $GL(2,3)$ had to be excluded from the consideration of the family of general linear groups $GL(n,q)$ as mentioned before. Also $SL(2,3)$ plays an analogous exceptional role within the family of special linear groups $SL(n,q)$ just as S_4 respectively A_4 does within the family of symmetric respectively alternating groups.

Now we report about the four theorems from the forthcoming paper [25].

So for our first group $G = GL(2,3)$ let us introduce some notation. We define $R = I(G)$ and $K = E(G)$. Let Z be the center of G. Finally we define

$$N = \{\alpha \in R \mid G\alpha \subseteq Z\} \, .$$

Then it is easy to show that

$$N = \{\alpha \in R \mid G\alpha \subseteq Z, \ Z\alpha = 1\} \, .$$

Concerning the group $G = GL(2,3)$ we prove in [25] the following two theorems about the nearrings

$$R = I(G) \subseteq A(G) \subseteq E(G) = K \ :$$

Theorem 5. *It is true that $R = I(G) = A(G)$. Furthermore N is an ideal of R such that $N^2 = 0$ and R/N is isomorphic to $I(S_4)$. The order of N is 2^{23} and the order of R is $2^{58} \cdot 3^3$.*

Theorem 6. *N is also an ideal of $E(G) = K$ such that K/N is isomorphic to $I(S_4) \times GF(2)$, the direct product of $I(S_4)$ and the field of order 2. The order of K is $2^{59} \cdot 3^3$. Thus K does not coincide with R.*

Note that the structure and the order of $R/N \simeq I(S_4) = A(S_4) = E(S_4)$ is known, see [5] and Theorem 1.

The other group we consider in that article is $S = SL(2,3)$, the special linear group of all matrices of G with determinant 1. So for our second group $S = SL(2,3)$ let us also introduce some notation. We define $R_S = I(S)$ and $K_S = E(S)$. Of course Z is also the center of S. Finally we define

$$J_S = \{\alpha \in R_S \mid S\alpha \subseteq Z, \ Z\alpha = 1\} \, .$$

Concerning the group $S = SL(2,3)$ we prove in [25] the following two theorems about the nearrings

$$R_S = I(S) \subseteq A(S) \subseteq E(S) = K_S \ :$$

Theorem 7. *J_S is an ideal of $R_S = I(S)$ such that $J_S{}^2 = 0$ and R_S/J_S is isomorphic to $I(A_4) \times GF(2)$, the direct product of $I(A_4)$ and the field of order 2. The order of J_S is 2^{11} and the order of R_S is $2^{22} \cdot 3$.*

Theorem 8. *It is true that* $A(S) = E(S) = K_S$. *Furthermore* J_S *is also an ideal of* K_S *such that* K_S/J_S *is isomorphic to* $A(A_4) \times GF(2)$, *the direct product of* $A(A_4)$ *and the field of order* 2. *The order of* K_S *is* $2^{28} \cdot 3$. *Thus* K_S *does not coincide with* R_S.

Note that the structure and the order of $I(A_4)$ and $A(A_4) = E(A_4)$ is known, see Theorem 3 and Theorem 4.

3   Coincidence theorems for linear and quasisimple groups

In this section we give four theorems about some coincidences of the three endomorphism nearrings for certain groups. The first two theorems deal with general linear groups and are planned to be published together in a paper entitled *"The nearring generated by the automorphisms of a finite general linear group"*. The theorem with the long proof is easy to state.

Theorem 9. *All general linear groups* $G = GL(n, q)$ *satisfy* $I(G) = A(G)$.

Here of course n is an arbitrary natural number and q is a prime power.

Note that in the proof only the cases $n \geq 3$ and the cases $n = 2$ and $q > 3$, i. e. only the non-solvable cases, have to be and will be considered, because the cases $n = 1$ are clear and the two exceptional cases $n = q = 2$ and $n = 2$, $q = 3$ are considered already in [11] and in Theorem 5 respectively.

The second theorem has a relatively short proof, if one uses material from other theorems.

Theorem 10. *Let* $G = GL(n, q)$ *be a general linear group with* $n \geq 2$. *Then it is true that* $A(G) = E(G)$ *if and only if* $q = 2$.

The other two theorems about coincidences of the three endomorphism nearrings deal with special linear groups and quasisimple groups and are planned to be published together in a paper entitled *"The nearring generated by the inner automorphisms of a finite quasisimple group"*.

For the special linear group $SL(n, q)$, the subgroup of $GL(n, q)$ whose elements have determinant 1, we give the following result.

Theorem 11. *All special linear groups* $S = SL(n, q)$ *except* $SL(2, 3)$ *satisfy* $I(S) = A(S) = E(S)$.

Note that also here only the non-solvable cases need to be considered and that the exception is treated in Theorem 8. We mention also that one direction of Theorem 10 is contained in Theorem 11, because $GL(n, 2) = SL(n, 2)$.

Actually Theorem 11 is a consequence of a more general theorem about quasisimple groups with cyclic center to be stated next. We remind the reader that a group S is called quasisimple if it is a perfect group, i. e. if it coincides with its derived group S' and if the central factor group $S/Z(S)$ is a simple non-abelian group. So by II.6.10 and II.6.13 in [9] a non-solvable special linear group $SL(n, q)$ is a quasisimple group with cyclic center. So Theorem 11 is essentially a consequence of the following general result.

Theorem 12. *All finite quasisimple groups S with cyclic center satisfy* $I(S) = A(S) = E(S)$.

We remark that the difficult to prove part in the last theorem is $I(S) = A(S)$ whereas the part $A(S) = E(S)$ is relatively easy to prove. Naturally this is analogous to the situation concerning the proofs of Theorem 9 and Theorem 10. Furthermore the easy to prove part is independent of the condition of the cyclicity of the center.

But we emphasize that the condition of the cyclicity of the center is essential for the proof of $I(S) = A(S)$. On the other hand, not so many quasisimple groups have non-cyclic centers. According to the classification of finite simple groups (see Theorem 4.236 in the book by D. Gorenstein [7]), if S is a quasisimple group with non-cyclic center, then its central factor group $S/Z(S)$ is isomorphic to one of the following groups: $D_l(q)$ with $l \geq 4$ even and q odd, $D_4(2)$, $PSL(3,4)$, $PSU(4,3^2)$, $PSU(6,2^2)$, $Sz(8)$, or $^2E_6(2)$, using the notation in [7] or in [8]. It seems (but it has to be checked more carefully) that in each of these cases S admits an automorphism which is not scalar on the center of S. At least, this has been proved explicitly for the group $Sz(8)$ by J. L. Alperin and D. Gorenstein in [1]. This would mean that the condition that the center of S is cyclic is necessary and sufficient for a quasisimple group S to satisfy $I(S) = A(S)$.

4   Nilpotent and metacyclic groups

A complete solution to the general problem to characterize all groups whose endomorphism nearrings coincide (or at least two of them coincide) seems to be far away. So it is very natural to restrict oneself to certain families of groups as others and also we did. We considered in the preceeding sections only classes of groups all of whose members are non-solvable except a few which we had to treat separately. To consider also some classes of solvable groups a first natural step would be to study metacyclic groups.

The class of metacyclic groups is not always defined by different authors to be the same. We call a group G metacyclic if it has a normal subgroup A such that both A and G/A are cyclic.

A satisfactory result concerning a proper subclass of the class of metacyclic groups is given by J. J. Malone and C. Maxon in [16]. There they show that all ZS-metacyclic groups G satisfy $I(G) = A(G) = E(G)$. ZS-metacyclic groups can be defined as groups all of whose Sylow subgroups are cyclic.

But there exist other metacyclic groups G than ZS-metacyclic groups that also satisfy $I(G) = E(G)$. For example the holomorph G of a cyclic group of order p^2 with p an odd prime satisfies $I(G) = E(G)$. By definition, the holomorph of a group A is the natural semidirect product of A and Aut A.

What we report now came out of our struggling to find necessary and sufficient conditions for a metacyclic group G to satisfy $I(G) = E(G)$. Under the heading of this section or something similar is planned some paper part of which will be the following two theorems.

Theorem 13. *Every finite metacyclic group G with $\mathrm{I}(G) = \mathrm{E}(G)$ contains a normal subgroup A with the following properties:*

(1) *A and G/A are cyclic.*

(2) *$[A, G] = A$.*

(3) *For every prime p dividing the order of A each Sylow p-subgroup A_p of A is also a Sylow p-subgroup of the centralizer $\mathrm{C}_G(A_p)$ of A_p in G.*

These conditions together are very close to be a sufficient condition for a finite group to satisfy $\mathrm{I}(G) = \mathrm{E}(G)$.

Let us consider a finite group G satisfying the conclusion of Theorem 13. Then G satisfies $\mathrm{I}(G) = \mathrm{A}(G)$. Assume now that $\mathrm{I}(G)$ is not equal to $\mathrm{E}(G)$ and that furthermore G is a group of minimal order under these restrictions. Then one can prove that the group G contains exactly one minimal normal subgroup. From this follows that A is a cyclic p-group for some odd prime number p and $\mathrm{C}_G(A) = A$. Therefore G is a subgroup of the holomorph of A. So, in some sense, the problem is now reduced to the question whether the holomorph of a cyclic p-group for an odd prime p satisfies $\mathrm{I}(G) = \mathrm{E}(G)$. This question has an unexpected answer.

Theorem 14. *Let p be an odd prime. Let G be the holomorph of a cyclic group of order p^n. Then G satisfies $\mathrm{I}(G) = \mathrm{E}(G)$ if and only if $n \leq p$.*

The detailed proof of Theorem 14 is not short and we would like to give here a brief outline of this proof as we are sure it has some general nature and can be applied in other situations.

So let G be the holomorph of a cyclic group A of order p^n, where p is an odd prime and $n \geq 1$. Then $G = A \cdot B$ is a semidirect product of two cyclic groups A and B of orders p^n and $p^{n-1}(p-1)$ respectively. Let F be the ring of integers modulo p^n. Roughly speaking, elements of $\mathrm{E}(G)$ are interpreted as mappings from the group of units W of this ring F into F. Let $F[x]$ be the ring of all polynomial functions. We consider two polynomials f, g from $F[x]$ to be equal if $f(u) = g(u)$ for all units u from F. Then $\mathrm{I}(G)$ corresponds to the ideal $(x - 1)F[x]$ of $F[x]$. On the other hand, $\mathrm{E}(G)$ can be interpreted as the ideal $pF[x]$. One can show that these two ideals have the same cardinality if and only if $n \leq p$. This completes the sketch of the proof of Theorem 14.

5 Non-nilpotent groups and the Peirce decomposition

Under the heading of this section we plan a paper which will include the following material.

Let G be a finite group. Assume that $G = A \cdot B$ is a semidirect product of a minimal normal abelian subgroup A and some subgroup B. Let ε be the projection of G onto B. Then $\varepsilon^2 = \varepsilon$, that is ε is an idempotent.

Set $R = \mathrm{I}(G)$ and assume for a moment that $\varepsilon \in R$. Then we have the Peirce decomposition of R relative to the idempotent ε which means we have

$$R = \varepsilon \circ R + \mathrm{Ann}_R(\varepsilon) \quad \text{with} \quad \varepsilon \circ R \cap \mathrm{Ann}_R(\varepsilon) = 0 \,, \tag{1}$$

where $\varepsilon_\circ R = \{\varepsilon_\circ r \mid r \in R\}$ and $\text{Ann}_R(\varepsilon) = \{r \in R \mid \varepsilon_\circ r = 0\}$.

As usual we define the annihilator of A as $\text{Ann}_R(A) = \{r \in R \mid Ar = 0\}$ and the annihilator of G/A as $\text{Ann}_R(G/A) = \{r \in R \mid Gr \subseteq A\}$.

Set $N = \text{Ann}_R(A) \cap \text{Ann}_R(G/A)$. Then N is an ideal of R such that $N^2 = 0$ and such that

$$N = (N \cap \varepsilon_\circ R) + (N \cap \text{Ann}_R(\varepsilon)) \text{ and } (N \cap \varepsilon_\circ R) \cap (N \cap \text{Ann}_R(\varepsilon)) = 0. \quad (2)$$

A natural way to study R is now to consider first the quotient nearring R/N and then to use the decomposition (2) of N to obtain the order of N and thus of R.

It turns out that the quotient nearring R/N is a subdirect product of the endomorphism ring $R/\text{Ann}_R(A)$ and the endomorphism nearring $\text{I}(G/A) \simeq \text{I}(B)$. The group B is smaller than G, therefore $\text{I}(B)$ can be assumed to be known. The ring $R/\text{Ann}_R(A)$ is a finite primitive ring, therefore it is isomorphic to the full matrix ring F_n for some $n \geq 1$ and some finite field F (this ring is simple!). In particular, if B is a cyclic group, this ring $R/\text{Ann}_R(A)$ is simply a field of the same order as A.

What can one expect when calculating the order of N by using the decomposition (2)? Every mapping $\alpha \in N \cap \varepsilon_\circ R$ induces a constant mapping $Ab \to A$ on each coset Ab and annihilates A. Therefore the order of $N \cap \varepsilon_\circ R$ is at most $|A|^{|B|-1}$.

Now let $\alpha \in N \cap \text{Ann}_R(\varepsilon)$. For any $b \in B$ the mapping $\alpha_b : A \to A$ is defined by $a\alpha_b = (ab)\alpha$ for $a \in A$. One can verify that α_b coincides with one of the endomorphisms induced by R on A. From this follows that

$$|N \cap \text{Ann}_R(\varepsilon)| \leq k^{|B|-1} \,,$$

where $k = |R/\text{Ann}_R(A)|$ is the order of the endomorphism ring induced by R on A.

The first question arising here is of course whether ε really belongs to R. For example, if G is a p-group for some prime p and A, B are both non-trivial, then ε can not lie in R. Due to this reason, the Peirce decomposition can not be used for nilpotent groups. It seems, one can prove that ε belongs always to R provided $C_G(A) = A$. From this follows that in this case R/N is the direct product of the endomorphism ring $R/\text{Ann}_R(A)$ and $\text{I}(B)$.

Now we shall apply the just outlined use of the Peirce decomposition when we will consider the so-called minimal non-nilpotent groups. Namely, we assume from now on that G is a finite group with the following

Property P. *The group G has a minimal normal subgroup A of order p^n for some prime p and $n \geq 1$ such that $C_G(A) = A$ and such that G/A is cyclic of order q prime to p.*

It follows that A is an abelian group, even an elementary abelian p-group. Therefore A can be considered as a vector space over $\text{GF}(p)$ of dimension n. It is easy to see that $G = A \cdot B$ is a semidirect product of A and a cyclic group B of order q. As the case $q = 1$ is trivial, we assume $q > 1$.

As above, we set $R = \mathrm{I}(G)$ and $N = \mathrm{Ann}_R(A) \cap \mathrm{Ann}_R(G/A)$. It can be shown that the quotient nearring $R/\mathrm{Ann}_R(A)$ is isomorphic to the field $\mathrm{GF}(p^n)$ of order p^n.

It can also be shown that the second quotient nearring $R/\mathrm{Ann}_R(G/A)$ is isomorphic to the ring Z_q of integers modulo q. Therefore $R/\mathrm{Ann}_R(A)$ and $R/\mathrm{Ann}_R(G/A)$ have relatively prime orders. Hence R/N is isomorphic to the direct product

$$\mathrm{GF}(p^n) \times Z_q . \tag{3}$$

We are able to prove the following result.

Theorem 15. *The order of R is $qp^{n(2q-1)}$.*

Sketch of a proof: Let ε be the projection of G on B. We show first that ε lies in R. In fact, let b be a generator of B, then $b - 1$ is an element of the group $GL(A)$ of all invertible linear transformations of A and $(b - 1)^m$ is the identical transformation of A for some integer $m \geq 1$. It is not difficult to verify that $\varepsilon = -(b - 1)^m + 1$. Therefore ε lies in R.

Now we calculate the order of N using the decomposition (2) of N. First we show that

$$|N \cap \varepsilon_\circ R| = p^{n(q-1)}. \tag{4}$$

We know already that the number on the right is an upper bound, i. e. we have $|N \cap \varepsilon_\circ R| \leq p^{n(q-1)}$. Consider the subset S of R consisting of all elements

$$\gamma = \varepsilon_\circ(a_1 + \cdots + a_q)$$

for various sequences $(a_1, ..., a_q)$ in A^q (here a_i is considered as the inner automorphism of G induced by the element a_i). It is easy to show that $S \subseteq N \cap \varepsilon_\circ R$. Moreover, the correspondence

$$(a_1, ..., a_q) \rightarrow \varepsilon_\circ(a_1 + \cdots + a_q) \tag{5}$$

is a group homomorphism from the direct product of q copies of A onto the additive group of the subnearring S.

We will use the following lemma from linear algebra.

Lemma. *Let V be a finite-dimensional vector space and B a cyclic irreducible subgroup of order $q > 1$ of $GL(V)$. Let $v_1, ..., v_q$ be vectors from V. Then the equality*

$$v_1 + v_2 b + \cdots + v_q b^{q-1} = 0$$

holds for each non-identity $b \in B$ if and only if $v_1 = \cdots = v_q$.

This lemma means that the kernel of the homomorphism (5) is the diagonal subgroup $\{(a, ..., a) \mid a \in A\}$ of order p^n. Therefore the range S of this homomorphism has order $p^{nq}/p^n = p^{n(q-1)}$ and the statement (4) follows.

Next we show that

$$|N \cap \mathrm{Ann}_R(\varepsilon)| = p^{n(q-1)}. \tag{6}$$

We know already that the number on the right is an upper bound, i. e. we have $|N \cap \text{Ann}_R(\varepsilon)| \leq p^{n(q-1)}$.

What follows now is the construction of exactly $p^{n(q-1)}$ different mappings in $N \cap \text{Ann}_R(\varepsilon)$. Let $\alpha \in N \cap \text{Ann}_R(\varepsilon)$, then

$$\alpha = (-r_1 + (1 - \varepsilon) \circ g_1 + r_1) + \cdots + (-r_k + (1 - \varepsilon) \circ g_k + r_k),$$

where $g_i \in \text{Inn}\, G$ and r_i is a sum of m_i inner automorphisms of G. For any $a \in A$ and $b \in B$ we have

$$a\alpha_b = a(g_1 b^{m_1} + \cdots + g_k b^{m_k}),$$

that is α_b is a polynomial in b with coefficients g_i taken from $\text{Inn}\, G$. As b has multiplicative order q, we need to consider only polynomials

$$\alpha_b = \gamma_0 + \gamma_1 b + \cdots + \gamma_q b^{q-1} \tag{7}$$

of degree less than q with coefficients γ_i from the endomorphism ring $R/\text{Ann}_R(A)$. Let P be the set of all such polynomials (7). Every polynomial in P corresponds to some $\alpha \in \text{Ann}_R(\varepsilon)$.

We have to use another lemma from linear algebra.

Lemma. *Let V be a finite dimensional vector space. Let B be a cyclic irreducible subgroup of order q of $GL(V)$ and K the subring of $\text{End}\, V$ generated by B. Assume that elements $k_1, ..., k_q$ from K satisfy*

$$k_1 + k_2 b + \cdots + k_q b^{q-1} = 0$$

for any $b \in B$. Then $k_1 = \cdots = k_q = 0$.

It follows from this lemma that the subnearring P contains exactly p^{nq} distinct elements. For $b = 1$ we can get an arbitrary endomorphism of A from the endomorphism ring $R/\text{Ann}_R(A)$. This implies $|N : N \cap P| = p^n$, therefore $N \cap P$ contains $p^{n(q-1)}$ distinct elements and the statement (6) is proved.

From (4) and (6) it follows that N has order $p^{2n(q-1)}$. Therefore by (3) we finally have $|R| = qp^{n(2q-1)}$. This completes the proof of Theorem 15.

Also, we are able to prove the following coincidence theorem.

Theorem 16. *Let G be a group satisfying the Property P. For $n = 1$ it is true that $\text{I}(G) = \text{A}(G) = \text{E}(G)$ whereas for $n \geq 2$ it is true that $\text{I}(G) \neq \text{A}(G) = \text{E}(G)$.*

Using arguments from the proof of Theorem 15 one can get the following structural theorem.

Theorem 17. *Let G be a group satisfying the Property P. Set $R = \text{I}(G)$. Then with $N_1 = \text{Ann}_R(G/A)$, $N_2 = \text{Ann}_R(A)$ and $N = N_1 \cap N_2$ it is true that:*

(1) *$N^2 = 0$ and the order of N is $p^{2n(q-1)}$.*

(2) *The quotient nearring $R/N = (N_1/N) \times (N_2/N)$ is the direct product of the two ideals N_1/N and N_2/N of R/N.*

(3) N_1/N *is the field of order p^n and the additive group $(N_1, +)$ is an elementary abelian group of order $p^{n(2q-1)}$.*

(4) N_2/N *is the ring Z_q of integers modulo q and the additive group $(N_2, +)$ is a semidirect product of the elementary abelian p-group N and some cyclic subgroup B_2 of order q. The normal subgroup N is the direct product of $2(q-1)$ copies of irreducible B_2-modules.*

(5) *The commutator subgroup of the additive group $(R, +)$ is N.*

References

[1] J. L. Alperin and D. Gorenstein. *The multiplicators of certain simple groups.* Proc. Amer. Math. Soc. **17** (1966), 515-519.

[2] J. R. Clay. *Nearrings: Geneses and Applications.* Oxford University Press. Oxford, New York, Tokyo 1992.

[3] Y. Fong. *The endomorphism near-rings of the symmetric groups.* Dissertation, Ph. D. University of Edinburgh, U. K. 1979.

[4] Y. Fong and J. D. P. Meldrum. *The endomorphism near-rings of the symmetric groups of degree at least five.* J. Austral. Math. Soc. **30A** (1980), 37-49.

[5] Y. Fong and J. D. P. Meldrum. *The endomorphism near-ring of the symmetric group of degree four.* Tamkang J. Math. **12** (1981), 193-203.

[6] A. Fröhlich. *The near-ring generated by the inner automorphisms of a finite simple group.* J. London Math. Soc. **33** (1958), 95-107.

[7] D. Gorenstein. *Finite simple groups: An introduction to their classification.* Plenum Press. New York 1982.

[8] R. L. Griess. *Schur multipliers of the known finite simple groups II.* Proc. Amer. Math. Soc **37** (1980), 279-282.

[9] B. Huppert. *Endliche Gruppen I.* Springer-Verlag. Berlin, Heidelberg, New York 1967.

[10] C. G. Lyons and J. J. Malone. *Endomorphism near-rings.* Proc. Edinburgh Math. Soc. **17** (1970), 71-78.

[11] C. G. Lyons and J. J. Malone. *Finite dihedral groups and d. g. near-rings I.* Compositio Mathematica **24** (1972), 305-312.

[12] C. G. Lyons and J. J. Malone. *Finite dihedral groups and d. g. near-rings II.* Compositio Mathematica **26** (1973), 249-259.

[13] C. G. Lyons and G. Mason. *Endomorphism near-rings of dicyclic and generalised dihedral groups.* Proc. Royal Irish Acad. **91A** (1991), 99-111.

[14] C. G. Lyons and J. D. P. Meldrum. *Reduction theorems for endomorphism near-rings.* Monatsh. Math. **89** (1980), 301-313.

[15] J. J. Malone. *Generalized quaternion groups and distributively generated near-rings.* Proc. Edinburgh Math. Soc. **18** (1973), 235-238.

[16] J. J. Malone and G. Mason. *ZS-metacyclic groups and their endomorphism near-rings.* To appear in Monatsh. Math.

[17] J. D. P. Meldrum. *On the structure of morphism near-rings.* Proc. Royal Soc. Edinburgh **81A** (1978), 287-298.

[18] J. D. P. Meldrum. *The endomorphism near-rings of finite general linear groups.* Proc. Royal Irish Acad. **79A** (1979), 87-96.

[19] J. D. P. Meldrum. *Near-rings and their links with groups.* Research Notes in Mathematics 134. Pitman Adv. Publ. Program. Boston, London 1985.

[20] G. Pilz. *Near-rings.* North-Holland/American Elsevier. Amsterdam 1977. Revised Edition 1983.

[21] J. J. Rotman. *The theory of groups.* Allyn and Bacon. Boston 1973.

[22] G. Saad, S. A. Syskin and M. J. Thomsen. *The coincidence of some nearrings defined by symmetric groups.* Intern. Conf. on Algebra. Book of Abstracts, p.432-433. Krasnoyarsk (Russia) 1993.

[23] G. Saad, S. A. Syskin and M. J. Thomsen. *The general linear group $GL(2,3)$ and related nearrings.* Intern. Conf. on Algebra. Book of Abstracts, p.433-434. Krasnoyarsk (Russia) 1993.

[24] G. Saad, S. A. Syskin and M. J. Thomsen. *On endomorphism nearrings of symmetric groups.* To appear in Results in Mathematics **26** (1994).

[25] G. Saad, S. A. Syskin and M. J. Thomsen. *Some linear groups and their endomorphism nearrings.* To appear.

ON THE STRUCTURE OF CERTAIN 2-TAME
NEAR-RINGS

S. D. Scott

Department of Mathematics & Statistics, The University of Ackland
Private Bag 92019, Auckland, New Zealand

Throughout this paper all near-rings are left distributive zero-symmetric and have an identity. Also all N-groups will be unitary. In this and other regards we shall be making use of conventions, notation and definitions from [17].

Chapter 11 of Meldrum's book [10] gives a very readable account of the structural properties (e.g. order) of the near-rings $I(G)$, $A(G)$ and $E(G)$ (N say) for certain well known finite groups G. The explanation in [10] does not go into details concerning how these properties have been derived. The original techniques due to the combined efforts of Fong, Lyons, Malone and Meldrum (see [2 to 7, 9, 11 and 12]) rest heavily on the group structure. Consequently, although these methods often display considerable insight and ingenuity, they necessarily remain quite ad hoc. There exist however, general techniques giving a structural characterization of N for large classes of finite groups. The examples dealt with in [10] always have abelian factors (i.e. there exist non-zero homomorphic images of N which are rings). In §8 of [17] a near-ring N is defined as ring-free if no non-zero homomorphic image is a ring. In this paper we characterize any 2-tame ring-free near-ring N, where $N/J(N)$ has $DCCR$. All such near-rings are finite and the method used allows an exact formulation of the order of the near-ring involved.

The above discussion summarizes the direction taken. Although the first section is taken up with preliminaries, the remainder of the paper contains much of interest. The relationship between tame N-groups and their annihilators is considered in section two. It can happen that the annihilator of certain factor N-groups decomposes into a direct sum. It is important to have information concerning how this happens.

In section three N-endomorphisms are used to investigate the situation where a ring-free near-ring N is 2-tame on some N-group V. If this is the case, then N-isomorphic minimal factors of V occur only under special conditions. This result is then used to show that, if V has N-isomorphic N-subgroups U and W, then $U = W$. A special case of this occurs when two elements u and w of V are such that $(0 : u) = (0 : w)$. In this situation it follows that $u = w$.

In the fourth section we obtain an estimate of the order of a near-ring N having a finite faithful N-group V with submodules U_i, $i = 1, 2$, such that $U_1 \cap U_2 = \{0\}$. When N is ring-free and 2-tame on V this estimate can be made precise. These delicate order arguments allow, in section five, the proof of the main theorem of this paper. This result is a full structural characterization of any 2-tame ring-free near-ring N, where $N/J(N)$ has $DCCR$. Although, by [16], such near-rings are

Y. Fong et al. (eds.), Near-Rings and Near-Fields, 239–256.
© 1995 *Kluwer Academic Publishers.*

finite the characterisation obtained is in fact far from trivial. Also results from
section four allow a precise formulation of the order of such a near-ring.

§1. Preliminaries

In this section we cover some basic material that will be required later. Although
most of the results stated are, in some sense known, they are included here for the
sake of clarity.

It is of value to briefly summarize some aspects of centrality. Central submodules of an N-group V are defined in [4]. Also elementary properties of these
submodules are given in 1.1, of [14]. It is further shown that the sum $Z(V)$ of all
such submodules is again central and, as a consequence, any N-subgroup H of V
is a central submodule if, and only if, $H \leq Z(V)$. The N-group V is said to be a
central sum of submodules V_i, $i = 1, 2$, if $V_1 + V_2 = V$, and $(v_1 + v_2)\alpha = v_1\alpha + v_2\alpha$
for all v_i, $i = 1, 2$, in V_i and α in N.

Proposition 1.1. *If the N-group V is a central sum of the submodules V_i, $i = 1, 2$,
then $V_1 \cap V_2 \leq Z(V)$ (see 1.3 of [14])*

In the case of a 2-tame N-group central sums arise from N-endomorphisms. In
section three we shall be using the result that follows to obtain information on
2-tame N-groups of a ring-free near-ring N.

Proposition 1.2. *If V is a 2-tame N-group and μ an N-endomorphism of V,
then $1 - \mu$ is an N-endomorphism of V and V is a central sum of $V\mu$ and $V(1-\mu)$
(see 1.4 of [15]).*

We now consider particular conditions under which centralizers exist. Suppose
V is an N-group where N is ring-free. If $U_1 \geq U_2$ are submodules of V then
$U_1 \cap U_2 = U_2$ and, by Zorn's lemma, there exists a maximal submodule H of V
such that $H \cap U_1 = U_2$. Let H_1 be any submodule of V such that $H_1 \cap U_1 = U_2$.
By a result of Betsch (see 3.24 of [10]) the lattice of $N/(0 : V)$-submodules of V
is distributive. Thus the lattice of submodules of V is distributive and we have
$[H_1 + H] \cap U_1 = U_2$. It follows from the maximality of H that $H_1 \leq H$, and the
submodule H is unique. We shall call H the *centralizer* of U_1/U_2 in V and denote
it by $C_V(U_1/U_2)$. The name centralizer is used since $C_V(U_1/U_2)$ is the unique
maximal submodule of V having the property that,

$$(u_1 + h)\alpha \equiv u_1\alpha + h\alpha \mod U_2$$

for all u_1 in U_1, h in $C_V(U_1/U_2)$ and α in N. It is clear that, $U_2 \leq C_V(U_1/U_2)$.
In the particular case where $U_2 = \{0\}$ we write $C_V(U_1/\{0\})$ as $C_V(U_1)$. In this
situation $C_V(U_1)$ is the unique maximal submodule H of V such that $H \cap U_1 = \{0\}$.

Only elementary details need be used to prove: -

Proposition 1.3. *If V is an N-group where N is ring-free and $U_1 \geq U_2 \geq K$ are submodules of V, then*

$$C_{V/K}(U_1/K)/(U_2/K) = C_V(U_1/U_2)/K.$$

A special case of the above definition occurs when $V = N$, U_1 is a right ideal ($= R$ say) of N and $U_2 = \{0\}$. It will be necessary to use centralizers of the form $C_N(R)$ in the proof of 4.2.

At this stage we look at some properties of tame N-groups, where N has a descending chain condition. In this area the concept of complete reducibility plays a quite useful role.

A submodule U of an N-group V will be called *completely reducible* in V if for every submodule $W \leq U$ of V we can find a submodule H of V, such that $W \cap H = \{0\}$, and $W \oplus H = U$. Basic properties of completely reducible submodules can be deduced in the same way as for ring modules. Thus we have that any submodule $W \leq U$ of V is completely reducible in V and also results similar to 15.3 of [1] hold.

We now define a *tame series* of a tame N-group V as a series

$$V_0 = \{0\} < V_1 < V_2 < \ldots < V_r = V$$

($r \geq 0$ an integer) of submodules of V, such that V_i/V_{i-1}, $i = 1, \ldots, r$, is completely reducible in V/V_{i-1}. The situation where N has $DCCR$ is of particular interest.

Theorem 1.4. *If V is a tame N-group and N has $DCCR$, then V has a tame series.*

Although 1.4 has been proved in [8] this theorem is initially of much older vintage (see p.90-91 of [15]). In the ring-free situation 1.4 can be considerably refined. It follows from 5.32 of [13] that: -

Proposition 1.5. *If N is a near-ring where $N/J_2(N)$ has $DCCR$, then the number of N-isomorphism types of type 2 N-groups is finite.*

A relatively elementary consequence of tameness, that is proved in [16], is the following.

Lemma 1.6. *If N is a ring-free near-ring and V a tame N-group with N-isomorphic minimal N-subgroups U_i, $i = 1, 2$, then $U_1 = U_2$*

It is now indicated how, if N (as in 1.4) is ring-free, then V has a composition series. The completely reducible factors of V are (see 15.3 of [1]) direct sums of minimal N-subgroups. However, by 1.5 and 1.6, these tame N-groups are finite direct sums. It follows that:-

Theorem 1.7. *If N is a ring-free near-ring with $DCCR$, then a tame N-group has a composition series.*

If the N-group of 1.7 is assumed to be 2-tame, then the result stated can be strengthened very much further (see [16]).

Theorem 1.8. *If the ring-free near-ring N has a faithful 2-tame N-group V and $N/J(N)$ has $DCCR$, then V and N are finite.*

The primary concern of this paper is to characterize all near-rings as in 1.8. Such near-rings and N-groups have already been the subject of detailed study. The main fact emerging from [16], is that, up to N-isomorphism, the N-group V is unique. However present undertakings basically evolved independently of [16]. There is therefore no need to refer further to this fertile source of information.

Our final results of this section take up the question of when a tame near-ring with $DCCR$ can be shown to be ring-free. Although the following theorem appears to be known the author has not seen a proof in print.

Theorem 1.9. *If N is tame on V and has $DCCR$, then a minimal N-group U is N-isomorphic to some minimal factor of V.*

Due to reasons stated above, we now quickly sketch how this theorem is proved. By considering $N/J(N)$, it is established that U is N-isomorphic to a minimal factor R_1/R_2 of N (see 5.32 of [13]). Take H/K as a minimal factor of N, N-isomorphic to R_1/R_2 and such that H is minimal. The minimality of H ensures that the right ideal K of N is uniquely maximal for the property that $K < H$. Since V is faithful, there exists v in V such that $(0 : v) \cap H < H$ and $(0 : v) \cap H \leq K$. Now consider the N-homomorphism λ of H onto vH/vK, defined by $\rho\lambda = v\rho + vK$, for all ρ in H. It can be shown, using the fact that $(0 : v) \cap H \leq K$, that H/K is N-isomorphic to vH/vK (i.e. $ker\lambda = K$). Thus vH/vK is N-isomorphic to U and 1.9 holds.

Theorem 1.9 allows us to answer the question formulated above.

Corollary 1.10. *Suppose the near-ring N has $DCCR$, is tame on the N-group V and V has a composition series*

$$V_0 = \{0\} < V_1 < V_2 < \ldots < V_r = V$$

($r \geq 0$ an integer). If no V_i/V_{i-1}, $i = 1, \ldots, r$, is a ring module then N is ring-free.

Corollary 1.10 is readily proved by resource to 1.9. As will be seen later the condition that V_i/V_{i-1} is not a ring module is frequently equivalent to the requirement that its additive group is non-abelian. These considerations complete our preliminary requirements.

§2. Annihilators.

In this section we shall present properties of the annihilators of certain tame N-groups. To assist with these developments we prove two results concerning sums of ideals.

Proposition 2.1. *If N is a near-ring and A, B and C ideals of N, such that $N = A + C = B + C$, then $N = A \cap B + C$.*

Proof. Since N has an identity

$$B = B \cdot N = B \cdot A + B \cdot C$$

and B is contained in $A \cap B + C$. However, $A \cap B + C$ contains C and therefore $A \cap B + C$ contains $B + C$. Thus $N = A \cap B + C$ and 2.1 holds.

Suppose N is a near-ring and S a non-empty subset of N. As is usual the subset S^r ($r \geq 1$ an integer) of N is simply the set of all products $\alpha_1 \alpha_2 \dots \alpha_r$ where α_i, $i = 1, \dots, r$, is in S. The ideal of N generated by any non-empty subset S of N is denoted by $Id(S)$.

Proposition 2.2. *If N is a near-ring and A and B ideals of N such that $N = A + B$, then $N = Id(A^r) + B$ for any integer $r \geq 1$.*

Proof. Since N has an identity

$$A = A \cdot N \subseteq A^2 + A \cdot B \subseteq A^2 + B.$$

Thus

$$N = A + B \subseteq A^2 + B + B = A^2 + B.$$

Similarly

$$A = A \cdot N = A \cdot (A^2 + B) \subseteq A^3 + B$$

and

$$N = A + B \subseteq A^3 + B + B = A^3 + B.$$

In this way we show that $N = A^r + B$ for any integer $r \geq 1$. Consequently $N = Id(A^r) + B$ and the proposition holds.

Theorem 2.3. *Suppose N is a near-ring with $DCCR$ and U and W are tame N-groups. We have $(0 : U) + (0 : W) = N$ if, and only if, no minimal factor of U is N-isomorphic to a minimal factor of W.*

Proof. Clearly we may assume $U \neq \{0\}$ and $W \neq \{0\}$. Suppose no minimal factor of U is N-isomorphic to one of W. By 1.5, the number of distinct N-isomorphism types of minimal factors of U is finite. Let $n(\geq 1)$ be this integer and

$m \geq 1$ correspondingly be the number of N-isomorphism types of minimal factors of W.

Let $\Delta_1, \ldots, \Delta_n$ be n minimal factors of U of distinct N-isomorphism type and $\Gamma_1, \ldots \Gamma_m$ be m minimal factors of W of distinct N-isomorphism type. Define,

$$A = \cap_{i=1}^{n}(0 : \Delta_i) \quad \text{and} \quad B = \cap_{j=1}^{m}(0 : \Gamma_j).$$

By 1.4, U has a tame series

$$U_0 = \{0\} < U_1 < U_2 < \ldots < U_r = U$$

of length r, where $r \geq 1$ is an integer. Since A annihilates all minimal factors of U it follows readily that $(U_{i+1}/U_i) \cdot A = \{0\}$ for $i = 0, \ldots, r - 1$. Thus

$$U \cdot A^r \subseteq U_{r-1} \cdot A^{r-1} \subseteq \ldots \subseteq U_1 \cdot A = \{0\}$$

and $A^r \leq (0 : U)$. Thus $Id(A^r) \leq (0 : U)$ and similarly there exists an integer $s \geq 1$, such that $Id(B^s) \leq (0 : W)$. If it is shown that $Id(A^r) + Id(B^s) = N$, then it will follow that $(0 : U) + (0 : W) = N$.

To this end we first prove that $A + B = N$. It will be shown that, for i in $\{1, \ldots, n\}$ and j in $\{1, \ldots, m\}$, $(0 : \Delta_i) \neq (0 : \Gamma_j)$. Suppose for some such i and j, $(0 : \Delta_i) = (0 : \Gamma_j)$. Now $(0 : \Delta_i)$ is a minimal $N/(0 : \Delta_i)$-group. Also since $(0 : \Delta_i) = (0 : \Gamma_j)$, Γ_j is a minimal $N/(0 : \Delta_i)$-group. Now the near-ring $N/(0 : \Delta_i)$ is primitive and has $DCCR$. By 7.9 of [10], Δ_i is $N/(0 : \Delta_i)$-isomorphic to Γ_j. Thus Δ_i is N-isomorphic to Γ_j contrary to our assumptions. Thus $(0 : \Delta_i) \neq (0 : \Gamma_j)$. The next step is to note that, by 7.9 of [10], $N/(0 : \Delta_i)$ is simple and $(0 : \Delta_i)$ is necessarily maximal. Clearly the same is true for $(0 : \Gamma_j)$. It now follows that for all i in $\{1, \ldots, n\}$, and j in $\{1, \ldots, m\}$, $(0 : \Delta_i) + (0 : \Gamma_j) = N$. Continued application of 2.1 yields the fact that, $A + (0 : \Gamma_j) = N$ for all j in $\{1, \ldots, n\}$. Again continued application of 2.1, gives us the fact that $A + B = N$. It now follows readily that $(0 : U) + (0 : W) = N$. Indeed, 2.2 implies $Id(A^r) + B = N$. Again, by 2.2., we conclude that $Id(A^r) + Id(B^s) = N$. From the previous paragraph $(0 : U) + (0 : W) = N$.

It remains to show that if $(0 : U) + (0 : W) = N$, U does not have a minimal factor N-isomorphic to a minimal factor of W. To obtain a contradiction assume some minimal factor Δ of U is N-isomorphic to the minimal factor Γ of W. If α is in $(0 : U)$, then $\Delta\alpha = 0$ and $(0 : U) \leq (0 : \Delta) = (0 : \Gamma)$. Similarly, $(0 : W) \leq (0 : \Gamma)$. Since $N \not\leq (0 : \Gamma)$, we have the desired contradiction. The proof of 2.3 is complete.

Theorem 2.3 can be used to obtain a direct decomposition of certain annihilators.

Theorem 2.4. *Suppose the ring-free near-ring N has $DCCR$ and is tame on V. If U_i, $i = 1, 2$, are submodules of V such that $U_1 \cap U_2 = \{0\}$, then*

$$(U_1 \oplus U_2 : V) = (U_1 : V) \oplus (U_2 : V).$$

Proof. It will first be shown that no minimal factor of U_1 is N-isomorphic to a minimal factor of U_2. To obtain a contradiction suppose that the minimal factor X_1/X_2 of U_1 is N-isomorphic to the minimal factor Y_1/Y_2 of U_2. If $H = X_2 + Y_2$, then $(X_1 + H)/H$ is N-isomorphic to

$$X_1/[X_1 \cap (X_2 + Y_2)] = X_1/X_2.$$

Similarly $(Y_1 + H)/H$ is N-isomorphic to Y_1/Y_2. Thus V/H has minimal N-isomorphic N-subgroups $(X_1 + H)/H$ and $(Y_1 + H)/H$. By 1.8, this can only happen if $X_1 + H = Y_1 + H$. Since this implies $X_1 + Y_2 = X_2 + Y_1$, we conclude that

$$X_1 = X_1 \cap [X_2 + Y_1] = X_2.$$

This contradiction to the nature of X_1/X_2 means no minimal factor of U_1 is N-isomorphic to a minimal factor of U_2.

By 2.3, we have that $(0 : U_1) + (0 : U_2) = N$ and 1 can be expressed in the form $e_1 + e_2$, where e_1 is in $(0 : U_1)$ and e_2 is in $(0 : U_2)$. If α is in $(U_1 \oplus U_2 : V)$, then $\alpha = \alpha e_1 + \alpha e_2$. However, for v in V, $v\alpha = u_1 + u_2$ where u_i is in U_i, $i = 1, 2$. Thus, $v\alpha e_1 = u_2 e_1$ and, since $u_2 = u_2 e_1 + u_2 e_2 = u_2 e_1$, we have $w\alpha e_1 = u_2$. It follows that αe_1 is in $(U_2 : V)$. Similarly αe_2 is in $(U_1 : V)$. Thus α is in $(U_1 : V) + (U_2 : V)$ and we conclude that

$$(U_1 \oplus U_2 : V) \leq (U_1 : V) + (U_2 : V).$$

Clearly the $(U_i : V)$, $i = 1, 2$, are contained in $(U_1 \oplus U_2 : V)$ and the reverse inclusion holds. The sum $(U_1 : V) + (U_2 : V)$ is direct since, if β is in $(U_1 : V) \cap (U_2 : V)$, then $V\beta \subseteq U_1 \cap U_2 = \{0\}$ and, from the faithfulness of V, we have $\beta = 0$. Thus

$$(U_1 \oplus U_2 : V) = (U_1 : V) \oplus (U_2 : V)$$

and the theorem is completely proved.

§3. Use of N-endomorphisms.

In this section we shall be using N-endomorphisms to obtain information on 2-tame N-groups of a ring-free near-ring N. The first theorem shows that N-isomorphic minimal factors of such N-groups occur only under special conditions. In material that follows we will be using this theorem to obtain other important structural restraints.

Theorem 3.1. *Let V be a 2-tame N-group and U_1/U_2 and W_1/W_2, N-isomorphic minimal factors of V. If N is ring-free, then $U_2 + W_1 \geq U_1$ and $U_1 + W_2 \geq W_1$.*

Proof. Suppose $U_2 + W_1 \not\geq U_1$. If $U_1 \cap W_1 \not\geq U_2$, then

$$U_2 + W_1 \geq U_2 + U_1 \cap W_1 = U_1.$$

Thus $U_1 \cap W_1 \leq U_2$ and the N-group

$$(U_1 + W_1)/(U_2 + W_1) = (U_1 + U_2 + W_1)/(U_2 + W_1)$$

is N-isomorphic to

$$U_1/[U_1 \cap (U_2 + W_1)] = U_1/U_2.$$

Thus with $X_1 = U_1 + W_1$ and $X_2 = U_2 + W_1$, we see that X_1/X_2 is a minimal factor of V, N-isomorphic (by δ say) to W_1/W_2 and also such that $W_1 \leq X_2$. Let κ be the natural N-homomorphism of X_1/W_2 onto

$$(X_1/W_2)/(X_2/W_2)$$

and λ be the natural N-isomorphism of this N-group onto X_1/X_2. Thus $\kappa\lambda\delta$ is an N-homomorphism of X_1/W_2 onto W_1/W_2 and is therefore an N-endomorphism of X_1/W_2. If 1 is the identity N-endomorphism on X_1/W_2 we have, by 1.2, that

$$X_1/W_2 = (X_1/W_2)\kappa\lambda\delta + (X_1/W_2)(1 - \kappa\lambda\delta)$$

and that this sum of submodules is central. Since $X_2 \geq W_1$, $(W_1/W_2)\kappa = \{0\}$ and $(W_1/W_2)(1 - \kappa\lambda\delta) = W_1/W_2$. Thus

$$(X_1/W_2)\kappa\lambda\delta \cap (X_1/W_2)(1 - \kappa\lambda\delta) \geq W_1/W_2$$

and, by 1.1, W_1/W_2 is a central submodule of X_1/W_2. This implies $N/(0 : W_1/W_2)$ is a ring and, since $W_1/W_2 \neq \{0\}$, we have a contradiction. It has been shown that $U_2 + W_1 \geq U_1$. An entirely similar argument shows $U_1 + W_2 \geq W_1$. The theorem is completely proved.

A proposition is now stated that allows us to prove an interesting consequence of 3.1.

Proposition 3.2. *Let V_i, $i = 1, 2$, be N-groups and Y/X a factor of V_1. If σ is an N-isomorphism of V_1 onto V_2 then $Y\sigma/X\sigma$ is a factor of V_2, N-isomorphic to Y/X.*

We come now to the statement and proof of the theorem hinted at above.

Theorem 3.3. *Let V be a 2-tame N-group and U and W, N-isomorphic N-subgroups of V. If N is ring-free, then $U = W$.*

Proof. Suppose that $U + W > W$ and let x be in $U + W$ but not in W. By Zorn's lemma, we can find an N-subgroup X of V maximal for the property that X excludes x and $W \leq X < U + W$. It follows that, with $Y = xN + W$, $Y \leq U + W$ and Y/X is a minimal factor of V. Now $(U + W)/W$ is N-isomorphic, by δ say, to $U/U \cap W$. Let Y_1 be an N-subgroup of U containing $U \cap W$ and such that $(Y/W)\delta = Y_1/U \cap W$, and let X_1 be an N-subgroup of U containing $U \cap W$ and such that $(X/W)\delta = X_1/U \cap W$. Clearly Y/X is N-isomorphic to $(Y/W)/(X/W)$. Also, by 3.2, $(Y/W)/(X/W)$ is N-isomorphic to

$$(Y_1/U \cap W)/(X_1/U \cap W),$$

which is N-isomorphic to Y_1/X_1. Thus the minimal factor Y/X of $U + W$ is N-isomorphic to the minimal factor Y_1/X_1 of U. Let λ be an N-isomorphism of U onto W. By 3.2, Y_1/X_1 is N-isomorphic to the minimal factor $Y_1\lambda/X_1\lambda$ of W. Thus Y/X is N-isomorphic to $Y_1\lambda/X_1\lambda$. Since $Y_1\lambda \leq W \leq X$ we have $Y_1\lambda + X \not\geq Y$ contrary to 3.1. Thus $U + W \leq W$. Similary $U + W \leq U$ and $U = V + W = W$. The proof of 3.3 is complete.

Theorem 3.3 is of quite general character and also facilitates the proof of the next theorem, which is of an equally general nature. The theorem to be proved will be used at a key point in proving 4.2.

Theorem 3.4. *Suppose V is a 2-tame N-group where N is ring-free. If u and w in V, are such that $(0 : u) = (0 : w)$, then $u = w$.*

Proof. The map δ taking α in N to $u\alpha$ in uN is an N-homomorphism of N onto uN. Since $\ker\delta = (0 : u)$, we have uN is N-isomorphic to $N/(0 : u)$. However, $N/(0 : u)(= N/(0 : w))$ is, by a similar argument, N-isomorphic to wN. Thus uN is N-isomorphic to wN and, by 3.3, $uN = wN$. Define a map λ of uN into $uN(= wN)$ by mapping $u\beta$, β in N, to $w\beta$. Indeed λ is well defined, for if $u\beta_1 = u\beta_2$, where β_i, $i = 1, 2$, are in N, then $\beta_1 - \beta_2$ is in $(0 : u)(= (0 : w))$ and $w\beta_1 = w\beta_2$. It is easily verified that λ is an N-automorphism of uN, since if $(u\gamma)\lambda = 0 = w\gamma$ for some γ in N, then $u\gamma = 0$. By 1.2, we have uN is the central sum of $uN\lambda$ and $uN(1 - \lambda)$. Since λ is an N-automorphism of uN, it follows, from 1.1, that $uN(1 - \lambda) \leq Z(uN)$. Thus $uN(1 - \lambda)$ is a ring module and $N/(0 : uN(1 - \lambda))$ is a ring. Since N is ring-free this can only happen if $uN(1 - \lambda) = 0$ and $\lambda = 1$. From this we conclude that $(u1)\lambda = u = w1 = w$. Thus $u = w$ and the proof of 3.4 is complete.

The final result of this section is a further application of 3.1 and is of a more specialized nature. This result is also required in the order arguments of the next section.

Theorem 3.5. *Suppose the ring-free near-ring N has $DCCR$ and is 2-tame on V and V has a unique minimal N-subgroup U. If v in V is such that $v(U:V) = \{0\}$, then $v = 0$.*

Proof. First it is shown that no minimal factor of V/U is N-isomorphic to U. To obtain a contradiction suppose Y_1/X_1 is a minimal factor of V/U, N-isomorphic to U. There exist submodules Y and X of V containing U and such that $Y/U = Y_1$, and $X/U = X_1$. Clearly Y/X is a minimal factor of V, N-isomorphic to U and, since $U + X \le X$ we have a contradiction to 3.1.

Now, by 2.3, $(0 : V/U) + (0 : U) = N$ and, since $(0 : V/U) = (U : V)$, we see that $(U : V) + (0 : U) = N$. However, $N(U : V) \subseteq (U : V)$ and $vN(U : V) = \{0\}$. If $vN \ne \{0\}$, then, by 1.8, vN has a minimal N-subgroup W. Since W is a minimal N-subgroup of V it follows, by the uniqueness of U, that $U \le vN$. From above, $(U : V) \le (0 : vN) \le (0 : U)$, contradicting the fact that $(U : V) + (0 : U) = N$. Thus $vN = \{0\}$ and, since V is unitary, $v = 0$. The theorem is entirely proved.

§4. Order Arguments

In this section we shall first obtain an estimate for the order of any near-ring having a finite faithful N-group with submodules U_i, $i = 1, 2$, such that $U_1 \cap U_2 = \{0\}$. It will then be shown how results of previous sections can be used, in the ring-free 2-tame situation, to make this estimate precise. This last result, allows us, in the next section, to present a characterisation of any ring-free 2-tame near-ring N where $N/J(N)$ has $DCCR$.

Theorem 4.1. *If V is a finite faithful N-group and U_i, $i = 1, 2$, are submodules of V such that $U_1 \cap U_2 = \{0\}$, then N is finite and*

$$|N| \le |U_1|^{|V/U_2|-1}|N/(U_1 : V)|.$$

Proof. Since V is finite $|M_0(V)| \le |V|^{|V|}$ and $M_0(V)$ is finite. Also the map taking α in N to the function that takes v in V to $v\alpha$ in V, is readily seen to be a near-ring embedding of N into $M_0(V)$. Thus $|N| \le |M_0(V)|$ and N is finite. Clearly

$$|N| = |(U_1 : V)||N/(U_1 : V)|$$

and it remains to show that

$$|(U_1 : V)| \le |U_1|^{|V/U_2|-1}.$$

Let $v_1, \dots, v_k$ ($k \ge 1$ an integer) be a set of distinct coset representatives of U_2 in V where $v_1 = 0$. If this collection consists of the single element $v_1 (= 0)$, then $U_2 = V$ and $U_1 = \{0\}$. In this case $|(U_1 : V)| = |(0 : V)| = 1 \le 0^{1-1}$. Thus we may assume $k \ge 2$. Let $\triangle = \{v_2, \dots v_k\}$ and $U_1^\triangle$ be the set of all functions of $\triangle$

into U_1. We shall show that there exists an injective function τ of $(U_1 : V)$ into U_1^Δ. If α is in $(U_1 : V)$, define $(\alpha)\tau$ to be the function given by $x(\alpha)\tau = x\alpha$ for all x in Δ. Since for x in Δ, $x\alpha$ is in U_1, $(\alpha)\tau$ is in U^Δ and τ is a function of $(U_1 : V)$ into U_1^Δ. We now show that τ is injective. Suppose $(\alpha_1)\tau = (\alpha_2)\tau$, where α_i, $i = 1, 2$, are in $(U_1 : V)$. This implies $v_j\alpha_1 = v_j\alpha_2$ for all j in $\{2, \dots, k\}$ and, since $0\alpha_1 = 0\alpha_2$, $v_j\alpha_1 = v_j\alpha_2$ for all j in $\{1, \dots, k\}$. However, if u is in U_2, then, because U_2 is a submodule of V

$$(v_j + u)\alpha_1 = v_j\alpha_1 + y,$$

where y is in U_2. Since α_1 is in $(U_1 : V)$, $v_j\alpha$, and $(v_j + u)\alpha_1$ are in U_1 and $y = 0$. Thus $(v_j + u)\alpha_1 = v_j\alpha_1$ and an entirely similar argument shows $(v_j + u)\alpha_2 = v_j\alpha_2$. Since any element v of V is of the form $v_j + u$, we conclude that $v(\alpha_1 - \alpha_2) = 0$, for all v in V. Thus $\alpha_1 = \alpha_2$ and τ is injective. It now follows that

$$|(U_1 : V)| \le |U_1|^{|\Delta|}$$

and, because Δ has $|V/U_2| - 1$ elements,

$$|(U_1 : V)| \le |U_1|^{|V/U_2|-1}.$$

The theorem is therefore proved.

The above estimate is, in some situations, an exact formulation of the order of N.

Theorem 4.2. *Suppose the ring-free near-ring N is 2-tame on the finite N-group V. If U is a minimal N-subgroup of V and $H = C_V(U)$, then*

$$|N| = |U|^{|V/H|-1}|N/(U : V)|.$$

As in the proof of 4.1 it follows that the N of 4.2 is finite. Before proving 4.2 we establish a lemma that will greatly simplify our calculations.

Lemma 4.3. *If 4.2 holds when $H = \{0\}$, then it holds generally*

Proof. The ring-free near-ring $N/(H : V)$ $(= N_1$ say) is 2-tame on V/H. Furthermore we have that $\bar{U}($ $= (U + H)/H)$ is a minimal N-subgroup of V/H. The centralizer $C_{V/H}(\bar{U})$ can be calculated. This submodule of V/H can be expressed in the form W/H, where $W \ge H$ is an N-subgroup of V. Since $\bar{U} \cap (W/H) = \{0\}$, we have that $(U + H) \cap W = H$ and consequently $U \cap W = \{0\}$. However, $W \ge H$ and it follows that $W = H$. Thus $C_{V/H}(\bar{U}) = \{0\}$. It follows that if 4.2 holds whenever $C_V(U) = \{0\}$, then it will hold for N_1 with V replaced by V/H, U replaced by $\bar{U}$ and H by $\{0\}$. Consequently

$$|N_1| = |\bar{U}|^{|V/H|-1}|N_1/A| \qquad \text{(a)},$$

where A is the ideal of N_1 consisting of all α_1 in N_1 such that $V/H \cdot \alpha_1 \subseteq \bar{U}$. Thus A consists precisely of those elements $\alpha + (H : V)$, $\alpha \in N$, of $N/(H : V)$ for which $V/H \cdot \alpha \subseteq \bar{U}$. This is equivalent to requiring α to be in $(U \oplus H : V)$. Therefore,

$$A = (U \oplus H : V)/(H : V).$$

Thus N_1/A can be written as

$$(N/(H : V))/((H \oplus U : V)/(H : V))$$

and this near-ring is isomorphic to $N/(U \oplus H : V)$. Since $\bar{U}$ is N-isomorphic to U, (a) reduces to

$$|N/(H : V)| = |U|^{|V/H|-1}|N/(U \oplus H : V)| \qquad \text{(b)}.$$

However, 2.4 implies

$$|(U \oplus H : V)| = |(U : V)||(H : V)|$$

and (b) reduces to

$$|N| = |U|^{|V/H|-1}|N/(U : V)|.$$

This clearly completes the proof of lemma 4.3.

Proof of 4.2. It follows from comments above that N is finite. Also we may assume, by 4.3, that $C_V(U) = \{0\}$. Since

$$|N| = |(U : V)||N/(U : V)|,$$

it remains to show that $|(U : V)| = |U|^{|V|-1}$. Clearly U is the unique minimal N-subgroup of V otherwise it would follow that $C_V(U) > \{0\}$. By 3.5, we have $v(U : V) \neq \{0\}$ for any $v \neq 0$ in V and $(U : V) \neq \{0\}$. It follows from 10.1 of [17], that $(U : V)$ is a finite direct sum $R_1 \oplus \ldots \oplus R_k$ ($k \geq 1$ an integer) of minimal right ideals R_i, $i = 1, \ldots, k$, of N which are N-isomorphic to U. Thus $|(U : V)| = |U|^k$ and it remains to show that $k = |V| - 1$. This will follow if we can establish the existence of a bijective function f of $V \backslash \{0\}$ onto $\{1, \ldots, k\}$. The function f will first be given as a relationship between the sets $V \backslash \{0\}$ and $\{1, \ldots, k\}$. For v in $V \backslash \{0\}$ and i in $\{1, \ldots, k\}$ we write vfi if $vR_i \neq \{0\}$.

For given y in $V \backslash \{0\}$ there exists n in $\{1, \ldots, k\}$ such that yfn. This follows since, if $yR_i = \{0\}$, for all $i = 1, \ldots, k$, then $y(U : V) = \{0\}$. Furthermore, since $C_V(U) = \{0\}$, U is the unique minimal N-subgroup of V and, by 3.5, we have the contradiction that $y = 0$.

We now establish that f is a function. If this were not so then there would exist x in $V \backslash \{0\}$ such that xfj and xfl, for distinct j and l in $\{1, \ldots k\}$. Since the sum $R_j + R_l$ is direct $xR_j + xR_l$ is a central sum of xR_j and xR_l (see §1 of [14]).

However, R_j and R_l are contained in $(U : V)$ and both xR_j and xR_l coincide with U. This would imply (see 1.1) that U is a ring module and $N/(0 : U)$ is a ring. This contradiction to N being ring-free yields the fact that f is a function.

It is easy to establish that f is surjective. Indeed if r is in $\{1, \ldots, k\}$, then, from the faithfulness of V, there exists z in V such that $zR_r \neq \{0\}$ and thus $zf = r$.

To show f is injective suppose that for some u and w in V, $uf = wf(= s$ say$)$. It follows that $uR_s \neq \{0\}$. Now the sum $R_s + C_N(R_s)$ is direct and consequently $uR_s + uC_N(R_s)$ is a central sum of uR_s and $uC_N(R_s)$. Since $\{0\} < uR_s \leq U$, $uR_s = U$. If $uC_N(R_s) \neq \{0\}$ then, since V is finite, the N-subgroup $uC_N(R_s)$ contains a minimal submodule K. Furthermore, if $K \neq U$, then $C_V(U) \geq K$ contrary to the assumption that $C_V(U) = \{0\}$. Thus, whenever $uC_N(R_s) \neq \{0\}$ it contains U, and U is in the intersection $uR_s \cap uC_N(R_s)$. By 1.1, this would imply U is a ring module and $N/(0 : U)(\neq \{0\})$ is a ring. We therefore conclude that $uC_N(R_s) = \{0\}$ and $C_N(R_s) \leq (0 : u)$. If $C_N(R_s) < (0 : u)$ then, from the definition of $C_N(R_s)$, we have $(0 : u) \cap R_s > \{0\}$. In this case $R_s \leq (0 : u)$ and $uR_s = \{0\}$. Consequently $C_N(R_s) = (0 : u)$. Similarly $C_N(R_s) = (0 : w)$. By 3.4, it now follows that $u = w$. Thus the function f is bijective. This completes the proof of 4.2.

§5. Structural and Other Consequences

Theorems 4.1 and 4.2 have several consequences. They allow any 2-tame ring-free near-ring N, where $N/J(N)$ has $DCCR$ to be fully characterized. As was seen in 1.8 these near-rings are in fact finite. Further understanding of the structure of such a near-ring is the main focus of this section.

If the near-ring N has a faithful N-group V, then N can be canonically embedded into $M_0(V)$ by mapping α in N to the function taking v in V to $v\alpha$ in V. This will be expressed by saying N is regarded as a subnear-ring of $M_0(V)$. It is easy to verify that: -

Proposition 5.1. *A faithful N-group V is 2-tame (compatible) if, and only if, V is 2-tame (compatible) when N is regarded as a subnear-ring of $M_0(V)$.*

In 5.1, the N-submodules of V and the submodules of V with N regarded as a subnear-ring of $M_0(V)$ in fact coincide. This relationship will however not be pursued further.

Suppose V is a group and S a non-empty collection of normal subgroups of V. The notation $D(S)$ will be used to denote the subnear-ring of $M_0(V)$ consisting of all functions α in $M_0(V)$ such that $(v + U)\alpha \subseteq v\alpha + U$, for all v in V and U in S. A particular adaption of this notation is used for an N-group V. The near-ring $D(V, N)$ is taken as $D(S)$, where S in the collection of all submodules of V. It follows immediately from the definition that: -

Proposition 5.2. *Suppose N_i, $i = 1, 2$, are near-rings and V an N_i-group for $i = 1, 2$. If the N_1 and N_2-submodules of V coincide, then $D(V, N_1) = D(V, N_2)$.*

Any near-ring with a faithful N-group V can be embedded in $D(V, N)$. Details required in the proof of 5.4 follow:-

Proposition 5.3. *If V is a faithful N-group where N is regarded as a subnear-ring of $M_0(V)$, then N is a subnear-ring of $D(V, N)$ and the N-submodules and $D(V, N)$-submodules of V coincide. Furthermore $D(V, N)$ is compatible on V.*

Proof. Since a function γ of N has the property that $(x + W)\gamma \subseteq x\gamma + W$, for all x in V and N-submodules W of V, N is a subnear-ring of $D(V, N)$.

Also any N-submodule of V is clearly a $D(V, N)$-submodule of V. Since N is a subnear-ring of $D(V, N)$, $D(V, N)$-submodules are N-submodules. Therefore N and $D(V, N)$ -submodules coincide.

Let u be in V, α in $D(V, N)$ and β be the function taking w in V to $(u+w)\alpha - u\alpha$. Since

$$(v + U)\beta = (u + v + U)\alpha - u\alpha \subseteq (u + v)\alpha + U - u\alpha = v\beta + U$$

for all v in V it follows that β is in $D(V, N)$ and $D(V, N)$ is compatible on V. The proof of the proposition is complete.

We are now ready to prove the main theorem of this section.

Theorem 5.4. *Suppose the near-ring N is ring-free and 2-tame on V. If $N/J(N)$ has $DCCR$ then, with N regarded as a subnear-ring of $M_0(V)$, we have $N = D(V, N)$.*

Proof. Since, by 1.8, $|V|$ is finite we can proceed by induction on $|V|$. If $|V| = 1$, then $V = \{0\}$ and $N = \{0\}$ and the result holds. Suppose V has a minimal N-group U. Now $N/(U : V)$ is ring-free and 2-tame on V/U and V/U is finite. Therefore, regarding $N/(U : V)$ as a subnear-ring of $M_0(V/U)$, we see $N/(U : V)$ is finite and by induction

$$|N/(U : V)| = |D(V/U, N/(U : V))| \tag{c}.$$

Let $H = C_V(U)$. Since, by 5.3, H and U are $D(V, N)$-submodules of V such that $H \cap U = \{0\}$ it follows, from 4.1, that

$$|D(V, N)| \leq |U|^{|V/H|-1}|D(V, N)/A| \tag{d},$$

where A is the ideal of $D(V, N)$ consisting of all α in $D(V, N)$ such that $V\alpha \subseteq U$. Now $D(V, N)/A$ is faithful on V/U and so, by 5.3,

$$|D(V, N)/A| \leq |D(V/U, D(V, N)/A)| \qquad \text{(e)}.$$

However, since by 5.3, the $D(V, N)$ and N-submodules of V coincide, the $D(V, N)$ and N-submodules of V/U again coincide. Considering the faithful action of these near-rings on V/U yields the fact that $D(V, N)/A$ and $N/(U : V)$ -submodules of V/U also coincide. It follows, by 5.2, that

$$|D(V/U, D(V, N)/A)| = |D\big(V/U, N/(U : V)\big)|.$$

Thus by (e) we have that

$$|D(V, N)/A| \leq |D\big(V/U, N/(U : V)\big)|.$$

Furthermore, by (c) and (d), it follows that

$$|D(V, N)| \leq |U|^{|V/H|-1}|N/(U : V)|.$$

By 4.2, we see that

$$|D(V, N)| \leq |N|$$

and since, by 5.3, $N \subseteq D(V, N)$ we have $N = D(V, N)$. The proof of 5.4 is complete.

The above theorem gives the completed picture for the structure of a 2-tame ring-free near-ring N, where $N/J(N)$ has $DCCR$. Before considering applications of this result we shall derive an expression for the order of N.

Theorem 5.5. *Suppose the near-ring N is ring-free and 2-tame on V. If $N/J(N)$ has $DCCR$ and*

$$\{0\} = V_0 < V_1 < \ldots < V_{r-1} < V_r = V$$

($r \geq 1$ an integer) is a composition series of V, then

$$|N| = \Pi_{i=1}^{r}|V_i/V_{i-1}|^{|V/H_i|-1},$$

where $C_V(V_i/V_{i-1}) = H_i$ for $i = 1, \ldots, r$.

Proof. The case where $V = \{0\}$ may be considered to be excluded. If $r = 1$ then, by 1.5 of [14], N is dense in $M_0(V)$. Thus if N is regarded as a subnear-ring of $M_0(V)$, $N = M_0(V)$. In this case $H_1 = C_{V_1}(V_1) = \{0\}$,

$$|N| = |V_1|^{|V/H_1|-1} = |V_1/V_0|^{|V/H_1|-1}$$

and the theorem holds. Suppose $r > 1$. For $i \geq 2$ we have, by 1.3, that

$$C_{V/V_1}\big((V_i/V_1)/(V_{i-1}/V_1)\big) = H_i/V_1$$

and therefore, by induction

$$|N/(V_1 : V)| = \Pi_{i=2}^{r}|(V_i/V_1)/(V_{i-1}/V_1)|^{|(V/V_1)/(H_i/V_1)|-1}.$$

However the isomorphism theorems imply

$$|N/(V_1 : V)| = \Pi_{i=2}^{r}|V_i/V_{i-1}|^{|V/H_i|-1}.$$

By 4.2, it now follows that

$$|N| = |V_1/V_0|^{|V/H_1|-1}\Pi_{i=2}^{r}|V_i/V_{i-1}|^{|V/H_i|-1}$$
$$= \Pi_{i=1}^{r}|V_i/V_{i-1}|^{|V/H_i|-1}$$

and the proof of 5.5 is complete.

Many applications of 5.4 can be made. The first allows the formulation of conditions, where the 2-tame assumption, implies compatibility.

Theorem 5.6. *If V is a 2-tame N-group, N is ring-free and $N/J(N)$ has $DCCR$, then the N-group V is compatible.*

Proof. Clearly it is sufficient to prove $N/(0 : V)$ is compatible on V. If $N/(0 : V)$ is regarded as a subnear-ring of $M_0(V)$, then, by 5.4 and 5.3, $N/(0 : V)$ is compatible on V. It now follows, from 5.1, that $N/(0 : V)$ must be compatible on V in any case. The theorem therefore holds.

One situation of interest is where V is an Ω-group and $P_0(V)$ is the near-ring of zero-fixing polynomial maps of V into V.

Theorem 5.7. *If V is a finite Ω-group and $P_0(V)$ is ring-free, then $P_0(V) = D(S)$, where S is the collection of all ideals of V.*

Proof. By 5.4, $P_0(V) = D(V, P_0(V))$. However, by 7.123 of [13] the submodules of V are precisely the ideals of V. Thus $P_0(V) = D(S)$ and 5.7 holds.

Theorem 5.7 has far greater applicability than it may at first appear. The condition that $P_0(V)$ is ring-free is satisfied very frequently. Some indication of this fact is provided by considering the situation where V is a finite simple zero-symmetric Ω-group (see §5 of [14]). According to 12.6 of [14], $P_0(V) = M_0(V)$ for a very wide selection of such Ω-groups. In this context 1.10 supplies us with precise information as to when any near-ring N, as in 5.4, is ring-free. We see that this property is, in fact, inherited from a relatively weak condition on V. This is particularly true for near-rings of the form $P_0(V)$ (V as in 5.7). Also theorem 5.5 yields a formula for the order of $P_0(V)$ (as in 5.7).

A finite group V is a faithful compatible N-group where N is taken as one of the near-rings $I(V)$, $A(V)$ or $E(V)$. Suppose

$$\{0\} = V_0 < V_1 < \ldots < V_r = V$$

($r \geq 0$ an integer) is a composition (i.e. N-composition) series of V. We shall call the N-group V *completely non-abelian* if there does not exist i in $\{1, \ldots r\}$ such that V_i/V_{i-1} is additively abelian. Clearly this is the same as requiring that no V_i/V_{i-1}, i in $\{1, \ldots, r\}$, is a ring module. Examples of completely non-abelian groups where $N = I(V)$ are those groups generated from finite non-abelian simple groups under the binary operations of direct sum (two components) and wreath products.

Theorem 5.8. *Let V be a finite group and N be one of the near-rings $I(V)$, $A(V)$ or $E(V)$. If the N-group V is completely non-abelian, then $N = D(S)$, where S is the collection of all normal, characteristic or fully invariant subgroups of V, respectively.*

Proof. By 1.10, N is ring-free. By 5.4, $N = D(V, N)$. The submodules of V are the normal, characteristic or fully invariant subgroups of V according to whether N coincides with $I(V)$, $A(V)$ or $E(V)$. Thus $N = D(S)$, where S is the collection indicated. Theorem 5.8 therefore holds true.

Clearly theorem 5.5 allows us to state the order of the near-ring $I(V)$, $A(V)$ or $E(V)$ (as in 5.8).

This paper concludes with the statement of a corollary of 5.8. A finite group V is called an IA group if $I(V) = A(V)$ and an AE group if $A(V) = E(V)$.

Corollary of 5.8. *Suppose V is a completely non-abelian finite group with respect to $I(V)$. It follows that V is an IA group if, and only if, all normal subgroups are characteristic. A similar statement can be made for AE groups.*

The author would like to take this opportunity to express appreciation to G. Mason for the talk he gave at the 1991 near-ring and near-field conference calling attention to this problem (i.e. when is a group V an IA group or an AE group). His talk resulted in a time of fruitful discussion that encouraged me to complete this paper.

My expression of appreciation extends to Marilyn Talamaivao for her effort in reproducing this material. Also special thanks should go to Miss Linda Buckthought who has been a constant source of encouragement.

References

1. C.W. Curtis and I.Reiner, *Representation Theory of Finite Groups and Associative Algebras*, Interscience Publishers, New York (1966).
2. Y. Fong, *The Endomorphism Near-rings of the Symmetric Groups*, Ph.D. Dissertation, Uni. of Edin. U.K. (1979).
3. Y.Fong and J.D.P. Meldrum, *The Endomorphism Near-rings of the Symmetric Groups of Degree at Least Five*, J. Austral. Math. Soc. **30A** (1980), 37–49.
4. Y.Fong and J.D.P. Meldrum, *The Endomorphism Near-ring of the Symmetric Group of Degree Four*, Tamkang J. Math. **12** (1981), 193-203.
5. C.G. Lyons and J.J. Malone, *Endomorphism Near-rings*, Proc.Edin. Math. Soc. **17** (1970), 71–78.
6. C.G. Lyons and J.J. Malone, *Finite Didedral Groups and d.g. Near-rings I*, Comp. Math. **24** (1972), 305–312.
7. C.G. Lyons and J.J. Malone, *Finite Dihedral Groups and d.g. Near-rings II*, Comp. Math. **26** (1973), 249–259.
8. C.G. Lyons and J.D.P. Meldrum, *N-series and Tame Near-rings*, Proc. Royal Soc. Edin. **86A** (1980), 153–163.
9. J.J. Malone, *Generalized Quaternion Groups and Distributively Generated Near-rings*, Proc. Edin. Math. Soc **18** (1973), 235–238.
10. J.D.P. Meldrum, *Near-rings and their Links with Groups*, Pitman, London (1985).
11. J.D.P. Meldrum, *The Endomorphism Near-rings of Finite Linear Groups*, Proc. Royal Irish Acad. **79A** (1979), 87–96.
12. J.D.P. Meldrum, *The Endomorphism Near-ring of an Infinite Dihedral Group*, Proc. Royal Soc. Edin. **76A** (1977), 311–321.
13. G. Pilz, *Near-rings*, North-Holland, Amsterdam (1983).
14. S.D. Scott, *Linear Ω-Groups, Polynomial Maps*, Contr. Gen. Alg. **8** (1991), 239-293.
15. S.D. Scott, *Near-rings and Near-ring Modules*, Ph.D. disseration, Austral. Nat. Uni. (1970).
16. S.D. Scott, *On the Finiteness and Uniqueness of Certain 2-tame N-groups*, To appear.
17. S.D. Scott, *Tame Near-rings and N-groups*, Proc. Edin. Math. Soc. **23** (1980), 275–296.

RINGS WHICH ARE A HOMOMORPHIC IMAGE OF A CENTRALIZER NEAR-RING

Kirby C. Smith

Department of Mathematics, Texas A & M University
College Station, Texas 77843, USA

I. Introduction.

In this work the near-rings under consideration will be exclusively centralizer near-rings $M_A(G)$ where G is a finite group and A is a group of automorphisms of G.

In [2] the following question was discussed: which rings are isomorphic to $M_A(G)$ for some pair (A, G), that is which rings are of "$M_A(G)$-type". This turned out to be a difficult question to answer, and only a partial answer was given in [2]. It was shown that any (finite) field is of $M_A(G)$-type as is the ring $GF(2) \oplus GF(3) \oplus GF(5)$. ($GF(q)$ denotes the Galois field of order $q = p^n$ where p is a prime.) Also various direct sums of fields of characteristic 2 and/or 3 are of $M_A(G)$-type, but not all of these. Since a centralizer near-ring is rarely distributive, it is not unexpected that $M_A(G)$ is seldom a ring.

In this paper we investigate the following related question. Which rings are obtainable as a homomorphic image of a centralizer near-ring $M_A(G)$? We will show that a necessary condition for a ring to be a homomorphic image of $M_A(G)$ is that it be a direct sum of fields. We will further show that an arbitrary direct sum of prime fields is a homomorphic image of $M_A(G)$ for a suitable pair (A, G). This will be accomplished by examining the structure of $M_A(G)$ where G is the symmetric group S_n and $A = Inn(S_n)$, the group of inner automorphisms of S_n. We will see that the structure of this near-ring has possible independent interest.

We begin by recalling some information about $M_A(G)$ as found in [3]. The automorphism group A partitions G into disjoint orbits $\{0\}, \theta_1, \theta_2, \ldots, \theta_k$ where 0 is the identity element of G and $\{0\}$ is the trivial A-orbit. The orbits $\theta_1, \theta_2, \ldots, \theta_k$ are nontrivial A-orbits. If θ is an A-orbit of G we will use the notation $\theta(v)$ to mean that θ is that orbit of G which contains the element v in G.

For $v \in G$ we recall that the stabilizer of v is $\operatorname{stab}(v) = \{\phi \in A \mid \phi(v) = v\}$, a subgroup of A. We define the following partial order on the set of all A-orbits of G:

$\{0\} < \theta$ for all nontrivial orbits θ and $\theta_i < \theta_j$ if there exist $v \in \theta_i$, $w \in \theta_j$ with $\operatorname{stab}(v) \supset \operatorname{stab}(w)$.

(For sets A, B the notation $A \subset B$ means proper containment whereas $A \subseteq B$ allows for possible equality.) Moreover if θ_i and θ_j are nontrivial orbits, define θ_i and θ_j to be equivalent $(\theta_i \sim \theta_j)$ if there is a v in θ_i and a w in θ_j with $\operatorname{stab}(v) = \operatorname{stab}(w)$. We will use the notation $\theta_i \leq \theta_j$ to mean that either $\theta_i < \theta_j$ or else $\theta_i \sim \theta_j$.

Recall that a function f in $M_A(G)$ must preserve A-orbits. That is if θ is an

Y. Fong et al. (eds.), Near-Rings and Near-Fields, 257–270.

A-orbit, so is $f(\theta)$ and moreover $f(\theta) \leq \theta$. Also if $v \in G^*$ (G^* denotes the nonzero elements of G), then $\text{stab}\,(f(v)) \supseteq \text{stab}\,(v)$.

Let $N = M_A(G)$. It is shown in the theory of near-rings that the J_2-radical of N is the intersection of all strictly maximal left ideals of N (see, for example, [6]). (A left ideal of N is strictly maximal if it is maximal not only as a left ideal but also as an N-subgroup.) For a nontrivial A-orbit θ of G let

$$L(\theta) = \{f \in M_A(G) \mid f(\theta) < \theta\}.$$

The set $L(\theta)$ need not be a left ideal of $M_A(G)$, but when it is, it is a strictly maximal left ideal. Conversely if L is a strictly maximal left ideal of $M_A(G)$, then $L = L(\theta)$ for some nontrivial A-orbit θ. (See [3])

These nice features of $M_A(G)$ continue. For if θ_i and θ_j are nontrivial A-orbits with $\theta_i \sim \theta_j$, then $L(\theta_j)$ is a left ideal if and only if $L(\theta_j)$ is. Suppose that $L(\theta)$ is a left ideal for some orbit θ (and so it is automatically strictly maximal) Let $\theta = \theta^{(1)}, \ldots, \theta^{(t)}$ be a complete list of those orbits equivalent to θ. Then $L(\theta^{(1)}), \ldots, L(\theta^{(t)})$ are left ideals and $I = L(\theta^{(1)}) \cap \cdots L(\theta^{(t)})$ is a maximal 2-sided ideal of $M_A(G)$.

Let $I_1, \ldots, I_s$ be the complete set of ideals of $N = M_A(G)$ obtained in the above manner, one ideal for each equivalence class of nontrivial A-orbits θ such that $L(\theta)$ is a left ideal. Then the J_2-radical of N is $J_2(N) = I_1 \cap \cdots I_s$ and $N/J_2(N) \cong N/I_1 \oplus \cdots \oplus N/I_s$, where each N/I_j is a simple near-ring.

So for the centralizer near-ring $M_A(G)$, the determination of the radical of $M_A(G)$ is essentially the problem of deciding which nontrivial A-orbits θ have the property that $L(\theta)$ is a left ideal.

Our first main result is due to S. Bagley ([1]) who showed that if a finite ring R is a nonzero homomorphic image of a generalized centralizer near-ring, then R is a direct sum of fields. We state and prove the result for the special case $N = M_A(G)$.

Theorem 1. *Let R be a finite ring which is a nonzero homomorphic image of the near-ring $M_A(G)$. Then R is a direct sum of fields.*

Proof. We denote the nontrivial A-orbits of G by $\theta_1, \ldots, \theta_k$. Let $e_j \in N = M_A(G)$ be the idempotent associated with the orbit θ_j. (So $e_j(v) = v$ if v belongs to θ_j and $e_j(v) - 0$ if v does not belong to θ_j.) We distinguish some subsets of N, namely $N_{ij} = e_i N e_j = \{e_i n e_j \mid n \in N\}$, where $i, j = 1, \ldots, k$. (The set N_{ij} consists of those functions f in N such that $f(\theta_s) = \{0\}$ if $s \neq j$ and either $f(\theta_j) = \theta_i$ or $f(\theta_j) = \{0\}$.) We recall that for every i, N_{ii}^* is a multiplicative group.

Let $n \mapsto \bar{n}$ denote the homomorphism of N onto the ring R. Then $\overline{N_{ij}}$ signifies the image of N_{ij} under our homomorphism. We show first that $\overline{N_{ij}} = \{\bar{0}\}$ for all i, j with $i \neq j$.

Suppose $\overline{e_j} = \bar{0}$. If $n_{ij} \in N_{ij}$, then $n_{ij} = n_{ij}e_j$ and so $\overline{n_{ij}} = \overline{n_{ij}e_j} = \overline{n_{ij}}\,\overline{e_j} = \bar{0}$. In this case $\overline{N_{ij}} = \{\bar{0}\}$ and we may assume $\overline{e_j} \neq \bar{0}$.

We show now that if $n_{ij} \in N_{ij}$ ($i \neq j$) and $n_{jj} \in N_{jj}^*$, then $n_{ij} + n_{jj}$ must belong to N_{jj}. We know that $n_{ij} + n_{jj}$ belongs to N_{sj} for some s. If $s \neq j$ then $e_j(n_{ij} + n_{jj}) - e_j n_{jj} - e_j n_{ij} = 0 - e_j n_{jj} - 0 = -n_{jj}$. Since $\overline{N} = R$ is a ring, then

$\overline{-n_{jj}} = \overline{0}$. This means $\overline{n_{jj}} = \overline{0}$ and hence $\overline{e_j} = \overline{n_{jj}n_{jj}^{-1}} = \overline{0}$, a contradiction. So $s = j$ as desired.

We now have $n_{ij} + n_{jj} \in N_{jj}$ for every $n_{ij} \in N_{ij}$ and all $n_{jj} \in N_{jj}^*$. This means $e_j(n_{ij} + n_{jj}) - e_jn_{jj} - e_jn_{ij} = n_{ij} + n_{jj} - n_{jj} = n_{ij}$. Again, since $\overline{N} = R$ is a ring then $\overline{n_{ij}} = \overline{0}$ and $\overline{N_{ij}} = \{\overline{0}\}$.

Since N_{jj} is a multiplicative group then $\overline{N_{jj}^*}$ is either $\{\overline{0}\}$ or a multiplicative group. We have $N = \sum_{i,j} N_{ij}$ (a group direct sum) and therefore $\overline{N} = \sum_j \overline{N_{jj}} = R$. Moreover $N_{ii}N_{jj} = \{0\}$ if $i \neq j$ means that R is a finite ring with no nonzero nilpotent elements. From ring theory, R is a direct sum of fields.

II. Observations on $M_A(G)$, $A = Inn(G)$.

In this section we will be dealing with the centralizer near-ring $M_A(G)$ where $A = Inn(G)$, the group of inner automorphisms of G. As is customary we will use additive notation for the binary operation in G even though G is not necessarily abelian.

Let v be an element of G^*. We will call v *special* if it has the following property: whenever $i > 0$ is an integer such that iv is not a generator of the cyclic group $\langle v \rangle$, then $stab\,(iv) \supset stab\,(v)$. (It is always the case that $stab(iv) \supseteq stab(v)$.) In the next section we will show that if G is the symmetric group S_n, then virtually all elements of G are special.

Lemma 1. *Let $v \in G$ be special. Suppose v has order mn where $m > 1$, $n > 1$ and $(m,n) = 1$. Then $L(\theta(v))$ is not a left ideal of $M_A(G)$.*

Proof. Since mv is not a generator of $\langle v \rangle$ then $stab\,(mv) \supset stab\,(v)$ and $\theta(mv) < \theta(v)$. Since $\langle v \rangle$ is cyclic of order mn with $m > 1$, $n > 1$, and $(m,n) = 1$, then it is a group theory exercise to see that there is a positive integer i such that $(mi+1)v$ does not generate $\langle v \rangle$, and so $\theta((mi+1)v) < \theta(v)$.

Let f be any function in $M_A(G)$ such that $f(v) = miv$ where i is as above. So f belongs to $L(\theta(v))$. Let e be the idempotent in $M_A(G)$ associated with the orbit $\theta(v)$. We have

$$(e(f + e) - ee)(v) = e(miv + v) - v = e((mi + 1)v) - v = 0 - v = -v.$$

Since $stab\,(-v) = stab(v)$, then $\theta(-v)$ is equivalent to $\theta(v)$. This means $e(f+e)-ee$ does not belong to $L(\theta(v))$ and $L(\theta(v))$ is not a left ideal.

We recall that if $v \in G$ then $C(v) = \{w \in G \mid w+v = v+w\}$ is the centralizer of v in G, a subgroup of G. With $A = Inn(G)$, then $stab\,(v)$ is essentially $C(v)$ and it is not necessary to distinguish the two. We denote by $Z(G)$ the center of the group G.

Lemma 2. *Let v be a nonzero element of G. Suppose $w \neq 0$ is in $Z(C(v))$ such that $stab\,(w) \supset stab\,(v)$ and $\theta(v + w) \neq \theta(v)$. Then $L(\theta(v))$ is not a left ideal of $M_A(G)$.*

Proof. Since $A = Inn(G)$ and w belongs to $Z(C(v))$, then $stab\,(v + w) \supseteq stab\,(v)$. This means $\theta(v + w) \leq \theta(v)$.

Since $\theta(v+w) \neq \theta(v)$ by assumption, the idempotent e corresponding to the orbit $\theta(v)$ has the property that $e(v) = v$ and $e(v+w) = 0$. There is an h in $M_A(G)$ such that $h(v) = v + w$ and an f in $L(\theta(v))$ with $f(v) = -w$. We have $(e(f+h) - eh)(v) = e(-w+v+w) - e(v+w) = e(v) - 0 = v$. So $e(f+h) - eh$ is not in $L(\theta(v))$ and $L(\theta(v))$ is not a left ideal.

Corollary. Let v be a special element of G and suppose v has order p^k, where p is a prime. If $Z(C(v))$ contains an element w whose order is not a power of p and such that w and $v+w$ are special, then $L(\theta(v))$ is not a left ideal.

Proof. Since $A = Inn(G)$ and w belongs to $Z(C(v))$, then stab$(w) \supseteq$ stab(v). Since the order of w is not a power of p and $v + w = w + v$, then the order of $v+w$ is not a power of p. This means $\theta(v+w) \neq \theta(v)$.

If stab$(w) \supset$ stab(v), we are done by Lemma 2. If stab$(w) =$ stab(v), then $\theta(w) \sim \theta(v)$ and yet $\theta(w) \neq \theta(v)$ since w and v have different orders. We have stab$(v+w) \supseteq$ stab(v) because w is in $Z(C(v))$.

Assume stab$(v+w) \supset$ stab(v). Let e be the idempotent corresponding to $\theta(v)$ and let f be in $M_A(G)$ such that $f(v) = v + w$. Then $f \in L(\theta(v))$ and $(e(f+e) - ee)(v) = e(v+w+v) - e(v) = e(2v+w) - v$. Since $2v+w$ has order different from v, then $\theta(2v+w) \neq \theta(v)$ and $e(2v+w) = 0$. This means $e(f+e) - ee$ is not in $L(\theta(v))$ and $L(\theta(v))$ is not a left ideal.

The remaining situation is stab$(v+w) =$ stab$(v) =$ stab(w). Since the order of w is not a power of p, then either w or $v+w$ satisfies the hypothesis of Lemma 1, namely has order mn where $m > 1$, $n > 1$ and $(m,n) = 1$. So from this either $L(\theta(w))$ or $L(\theta(v+w))$ is not a left ideal. This means $L(\theta(v))$ is not a left ideal.

The next result is a positive one. It gives a situation in which $L(\theta)$ is a left ideal, in fact an ideal

Lemma 3. Assume v is a special element of G and suppose v has order p^k, where p is a prime and $k > 0$. Assume further that $\theta(v)$ contains every generator of the cyclic group $\langle v \rangle$. Finally, assume that if $x \in G$ has the property that stab$(x) \supseteq$ stab(v), then $x = iv$ for some integer i. Then $L(\theta(v))$ is an ideal of $M_A(G)$.

We note that since $A = Inn(G)$, the assumption about x is equivalent to saying that $Z(C(v)) = \langle v \rangle$.

Proof of Lemma 3. It is clear that $(L(\theta(v)), +)$ is a group.

Let g and h be in $M_A(G)$ and let f be in $L(\theta(v))$. There are integers i, j, k with $f(v) = ipv$, $g(v) = jv$ and $h(v) = kv$. So $(h(f+g) - hg)(v) = h(f(v) + g(v)) - h(g(v)) = h(ipv+jv) - h(jv)$, an element in $\langle v \rangle$, which we will conveniently call w.

(i) If jv does not generate $\langle v \rangle$ then p divides j and thus $ipv + jv$ does not generate $\langle v \rangle$. This makes w the difference of two elements in $\langle v \rangle$ that do not generate $\langle v \rangle$, so w does not generate $\langle v \rangle$. In this case $h(f+g) - hg$ belongs to $L(\theta(v))$.

(ii) Assume jv does generate $\langle v \rangle$. Then p does not divide j and p does not divide $ip+j$. This means $ipv + jv$ generates $\langle v \rangle$. Since $\theta(v)$ contains all generators of $\langle v \rangle$, then there exist α, β in A with $\alpha v = jv$ and $\beta v = ipv + jv$. We have

$$w = h(\beta v) - h(\alpha v) = \beta h(v) - \alpha h(v) = \beta kv - \alpha kv$$
$$= k(\beta v - \alpha v) = k(ipv + jv - jv) = klpv.$$

So the element w belongs to $\langle v \rangle$ since it is an integer multiple of v, and w is not a generator of $\langle v \rangle$. This means $\theta(w) < \theta(v)$, $h(f + g) - hg$ belongs to $L(\theta(v))$ and $L(\theta(v))$ is a left ideal. The right ideal property is easy to verify.

We note that if θ is any minimal nontrivial A-orbit of G under the partial ordering "$<$", then $L(\theta)$ is a left ideal. In this case $L(\theta)$ is the annihilator of the orbit θ.

III. The Near-ring $M_A(S_n)$, $A = Inn(S_n)$.

We are interested in near-rings N which have a nonzero ring as a homomorphic image. We will find that $M_A(S_n)$ is such a near-ring where S_n is the symmetric group on n elements and A is the group on inner automorphisms of S_n.

The following result due to A. G. Myasnikov [5] indicates why $A = Inn(G)$ is a natural choice for an automorphism group.

Theorem (Myasnikov). *Let G be a group and let $A = Inn(G)$. Then $(M_A(G), +)$ is abelian.*

Proof. For x, y in G let $[x, y] = x + y - x - y$. If $f \in M_A(G)$, then f commutes with every inner automorphism of G. From this it is easy to verify that if $[x, y] = 0$, then $[f(x), y] = 0$. In particular $[f(x), x] = 0$ and so $f(x) + x = x + f(x)$ for all x in G and all f in $M_A(G)$.

Let f, g be in $M_A(G)$ and $x, y \in G$. Using the above we have that $[x, y] = 0$ implies $[f(x), g(y)] = 0$. Since we always have $[x, x] = 0$, then $[f(x), g(x)] = 0$. This means $f(x) + g(x) = g(x) + f(x)$ for all f, g in $M_A(G)$ and all x in G. In other words $f + g = g + f$. This shows $(M_A(G), +)$ is abelian.

We note that if $A = Inn(G)$, Myasnikov's theorem implies that every homomorphic image of $M_A(G)$ has an abelian group structure.

We now begin our study of the near-ring $M_A(S_n)$ where $A = Inn(S_n)$. As is well known, the nontrivial A-orbit of S_n containing the permutation σ consists of those permutations having the same "cycle structure" as that of σ. For $\sigma \in S_n$, let Fix(σ) denote the set of elements left fixed by the permutation σ. We will use the more comfortable multiplicative notation for S_n in the following.

Lemma 4. *Let $G = S_n$ and $A = Inn(S_n)$. Then $\sigma \in S_n$ is not special if and only if σ has all of the following properties:*
(i) *$|\text{Fix}(\sigma)| = 0$, and*
(ii) *$\sigma = \sigma_1 \sigma_2 \cdots \sigma_r \tau$ (σ written as a product of disjoint cycles) where $r \geq 1$, each cycle σ_i has odd order, and τ is a transposition.*

Proof. Suppose σ^i does not generate $\langle \sigma \rangle$. We want to determine whether or not stab$(\sigma^i) \supset$ stab(σ). Clearly we may assume $\sigma^i \neq (1)$.

Assume σ is a cycle. Since σ^i does not generate $\langle\sigma\rangle$, then σ^i is not a cycle, it is a product of disjoint cycles $\beta_1, \ldots, \beta_r$ each having length less than the length of σ. Each β_j commutes with σ^i and yet β_j does not commute with σ. So $\mathrm{stab}\,(\sigma^i) \supset \mathrm{stab}\,(\sigma)$ in this case, and every cycle is special.

Assume σ is not a cycle. Write $\sigma = \sigma_1 \cdots \sigma_r$, a product of disjoint cycles. Then $\sigma^i = \sigma_1^i \cdots \sigma_r^i$. Since the order of σ^i is less than that of σ, there is at least one index j such that the order of σ_j^i is less than the order of σ_j. If σ_j^i is not the identity permutation, we use the fact that σ_j is a cycle and the above argument to find a permutation β that commutes with σ_j^i but not with σ_j. This β will commute with σ^i but not with σ. Again $\mathrm{stab}\,(\sigma^i) \supset \mathrm{stab}\,(\sigma)$.

We may now assume that $\sigma_j^i = (1)$ for every index j such that the order of σ_j^i is less than the order of σ_j. If σ_j is not a transposition, let β be any transposition that moves two elements that σ_j moves. Then β trivially commutes with $\sigma_j^i = (1)$ and therefore with σ^i. But β does not commute with σ.

We are now left with the situation where σ_j is a transposition and $\sigma_j^i = (1)$ for all j such that σ_j^i does not generate $\langle\sigma_j\rangle$. If $|\mathrm{Fix}\,(\sigma^i)| \geq 3$ (which will be the case if either $|\mathrm{Fix}\,(\sigma)| \neq 0$ or else $|\mathrm{Fix}\,(\sigma)| = 0$ and there is more than one transposition), let $\sigma_j = (s\,t)$ and let u be a third symbol which is fixed by σ^i Then $\beta = (s\,t\,u)$ is a 3-cycle which commutes with σ^i, but not with σ.

The remaining situation is $\sigma = \sigma_1\sigma_2 \cdots \sigma_{r-1}\tau$ where each cycle σ_k has odd order $n_k > 2$, τ is a transposition and $|\mathrm{Fix}\,(\sigma)| = 0$. One now verifies that $\mathrm{stab}\,(\sigma) = \mathrm{stab}\,(\sigma^2)$ and σ^2 does not generate $\langle\sigma\rangle$. This means σ is not special.

We note that in S_2, S_3 and S_4 no permutation can have the form of Lemma 4 and so every element is special. In S_5 we have our first non-special elements, namely those in the orbit of $(123)(45)$. All elements of S_6 are special. Those non-special elements of S_7 belong to the orbit of $(12345)(67)$.

Our next goal is to describe the radical of $M_A(S_n)$. Toward this end we need to determine which sets $L(\theta)$ are left ideals of $M_A(S_n)$.

Lemma 5. *Let σ be a permutation in S_n, where $n \geq 5$ and $|\mathrm{Fix}\,(\sigma)| = 2$. Then $L(\theta(\sigma))$ is not a left ideal.*

Proof. Let $\mathrm{Fix}\,(\sigma) = \{i, j\}$ and let τ be the transposition $(i\,j)$. If the permutation ω in S_n commutes with σ then $\omega(\mathrm{Fix}\,(\sigma)) = \mathrm{Fix}\,(\sigma)$, i.e. $\mathrm{Fix}\,(\sigma)$ is invariant under ω. This means ω commutes with τ since $|\mathrm{Fix}\,(\sigma)| = 2$. Hence $\tau \in Z(C(\sigma))$.

Since σ and $\sigma\tau$ have different cycle structures, we have $\theta(\sigma) \neq \theta(\sigma\tau)$. Since σ moves at least three elements ($n \geq 5$), then $\mathrm{stab}\,(\tau) \supset \mathrm{stab}\,(\sigma)$, and $\theta(\tau) < \theta(\sigma)$. Moreover σ is special and so Lemma 2 applies to show that $L(\theta(\sigma))$ is not a left ideal.

The restriction $n \geq 5$ in Lemma 5 is needed. For let $\sigma = (12)$ in S_4. Then one verifies that $L(\theta(\sigma))$ is a left ideal.

Corollary. *If σ in S_n, $n \geq 5$, is not special, then $L(\theta(\sigma))$ is not a left ideal.*

Proof. If σ is not special then $|\mathrm{Fix}\,(\sigma^2)| = 2$ and $L(\theta(\sigma^2))$ is not a left ideal by Lemma 5. Since $\theta(\sigma^2) \sim \theta(\sigma)$, then $L(\theta(\sigma))$ is not a left ideal .

Lemma 6. *If σ is a permutation in S_n, $n \geq 5$, which does not have prime power order, then $L(\theta(\sigma))$ is not a left ideal.*

Proof. If σ is special, then Lemma 1 applies and we are done. If σ is not special, then by the corollary to Lemma 5, $L(\theta(\sigma))$ is not a left ideal.

For $\sigma \neq (1)$ write σ as a product of disjoint cycles, $\sigma = \sigma_1\sigma_2\cdots\sigma_r$. We will call σ homogeneous if each of the cycles σ_i has the same length.

Lemma 7. *If σ is a permutation in S_n, $n \geq 5$, which is not homogeneous then $L(\theta(\sigma))$ is not a left ideal.*

Proof. We may assume σ is special by the corollary to Lemma 5. Assuming $\sigma = \sigma_1\sigma_2\cdots\sigma_r$ is not homogeneous, then there are two cycles having different lengths, say σ_1 and σ_r. Let $\sigma_1, \sigma_2, \ldots, \sigma_s$ be the cycles having the same length as σ_1. Let $\omega = \sigma_1^{-1}\sigma_2^{-1}\cdots\sigma_s^{-1}$. Then $\text{stab}(\omega) \supset \text{stab}(\sigma)$ and $\theta(\omega) < \theta(\sigma)$. Since $\sigma\omega$ and σ have different cycle structures, then $\theta(\sigma\omega) \neq \theta(\sigma)$. By Lemma 2, $L(\theta(\sigma))$ is not a left ideal.

We put all of this together in the following theorem.

Theorem 2. *Let σ be a non-identity permutation in S_n, $n \geq 5$. Then $L(\theta(\sigma))$ is a left ideal of $M_A(S_n)$, $A = \text{Inn}(S_n)$, if and only if σ has all of the following properties:*
(a) $|\text{Fix}(\sigma)| \neq 2$,
(b) *σ is homogeneous, and*
(c) *σ has prime power order.*

Proof. Lemmas 5–7 establish the necessity of properties (a)–(c). It remains to prove that if σ has these properties, then $L(\theta(\sigma))$ is a left ideal. We will do this by showing that the conditions of Lemma 3 are satisfied by σ.

Since σ is homogeneous, then σ is special. Also, every generator of the cyclic group $\langle\sigma\rangle$ has the same cycle structure as σ and so they all belong to $\theta(\sigma)$.

We show now that if β is in S_n such that $\text{stab}(\beta) \supseteq \text{stab}(\sigma)$, then $\beta = \sigma^i$ for some integer i. Since σ is homogeneous, there exists a cycle γ and an integer $r > 0$ such that $\gamma^r = \sigma$. Since $\gamma\sigma = \sigma\gamma$, then $\gamma\beta = \beta\gamma$. The fact that γ is a cycle means $\beta = \gamma^s\omega$ for some integer s and some permutation ω on $\text{Fix}(\gamma) = \text{Fix}(\sigma)$ which is disjoint from σ. Since $\beta = \gamma^s\omega$ commutes with every permutation disjoint from σ and $|\text{Fix}(\sigma)| \neq 2$, then $\omega = (1)$ and $\beta = \gamma^s$.

The permutations γ, $\sigma = \gamma^r$ and $\beta = \gamma^s$ belong to the cyclic group $\langle\gamma\rangle$ and every element of this group is special. Even though σ has prime power order, say p^k, $k > 0$, γ could have composite order. However the fact that $\text{stab}(\beta) \supseteq \text{stab}(\sigma)$ implies that β must have prime power order p^s for some $s > 0$. (For if the order of β is divisible by a prime $q \neq p$, then one can find a p-cycle that commutes with σ but does not commute with β.) Since σ has order p^k and β has order p^s, and since $\langle\gamma\rangle$ is cyclic, then either $\beta \in \langle\sigma\rangle$ or else $\sigma \in \langle\beta\rangle$. The fact that $\text{stab}(\beta) \supseteq \text{stab}(\sigma)$ and both β and σ are special implies $\beta \in \langle\sigma\rangle$.

Hence the only orbits θ with $\theta \leq \theta(\sigma)$ are those of the form $\theta(\sigma^i)$. By Lemma 3, $L(\theta(\sigma))$ is a left ideal of $M_A(S_n)$.

We note that in S_2, S_3 and S_4 one can verify that $L(\theta)$ is a left ideal for all orbits θ.

For convenience we will call an element σ in S_n, $n \geq 5$, a *radical* element if it has properties (a)-(c) of Theorem 2, and call its orbit $\theta(\sigma)$ a *radical* orbit.

The proof of the following corollary follows from the discussion of the radical in the introduction and Theorem 2.

Corollary 1. *Let $N = M_A(S_n)$ where $n \geq 5$ and $A = Inn(S_n)$. The radical of N consists of those functions f in N such that $f(\theta(\sigma)) < \theta(\sigma)$ for all radical elements σ.*

Corollary 2. *Let $N = M_A(S_n)$ where $n \geq 2$ and $A = Inn(S_n)$. If $L(\theta)$ is a left ideal of N, then it is a 2-sided ideal.*

Sketch of proof. In S_2, S_3 and S_4 we have $\text{stab}(\beta) = \text{stab}(\sigma)$ if and only if $\theta(\beta) = \theta(\sigma)$. So each orbit of S_i, $i = 2,3,4$, is in an equivalence class by itself.

In S_n, $n \geq 5$, $L(\theta(\sigma))$ is a left ideal precisely when σ is a radical element. In this case one verifies that $\theta(\sigma)$ is also in an equivalence class by itself.

If $L(\theta)$ is a left ideal and $\{\theta\}$ is an equivalence class, then $L(\theta)$ is an ideal.

Corollary 2 says that every strictly maximal left ideal L of $N = M_A(S_n)$ is a maximal ideal. So N/L is a simple near-ring. Which simple near-ring?

Lemma 8. *Let $N = M_A(S_n)$ where $n \geq 2$ and $A = Inn(S_n)$. If $L(\theta(\sigma))$ is a left ideal then the factor near-ring $N/L(\theta(\sigma))$ is the prime field $GF(p)$, where σ has prime power order p^k for some integer k.*

Proof. By Myasnikov's theorem $(N/L(\theta(\sigma)), +)$ is an abelian group. The ideal $L(\theta(\sigma))$ consists of those functions f in N such that $f(\sigma) = \sigma^i$ belongs to the subgroup H of nongenerators of $\langle \sigma \rangle$, where $|\langle \sigma \rangle| = p^k$. In other words $L(\theta(\sigma)) = \{f \in N \mid f(\sigma) \in H\}$.

Define $\phi : N \to GF(p)$ as follows: if $f \in N$, then $f(\sigma) = \sigma^j$ for some integer j, let $\phi(f) = \bar{j} \in GF(p)$. The function ϕ is a homomorphism of N with kernel $L(\theta(\sigma))$. Hence $N/L(\theta(\sigma)) \cong GF(p)$.

Theorem 3. *Let $N = M_A(S_n)$ where $n \geq 2$ and $A = Inn(S_n)$. Let $\text{rad}(N)$ denote the J_2-radical of N Then*
(a) *if $n = 2$, $\text{rad}(N) = \{0\}$ and $N \cong GF(2)$,*
(b) *if $n = 3$, $\text{rad}(N) = \{0\}$ and $N \cong GF(2) \oplus GF(3)$,*
(c) *if $n = 4$, $N/\text{rad}(N) \cong GF(2) \oplus GF(2) \oplus GF(2) \oplus GF(3)$,*
(d) *if $n \geq 5$, $N/\text{rad}(N)$ is isomorphic to a direct sum of prime fields, one prime field $GF(p_j)$ for each prime factor (with multiplicity) of $\frac{n!}{n-2}$.*

We illustrate part (d) with the near-ring $N = M_A(S_7)$. Since $\frac{7!}{7-2} = 2^4 3^2 7$, then $N/\text{rad}(N) \cong GF(2) \oplus GF(2) \oplus GF(2) \oplus GF(2) \oplus GF(3) \oplus GF(3) \oplus GF(7)$.

Proof of Theorem 3. We leave the cases $n = 2, 3$ and 4 as exercises and prove only (d).

Let $S = \{n, n-1, n-3, \ldots, 2\}$. With each simple summand N_i of $N/\text{rad}(N)$ there corresponds a unique radical orbit $\theta(\sigma)$ where $N_i \cong N/L(\theta(\sigma))$. Let $|\text{Fix}(\sigma)|$

$= n - k$. Since $|\text{Fix}(\sigma)| \neq 2$, then $k \neq n - 2$ and k belongs to S. Since σ is homogeneous of prime power order, say p^t, there is an integer q with $k = qp^t$. We associate the simple summand $N_i \cong GF(p)$ with the tth appearance of the prime p in the prime factorization of $k \in S$. In this way we associate each N_i with a unique prime in the factorization of $\frac{n!}{n-2}$.

Conversely, let k be in S and let p be a prime such that p^r ($r \geq 1$) divides k but p^{r+1} does not. If t is an integer with $1 \leq t \leq r$, then there is an integer q with $k = qp^t$. Let σ be any permutation in S_n which is the product of q disjoint cycles of length p^t. Then $|\text{Fix}(\sigma)| = n - k \neq 2$, σ is a radical element and there is a simple component N_i of $N/\text{rad}(N)$ with $N_i \cong N/L(\theta(\sigma)) \cong GF(p)$.

In this way we have a one-to-one correspondence between the prime divisors p of $\frac{n!}{n-2}$ and the direct summands $GF(p)$ of $N/\text{rad}(N)$.

Our main result of this section is now a corollary of Theorem 3.

Corollary. *Let R be a finite ring which is a direct sum of prime fields. Then there exists a positive integer n such that R is a homomorphic image of the near-ring $M_A(S_n)$, $A = Inn(S_n)$.*

Proof. Let $m = |R|$. Choose n large enough so that m divides $\frac{n!}{n-2}$. Since R is a direct sum of prime fields and since m divides $\frac{n!}{n-2}$, then R may be imbedded as a direct summand of $N/\text{rad}(N)$, where $N = M_A(S_n)$. This means R is a homomorphic image of $N/\text{rad}(N)$ by way of a projection map. Since $N/\text{rad}(N)$ is a homomorphic image of N, then R is a homomorphic image of N.

IV. Further Results on the Structure of $M_A(S_n)$, $A = Inn(S_n)$.

A. A decomposition of $M_A(S_n)$.

It is the goal of this subsection to show that the near-ring $M_A(S_n)$ decomposes into a direct sum of near-rings in a natural way. To find this decomposition, we need to define an equivalence relation on the set (A, S_n) of nonzero A-orbits of S_n. The equivalence classes will then give rise to a decomposition of $M_A(S_n)$.

For two nonzero orbits θ_1 and θ_2 in (A, S_n), define θ_1 r-related to θ_2 denoted $\theta_1 r \theta_2$, if either $\theta_1 \leq \theta_2$ or $\theta_2 \leq \theta_1$. Further, define θ_1 and θ_2 to be R-related, denoted $\theta_1 R \theta_2$, if there is a finite sequence of nonzero orbits $\theta_1 = \beta_1, \beta_2, \ldots,$ $\beta_k = \theta_2$ such that $\beta_1 r \beta_2, \beta_2 r \beta_3, \ldots, \beta_{k-1} r \beta_k$. This R-relation is an equivalence relation on the set (A, S_n) of nonzero A-orbits.

We recall that an orbit θ in (A, S_n) is a minimal orbit if θ has the property that if β is an A-orbit such that $\beta \leq \theta$, then either $\beta = \{0\}$ or else $\beta \sim \theta$. Every R-equivalence class of (A, S_n) must contain a minimal orbit. Moreover, the R-relation on (A, S_n) induces an equivalence relation on the set Θ of minimal A-orbits, namely minimal A-orbits α and β are related if and only if they are R-related in (A, S_n). Since every R-equivalence class of (A, S_n) must contain a minimal orbit, then the number of R-equivalence classes of (A, S_n) equals the number of equivalence classes of Θ.

We need a description of the minimal A-orbits of S_n. This is given by the following lemma.

Lemma 9. *The minimal A-orbits of S_n, $A = \mathrm{Inn}(S_n)$, are of the form $\theta(\sigma)$, where σ has prime order with $|\mathrm{Fix}(\sigma)| \neq 2$.*

Sketch of proof. It is easy to see that if $\theta(\sigma)$ is minimal, then σ must have prime order. Moreover the following are true:

(a) If σ has prime order $p \neq 2$ with $|\mathrm{Fix}(\sigma)| = 2$, then $\theta(\sigma)$ is not a minimal orbit (for example in S_5 let $\sigma = (123)$ and $\beta = (45)$, then $\theta(\beta) < \theta(\sigma)$).

(b) If σ has order 2 with $|\mathrm{Fix}(\sigma)| = 2$ and if $n \neq 4$, then $\theta(\sigma)$ is not minimal (for example in S_6 let $\sigma = (14)(25)$ and $\beta = (14)(25)(36)$, then $\theta(\beta) < \theta(\sigma)$).

(c) If $n = 4$, then $\theta(12)$ is not minimal since $\theta(12)(34) < \theta(12)$.

(d) If σ has prime order with $|\mathrm{Fix}(\sigma)| \neq 2$, then $\theta(\sigma)$ is a minimal orbit.

Our immediate goal is to describe the R-equivalence classes of (A, S_n). The situations $n = 3$ and $n = 4$ can be computed individually.

Assume $n \geq 5$. Let $\theta(\sigma)$ be a minimal A-orbit of S_n where $|\mathrm{Fix}(\sigma)| > 2$. We will show that $\theta(\sigma)$ is R-related to $\theta(\tau)$, the minimal orbit of transpositions.

Assume σ has prime order $p \neq 2$. Let τ be a transposition moving two elements of $\mathrm{Fix}(\sigma)$. Then $\sigma\tau = \tau\sigma$, $(\sigma\tau)^2 = \sigma^2$, $(\sigma\tau)^p = \tau$ imply that $\theta(\sigma\tau) > \theta(\sigma)$ and $\theta(\sigma\tau) > \theta(\tau)$. This shows $\theta(\sigma)$ and $\theta(\tau)$ are R-related.

Assume σ has prime order $p = 2$. Then σ is a product of transpositions. Since $|\mathrm{Fix}(\sigma)| > 2$, there is a 3-cycle β which moves three elements of $\mathrm{Fix}(\sigma)$. We have $\sigma\beta = \beta\sigma$, $(\sigma\beta)^3 = \sigma$, $(\sigma\beta)^2 = \beta^2$ and so $\theta(\sigma\beta) > \theta(\sigma)$, $\theta(\sigma\beta) > \theta(\beta)$. This shows $\theta(\sigma)$ and $\theta(\beta)$ are R-related. If $n > 5$, then $|\mathrm{Fix}(\beta)| > 2$ and $\theta(\beta)$ is R-related to $\theta(\tau)$ and therefore $\theta(\sigma)$ is too. If $n = 5$, then σ must be a transposition and there is nothing to show.

We have now demonstrated that if $n \geq 5$, then $\{\theta(\sigma) \mid |\mathrm{Fix}(\sigma)| > 2\}$ is a subset of the R-equivalence class of (A, S_n) which contains the orbit $\theta(\tau)$ of transpositions.

The following lemmas will be helpful in the situation where $\theta(\sigma)$ is a minimal orbit with $|\mathrm{Fix}(\sigma)| < 2$.

Lemma 10. *Let α be an element of S_n. Let M be the set of elements moved by α and let $T = \mathrm{Fix}(\alpha)$. If $\beta \in S_n$ is such that $\alpha\beta = \beta\alpha$, then both M and T are setwise invariant under β.*

Proof. If $i \in T$, then $\alpha\beta(i) = \beta\alpha(i) = \beta(i)$ means $\beta(i) \in T$. Since β permutes $M \cup T$ with $M \cap T = \emptyset$, and since $\beta(T) = T$, then $\beta(M) = M$.

Lemma 11. *Let α be an r-cycle in S_n, let M be the set of elements moved by α and let $T = \mathrm{Fix}(\alpha)$. Then $\mathrm{stab}(\alpha) = \{\alpha^i \omega \mid i \in Z$ and ω is any permutation disjoint from $\alpha\}$.*

Proof. Since every $\alpha^i \omega$ commutes with α, then $\mathrm{stab}(\alpha) \supseteq \{\alpha^i \omega\}$. Let $\beta \in \mathrm{stab}(\alpha)$. By Lemma 10, $\beta(T) = T$. This means we may write β in the form $\beta = \gamma\omega$ where $\omega(T) = T$, γ is the identity on T and γ, ω are disjoint. Since $\alpha\beta = \beta\alpha$ and $\omega\alpha = \alpha\omega$, then $\gamma\alpha = \alpha\gamma$.

Since $|M| = r$, we may regard α and γ as members of S_r. In S_r we have $|\theta(\alpha)| = r!/r = (r-1)!$. This implies that the order of stab (α) in S_r is $r!/(r-1)! = r$. Thus in S_r, we have stab $(\alpha) = \langle \alpha \rangle$. This means $\gamma = \alpha^i$ for some i. So in S_n we have $\beta = \alpha^i \omega$ as desired. This shows stab $(\alpha) = \{\alpha^i \omega\}$.

Lemma 12. *Suppose n is even with $n \geq 6$. Let $\gamma \in S_n$ be a product of disjoint transpositions with $|\text{Fix}(\gamma)| = 0$. Then $\theta(\gamma)$ is R-related to $\theta(\tau)$, the orbit of transpositions.*

Proof. Let $\gamma = \tau_1 \cdots \tau_{s-1}\tau_s$, a product of disjoint transpositions. Let $\beta = \tau_1 \cdots \tau_{s-1}$. Then $|\text{Fix}(\beta)| = 2$ since $|\text{Fix}(\gamma)| = 0$. Lemma 10 implies stab $(\beta) \subset$ stab (γ), so $\theta(\beta) > \theta(\gamma)$. Lemma 10 also implies $\theta(\beta) > \theta(\tau_s)$. So $\theta(\gamma)$ is R-related to $\theta(\tau_s) = \theta(\tau)$.

Let $\theta(\sigma)$ be a minimal A-orbit of S_n with $|\text{Fix}(\sigma)| \leq 1$. Since $\theta(\sigma)$ is minimal, then σ has prime order p and $\sigma = \sigma_1 \cdots \sigma_t$, a product of disjoint cycles of length p.

Suppose $t = 1$. Then σ is a cycle of length p with $|\text{Fix}(\sigma)| = 0$ or 1. (This situation occurs when n or $n-1$ is prime.) If $\theta(\gamma) \geq \theta(\sigma)$, then Lemma 11 implies $\gamma = \sigma^i$ for some i. This means $\theta(\gamma) = \theta(\sigma)$ or else $\theta(\gamma) = \{(1)\}$. So in the case $t = 1$, $\{\theta(\sigma)\}$ is an R-equivalence class.

Suppose $t = 2$ and $|\text{Fix}(\sigma)| = 0$. Then $\sigma = \sigma_1\sigma_2$ and there is a 2p-cycle γ such that $\gamma^2 = \sigma$. Also γ^p is a product of p transpositions with $|\text{Fix}(\gamma^p)| = 0$. By Lemma 12, $\theta(\gamma^p)$ is R-related to $\theta(\tau)$, the orbit of transpositions. Since $\theta(\sigma) < \theta(\gamma)$ and $\theta(\gamma^p) < \theta(\gamma)$, then $\theta(\sigma)$ is R-related to $\theta(\tau)$.

Suppose $t = 2$, $|\text{Fix}(\sigma)| = 1$ and $p \neq 2$. Let γ be the 2p-cycle as above. Then γ^p has order 2 and is a product of p transpositions, say $\gamma^p = \tau_1 \cdots \tau_{p-1}\tau_p$. There exists a cycle β and an integer k such that $\beta^k = \tau_1 \cdots \tau_{p-1}$. Let $\alpha = \beta\tau_p$. Since $p \neq 2$, then β is not a 2-cycle. So $|\text{Fix}(\sigma)| = 1$ implies stab $(\alpha) = \{\beta^i\tau_p^j \mid i, j \in \mathbf{Z}\}$. This means stab $(\alpha) \subseteq$ stab (γ^p), so $\theta(\gamma^p) \geq \theta(\alpha)$. Since $\alpha^2 = \beta^2$, then $\theta(\alpha) < \theta(\beta)$. Finally, $\theta(\beta^k) < \theta(\beta)$. Since $|\text{Fix}(\beta^k)| = 3$, then $\theta(\beta^k)$ is R-related to the orbit of transpositions, $\theta(\tau)$, and so are $\theta(\gamma^p)$ and $\theta(\gamma)$. Since $\theta(\sigma) < \theta(\gamma)$, then $\theta(\sigma)$ is R-related to $\theta(\tau)$.

If $t = 2$, $|\text{Fix}(\sigma)| = 1$ and $p = 2$, then we are in S_5, with σ being a product of two transpositions. In this case $\{\theta(1234), \theta(13)(24)\}$ is an equivalence class.

We have now established the following theorem.

Theorem 4. *Let $N = M_A(S_n)$, where $A = \text{Inn}(S_n)$. Let (A, S_n) denote the set of nontrivial A-orbits of S_n, and let $C(\theta(\sigma))$ denote the R-equivalence class containing the orbit $\theta(\sigma)$.*

(a) *The R-equivalence classes of (A, S_3) are $C(\theta(12))$ and $C(\theta(123))$.*

(b) *The R-equivalence classes of (A, S_4) are $C(\theta(12))$ and $C(\theta(123))$.*

(c) *The R-equivalence classes of (A, S_5) are $C(\theta(12345))$, $C(\theta(1234))$ and $C(\theta(12))$.*

(d) *If $n \geq 6$ and if n is prime, then the R-equivalence classes of (A, S_n) are $C(\theta(12))$ and $C(\theta(\sigma))$ where σ is a cycle of length n.*

(e) *If $n \geq 6$ and if $n-1$ is prime, then the R-equivalence classes of (A, S_n) are $C(\theta(12))$ and $C(\theta(\sigma))$ where σ is a cycle of length $n-1$.*

(f) *If $n \geq 6$ and if neither n nor $n - 1$ is prime, then (A, S_n) has only one equivalence class, namely $C(\theta(12))$.*

Lemma 13. *Let $C(\theta(\sigma))$ be an R-equivalence class of (A, S_n). Let $N_\sigma = \{f \in M_A(S_n) \mid f(\theta(\beta)) = \{(1)\}$ for all $\theta(\beta) \notin C(\theta(\sigma))\}$. Then N_σ is an ideal of $M_A(S_n)$.*

Proof. Since N_σ is an annihilator of a set of elements in S_n, then it is a left ideal of $M_A(S_n)$. Since $C(\theta(\sigma))$ is an R-equivalence class, then for every g in $M_A(S_n)$ and for every orbit θ in $C(\theta(\sigma))$, we have $f(\theta) \in C(\theta(\sigma))$. This implies N_σ is a right ideal.

In the following theorem, N_σ denotes the ideal of $M_A(S_n)$ corresponding to the equivalence class $C(\theta(\sigma))$ of (A, S_n), as in Lemma 13 above. The proof follows easily from the preceding work.

Theorem 5.
(a) $M_A(S_3) = N_{(12)} \oplus N_{(123)}$.
(b) $M_A(S_4) = N_{(12)} \oplus N_{(123)}$.
(C) $M_A(S_5) = N_{(12345)} \oplus N_{(1234)} \oplus N_{(12)}$.
(d) *If $n \geq 6$ and n is prime, then $M_A(S_n) = N_{(12)} \oplus N_\sigma$, where σ is an n-cycle.*
(e) *If $n \geq 6$ and $n - 1$ is prime, then $M_A(S_n) = N_{(12)} \oplus N_\sigma$, where σ is a cycle of length $n - 1$.*
(f) *If $n \geq 6$ and neither n nor $n - 1$ is prime, then $M_A(S_n) = N_{(12)}$.*

B. On the Wedderburn Principal Theorem.

In the theory of rings, especially rings which are finite dimensional algebras, it is often the case that a ring R has a subring T such that $R = \mathrm{rad}(R) + T$, $\mathrm{rad}(R) \cap T = \{0\}$ (so necessarily $R/\mathrm{rad}(R) \cong T$). This is the Wedderburn Principal Theorem (see [4]). We show in this subsection that the obvious near-ring analogy to the Wedderburn Principal Theorem holds in $M_A(S_n)$ only when $n = 2$, 3 and 6.

The cases $n = 2$ and 3 are trivial since $M_A(S_2)$ and $M_A(S_3)$ are simple and semi-simple respectively.

Let $N = M_A(S_4)$. We will show that there is no subnear-ring T of N such that $N = \mathrm{rad}(N) + T$ and $\mathrm{rad}(N) \cap T = \{0\}$. Suppose, on the contrary, that there is such a T. Let t be an element of T, then $t = t1 = te_1 + te_2 + te_3 + te_4$, where the e_j's are idempotents associated with each of the four nontrivial A-orbits of S_4. Let e_1 be the idempotent corresponding to the orbit of 4-cycles. Since $(N, +)$ is abelian, we have

$$6t = 6te_1 + 6te_2 + 6te_3 + 6te_4 = 2te_1.$$

This means $6t = 2te_1$ belongs to T since T is a subnear-ring of N. Moreover, $2te_1(1234)$ is not a 4-cycle, so it is either (1) or (13)(24). Corollary 1 to Theorem 2 implies that $2te_1$ belongs to $\mathrm{rad}(N)$. We now have $2te_1$ in $\mathrm{rad}(N) \cap T = \{0\}$, so $2te_1 = 0$. Since a 4-cycle has order 4 and $2te_1 = 0$, then $te_1 \in \mathrm{rad}(N)$. We have now shown that for every t in T, te_1 belongs to $\mathrm{rad}(N)$.

Since $e_1 \in N$ and $N = \text{rad}(N) + T$, then there exist $r \in \text{rad}(N)$ and $t \in T$ such that $e_1 = r + t$. We have $e_1 = (r + t)e_1 = re_1 + te_1$. Since r and te_1 belong to $\text{rad}(N)$, then e_1 belongs to $\text{rad}(N)$. But in $N = M_A(S_4)$, one may verify that $\text{rad}(N)$ is nilpotent and so it does not contain e_1. This contradiction shows T does not exist.

We show now that $N = M_A(S_6)$ does satisfy the Wedderburn Principal Theorem. In $M_A(S_6)$ the only radical orbits are the minimal orbits. So $\text{rad}(N) = \{f \in N \mid f(\theta) = \{(1)\}$ for every minimal A-orbit $\theta\}$. Let $T = \{f \in N \mid f(\theta) = \{(1)\}$ for every non-minimal A-orbit $\theta\}$. Then T is a subnear-ring of N, $\text{rad}(N) + T = N$ and $\text{rad}(N) \cap T = \{0\}$.

Theorem 6. *The near-ring $N = M_A(S_n)$, $A = Inn(S_n)$, satisfies the Wedderburn Principal Theorem if and only if $n \in \{2, 3, 6\}$.*

Proof. It remains to show that if $n = 5$ or $n > 6$, then N does not satisfy the Wedderburn Principal Theorem.

Let $\sigma \in S_n$ be a cycle whose length r is a power of 2, $r = 2^k$, and such that $r \neq n - 2$ and r is maximal with these properties. Since $n = 5$ or $n > 6$, then $k > 1$. (For examples, if $n = 5$ or $n = 7$, then $r = 2^2$. If $n = 8$, then $r = 2^3$. If $n = 10$, then $r = 2^2$.) We note that the orbit $\theta(\sigma)$ is a radical orbit.

Suppose T is a subnear-ring of N such that $N = \text{rad}(N) + T$ with $\text{rad}(N) \cap T = \{0\}$. If $t \in T$, then $t = \sum_i te_i$ where e_i is the idempotent associated with the nontrivial A-orbit θ_i of S_n. If e_i belongs to $\text{rad}(N)$ (which will be the case if θ_i is not a radical orbit), then te_i belongs to $\text{rad}(N)$. If e_i does not belong to $\text{rad}(N)$, then e_i is associated with a radical orbit $\theta_i(\alpha)$, where α is homogeneous of prime power order and $|\text{Fix}(\alpha)| \neq 2$. Let e_1 be associated with the orbit $\theta(\sigma)$ where σ is defined as above.

Select the integer $m > 0$ large enough so that 2^{k-1} divides m, but 2^k does not, and such that $me_i = 0$ for all $e_i \notin \text{rad}(N)$ where e_i is associated with θ_i whose elements have prime power order p^s, $p \neq 2$. Since $(N, +)$ is abelian we have $mt = \sum_i mte_i$.

Since $mte_1(\theta(\sigma)) < \theta(\sigma)$, then $mte_1 \in \text{rad}(N)$. This means mte_j is in $\text{rad}(N)$ for all j and therefore $mt \in \text{rad}(N)$. Since $mt \in T$, then $mt \in \text{rad}(N) \cap T = \{0\}$ and so $mt = 0 = \sum_i mte_i$. This forces $mte_i = 0$ for every i. In particular $mte_1 = 0$, so $te_1(\theta(\sigma)) < \theta(\sigma)$ and te_1 belongs to $\text{rad}(N)$ for all $t \in T$.

Since $N = \text{rad}(N) + T$ then there exist elements $r \in \text{rad}(N)$ and $t \in T$ such that $e_1 = r + t$. We have $e_1 = (r + t)e_1 = re_1 + te_1$. Since both summands belong to $\text{rad}(N)$, then e_1 belongs to $\text{rad}(N)$, a contradiction since e_1 does not belong to $\text{rad}(N)$. So T cannot exist.

C. The distributor ideal of $M_A(S_n)$, $A = Inn(S_n)$.

For a near-ring N (with identity) let $S = \{a(b+c) - ac - ab \mid a, b, c \in N\}$, the set of distributors of N. Let $D(N)$ be the ideal generated by S. The ideal $D(N)$ is called the distributor ideal of N and it is the smallest ideal I of N such that N/I is a ring. In other words, $N/D(N)$ is a ring and if I is an ideal of N such that N/I is a ring, then $D(N) \subseteq I$. Now let $N = M_A(S_n)$. Since $N/\text{rad}(N)$ is a ring, then $D(N) \subseteq \mathrm{rad}\,(N)$. S. Bagley showed in [1] that for a centralizer near-ring one always has $\mathrm{rad}\,(N) \subseteq D(N)$. This gives the following result.

Theorem 7. Let $N = M_A(S_n)$, $A = Inn(S_n)$. Then $D(N) = \mathrm{rad}\,(N)$, that is $\mathrm{rad}\,(N)$ is the smallest ideal I of N such that N/I is a ring.

References

[1] S. Bagley, *Polynomial near-rings, distributor and J_2 ideals of generalized centralizer near-rings*, Ph.D. thesis, Texas A & M University, 1993.

[2] C. J. Maxson, M. R. Pettet and K. C. Smith, On semi-simple rings that are centralizer near-rings, *Pacific Journal of Mathematics* **101** (1982), 451–461.

[3] C. J. Maxson and K. C. Smith, The centralizer of a set of group automorphisms, *Communications in Algebra* **8** (1980), 211–230.

[4] B. R. McDonald, *Finite Rings With Identity*, Marcel Dekker, Inc., New York, 1974.

[5] A. Myasnikov, Centroid of a group and its links with endomorphisms and rings of scalars, preprint.

[6] G. Pilz, *Near-rings*, North Holland, Amsterdam, 1983

HOMOGENEOUS MAPS OF FREE RING MODULES

Andries B. van der Merwe

Department of Mathematics, Texas A & M University
College Station, Texas 77843-3368, U. S. A.

ABSTRACT

Given an R-module V, the near-ring of homogeneous maps $\mathcal{M}_R(V)$ is the set of maps $\{f : V \to V \mid (rv)f = r(vf)$ for all $r \in R$ and $v \in V\}$ under point-wise addition and composition of functions. When R is an integral domain and V is a finitely generated free module, the set $\{(v)f \mid f \in \mathcal{M}_R(V)\}$ will be described for any fixed element v in V.

1. Introduction

In this note, R will be a commutative integral domain with identity, and we will also assume that $R \neq \mathbb{Z}_2$. The set of maps $\mathcal{M}_R(R^n) := \{f : V \to V \mid (rv)f = r(vf)$ for all $r \in R$ and $v \in R^n\}$ is a left near-ring under point-wise addition and composition of functions, and the elements are called *homogeneous maps*. We will write the functions of $\mathcal{M}_R(R^n)$ on the right of the elements on which they act, therefore $\mathcal{M}_R(R^n)$ will satisfy the left distributive law. If S is a semigroup of endomorphisms of a group G, then the near-ring $\mathcal{M}_S(G) := \{f : G \to G \mid s(gf) = (sg)f$, for all $g \in G, s \in S\}$ has been investigated by several people (see for example [4], p. 62), and is known as a centralizer near-ring. Thus, $\mathcal{M}_R(V)$ is a specific example of a centralizer near-ring. In order to prove structure theorems for $\mathcal{M}_R(R^n)$, and to answer, for example, questions on whether or not it is possible to represent the homogeneous maps in $\mathcal{M}_R(R^n)$ locally by endomorphisms (see [3], section 5), it is essential to have a workable description of the sets $\{(v)f \mid f \in \mathcal{M}_R(V)\}$. We will obtain an unexpected relationship between these sets and divisorial ideals (see [2]), which arose in the investigations of Van der Waerden, Artin, and Krull in the 1930's. This will enable us to translate questions about homogeneous maps, to questions about divisorial ideals and vice versa.

2. General results

In this section we will generalize the concept of a centralizer near-ring, and then prove a generalization of Betsch's lemma (see [4], lemma 3.30). This will then be used to describe the sets $\{(v)f \mid f \in \mathcal{M}_R(R^n)\}$.

Let H be a subgroup of the group G, and let S be a semigroup of endomorphisms of G. By abuse of notation, we will denote by $\mathcal{M}_S(H)$ the near-ring $\{f : H \to H \mid s \in S$ and $s(h) \in H$ implies that $(sh)f = s(hf)\}$. For the remaining part of this section we will assume that $S = A$, where A is a group of automorphisms of G. Most of the useful concepts used in the study

Y. Fong et al. (eds.), Near-Rings and Near-Fields, 271–273.

of centralizer near-rings can be generalized in order to facilitate this more general definition. We can, for example, define an equivalence relation $\sim$ on H, by $h_1 \sim h_2$ if there exists an $a \in A$ such that $a(h_1) = h_2$, and refer to the non-zero equivalence classes as orbits. If $\overline{A} = \{a \in A \mid s(H) \subseteq H\}$, then $\mathcal{M}_{\overline{A}}(H) \supseteq \mathcal{M}_A(H)$, but we do not have equality in general. To see this, let G be the rationals under addition, $H = \mathbf{Z}$, $A = \{(\frac{2}{3})^k \mid k \in \mathbf{Z}\}$, and assume A acts on G by multiplication. Then $\mathcal{M}_{\overline{A}}(H) = \mathcal{M}_{\{1\}}(H) \neq \mathcal{M}_A(H)$, since if $f \in \mathcal{M}_A(H)$, then $(2)f = (\frac{2}{3} \cdot 3)f = \frac{2}{3}(3f)$. The stabilizer of $g \in G$ in A is defined by $St_A(g) := \{a \in A \mid a(g) = g\}$. Also, for $h \in H$, we will denote the set $\{a \in A \mid a(h) \in H\}$ by $St_A(h, H)$. Now we have all the appropriate definitions to state and prove a generalized Betsch's lemma.

Lemma 2.1. *Suppose $(G, +)$ is a group (written additively but not necessarily abelian), and let H be a subgroup of G, $h_1 \in H \setminus \{0\}$ and $h_2 \in H$. Then there exists $f \in \mathcal{M}_A(H)$ such that $(h_1)f = h_2$ if and only if $St_A(h_1) \subseteq St_A(h_2)$ and $St_A(h_1, H) \subseteq St_A(h_2, H)$.*

Proof: Let $f \in \mathcal{M}_A(H)$ be such that $(h_1)f = h_2$. If $a \in St_A(h_1, H)$, then $(ah_1)f = ah_2$ implies that $a \in St_A(h_2, H)$. Conversely, if $St_A(h_1) \subseteq St_A(h_2)$ and $St_A(h_1, H) \subseteq St_A(h_2, H)$, define $f : H \to H$ by

$$(ah_1)f = ah_2 \quad \text{for all } a \in St_A(h_1, H),$$
$$(k)f = 0 \quad \text{for all } k \in H \setminus St_A(h_1, H)h_1.$$

See [4], Lemma 3.30 for the remaining details.

3. Homogeneous maps of free ring modules

In this section we will use Lemma 2.1, with $G = Q^n$, $H = R^n$ and $A = Q^* = Q \setminus \{0\}$, where Q is the quotient field of R. Also, if $q \in Q^*$ and $v = (x_1, \ldots, x_n) \in Q^n$, then we will assume $qv = (qx_1, \ldots, qx_n)$. Since $St_A(h)$ is $\{1\}$ for all non-zero $h \in R^n$, we have that there exists $f \in \mathcal{M}_{Q^*}(R^n)$ such that $(h_1)f = h_2$ if and only if $St_A(h_1, H) \subseteq St_A(h_2, H)$. But now the key observation is that $\mathcal{M}_{Q^*}(R^n)$ equals $\mathcal{M}_R(R^n)$. To see this, first note that $\mathcal{M}_Q(R^n) = \mathcal{M}_{Q^*}(R^n)$, since $\mathcal{M}_{Q^*}(R^n)$ is zero-symmetric by the assumption that R has more than one non-zero element. So we have trivially that $\mathcal{M}_{Q^*}(R^n) \subseteq \mathcal{M}_R(R^n)$. Conversely, if $f \in \mathcal{M}_R(R^n)$, $v \in R^n$, $\frac{r}{s} \in Q^*$ and $\frac{r}{s}v \in R^n$, then $s(\frac{r}{s}vf) = r(vf)$, implies that $(\frac{r}{s}v)f = \frac{r}{s}(vf)$, so we get the following result.

Theorem 3.1. *Let $v \in R^n \setminus \{0\}$, then $\{(v)f \mid f \in \mathcal{M}_R(R^n)\} = \{w \in R^n \mid St_{Q^*}(v, R^n) \subseteq St_{Q^*}(w, R^n)\}$.*

Now we will see how this theorem relates homogeneous maps to divisorial ideals. For $v = (x_1, \ldots, x_n) \in R^n \setminus \{0\}$, let $I(x_1, \ldots, x_n) = \{y \in R \mid (v)f = (y, 0, \ldots, 0)$ for some $f \in \mathcal{M}_R(R^n)\}$. By using the fact that $End_R(R^n) \subseteq \mathcal{M}_R(R^n)$, we see that $\{(v)f \mid f \in \mathcal{M}_R(R^n)\} = \{(y_1, \ldots, y_n) \in R^n \mid y_i \in I(x_1, \ldots, x_n)\}$. It is also not too hard to see that $I(x_1, \ldots, x_n)$ is an ideal of R, and that it contains the ideal generated by $x_1, \ldots, x_n$. So from Theorem 3.1 and the definition of $I(x_1, \ldots, x_n)$, we have that $I(x_1, \ldots, x_n) = \{y \in R \mid q \in Q \text{ and } qx_i \in R, i = 1, 2, \ldots, n, \text{ imply that } qy \in R\}$. Recall from [2], that a non-zero fractional ideal F of an integral domain D with quotient field K, is *divisorial* if there is a non-zero fractional ideal E of D such that $F = E^{-1} = D : E = \{x \in K \mid xE \subseteq D\}$. Let $F = J^{-1}$, where J is the ideal generated by $x_1, \ldots, x_n$, then since $F = \{q \in Q \mid qx_i \in R, i = 1, 2, \ldots, n\}$, we have that $I(x_1, \ldots, x_n) = F^{-1} = (J^{-1})^{-1}$, and thus $I(x_1, \ldots, x_n)$ is an integral divisorial ideal. So we see that we can identify $\{(v)f \mid f \in \mathcal{M}_R(R^n)\}$ with the divisorial ideal $I(x_1, \ldots, x_n)$. We also have a mapping that takes an ideal that can be generated by n elements, say $z_1, \ldots, z_n$, to the divisorial ideal $I(z_1, \ldots, z_n)$. These concepts are closely related to $\star$-Operations and v-Operations (see chapter 5 in [1]). This connection established above, can be used to translate results obtained by Heinzer into theorems about homogeneous maps.

References

[1] R. Gilmer, *Multiplicative ideal theory*, Marcel Dekker, Inc., Pure and Applied Mathematics, vol. 12, New York, 1972.

[2] W. Heinzer, Integral domains in which each non-zero ideal is divisorial, *Mathematica* **15** (1968), 164–170.

[3] C. J. Maxson and A. P. J. van der Walt, Homogeneous maps as piecewise endomorphisms, *Comm. in Alg.* **20** (1992), 2755–2776.

[4] J. D. P. Meldrum, *Near-rings and their links with groups*, Pitman Research Notes Series, No. 134, 1985.

A DECODING STRATEGY FOR EQUAL WEIGHT CODES FROM FERRERO PAIRS *

Gerhard G. Wagner

Institut für Mathematik, Johannes Kepler Universität Linz
Altenbergerstr. 69, A-4040 Linz, Austria

ABSTRACT

Ferrero pairs (N, Φ), where $(N, +)$ is a group and Φ is a group of fixed point free automorphisms on N, can be used to construct equal weight codes. In the sequel, a decoding strategy for this kind of codes is discussed.

1 Basic Definitions

Definition 1 *[2] A (finite) Ferrero pair is an ordered pair (N, Φ), where*

- $(N, +)$ *is a finite group;*

- $\Phi \leq Aut(N)$ *with $\Phi \neq \{id_N\}$, and every $\varphi \in \Phi \setminus \{id_N\}$ is fixed point free, i.e.,*

$$\forall x \in N : \quad \varphi(x) = x \Longrightarrow x = 0. \tag{1}$$

Definition 2 *[1] Let X be a set of v elements ("points") and let $\mathcal{B}$ be a proper subset of $\mathcal{P}(X)$, the power set of X, with b elements ("blocks"). The pair $(X, \mathcal{B})$ is called balanced incomplete block design (BIB-design, BIBD), if all the following conditions are fulfilled:*

- *Every point $x \in X$ is contained in exactly r blocks of $\mathcal{B}$.*

- *Every block $B \in \mathcal{B}$ consists of precisely k points with $k \geq 2$.*

- *There exists $\lambda \in \mathbb{N}$ such that any two points $x, y \in X$ with $x \neq y$ occur in exactly λ blocks.*

The numbers v, b, r, k and λ are the parameters of $(X, \mathcal{B})$.

2 Constructing an Equal Weight Code via Ferrero Pair and BIB-Design

Let (N, Φ) be a Ferrero pair and define

$$\mathcal{B}^* := \{\Phi(a) + b \mid a \in N \setminus \{0\}, \ b \in N\} \tag{2}$$

---

*Supported by the Austrian Fonds zur Förderung der wissenschaftlichen Forschung, Projekt P9111-PHY

Y. Fong et al. (eds.), Near-Rings and Near-Fields, 275–278.
© 1995 *Kluwer Academic Publishers.*

with $\Phi(a) + b := \{\varphi(a) + b \mid \varphi \in \Phi\}$ for fixed a and b. Then the following holds:

Theorem 1 *[2] $(N, \mathcal{B}^*)$ is a balanced incomplete block design with the following parameters:*

$$v = |N|, \ k = |\Phi|, \ b = v(v-1)/k, \ r = v-1, \ \lambda = k-1.$$

The sets $\Phi(a)$ are usually called "basic blocks" of the design.

Let us now denote the elements of N by $x_1, x_2, ..., x_v$. Every block $B \in \mathcal{B}^*$ induces a codeword c of the desired code in the following way:

$$c := (c_1, c_2, ...c_v) \in (\mathbf{Z}_2)^v \text{ with } c_i = \begin{cases} 1, & \text{if } x_i \in B; \\ 0, & \text{otherwise.} \end{cases} \quad (3)$$

This yields a $(v, b, 2(k - \mu))$-code $\mathcal{C}$, where

$$\mu = max\{|B_1 \cap B_2| : B_1, B_2 \in \mathcal{B}^*, \ B_1 \neq B_2\} \quad (4)$$

(cf. [4]). The weight w of all codewords in $\mathcal{C}$, i.e., the number of 1s in every codeword, obviously equals the parameter k of $(N, \mathcal{B}^*)$. Such codes are called *equal weight codes*.

3   Decoding the Resulting Codes

Suppose, a word $\mathcal{W}$ is coming in. If no error has been made during transmission, then $\mathcal{W}$ corresponds to a block B belonging to the underlying BIB-design. Otherwise, $\mathcal{W}$ induces a set $W := \{x_1, x_2, ..., x_z\}$ in the obvious way with $x_1, x_2, ..., x_z \in N$, and $z \in \mathbf{N}$. If $z > 2k - \mu - 1$ or $z < \mu + 1$, one can show (cf. [5]) that more than $k - \mu - 1$ errors must have occurred, which means that we cannot decode correctly. Thus, let us assume $2k - \mu - 1 \geq z \geq \mu + 1$ for the subsequent considerations. With $\Phi := \{\varphi_1, \varphi_2, ..., \varphi_k\}$, we get the following system of equations, where $a, b \in N$ are unknown:

$$\left\{ \begin{array}{ccc} x_1 & = & \varphi_{i_1}(a) & + & b \\ x_2 & = & \varphi_{i_2}(a) & + & b \\ \vdots & & \vdots & \\ x_z & = & \varphi_{i_z}(a) & + & b \end{array} \right\} \quad (5)$$

If errors in transmission have occurred, (5) may be unsolvable, but one can prove:

Proposition 1 *[5] There exists a subsystem of (5) with at least $\left\lceil \frac{z+\mu+1}{2} \right\rceil$ equations, which has a unique solution for a and b. (By $\lceil x \rceil$ we mean the smallest integer greater or equal to x for any $x \in \mathbf{Z}$.)*

In other words, Proposition 1 tells us that there must exist a subset W' of W with $|W'| \geq \left\lceil \frac{z+\mu+1}{2} \right\rceil$ and a uniquely determined pair $(a, b) \in (N \setminus \{0\} \times N)$ such that W' is contained in $\Phi(a) + b$. This can also be written as

$$(\exists!(a, b) \in (N \setminus \{0\} \times N)): \quad |W \cap (\Phi(a) + b)| \geq \left\lceil \frac{z + \mu + 1}{2} \right\rceil, \quad (6)$$

which eventually gives the following result:

Proposition 2

$$(\exists!(a, b) \in (N \setminus \{0\} \times N)) : \quad |(W - b) \cap \Phi(a)| \geq \left\lceil \frac{z + \mu + 1}{2} \right\rceil. \tag{7}$$

Proposition 2 leads us to a first attempt for decoding:

Algorithm 1 *Try to subtract elements $n \in N$ until one of them fulfils*

$$|(W - n) \cap \Phi(m)| \geq \left\lceil \frac{z + \mu + 1}{2} \right\rceil \tag{8}$$

for some basic block $\Phi(m)$. If we find such an n, we are done because of the uniqueness statement in Proposition 2 and we decode W to $\Phi(m) + n$.

As input for this algorithm we need W and $\{\Phi(a) \mid a \in N \setminus \{0\}\}$, which is not very much, and therefore good.

The problem is, that in whatever ordering one takes those n and $\Phi(m)$, it always can happen that we find the right n and m to decode only in the last intersection. Thus, we cannot improve the complexity of the worst case. Nevertheless, one observes that none of the basic blocks contains 0, the identity of $(N, +)$. Hence, it will be reasonable to extend Algorithm 1 as follows:

Algorithm 2
Step 1: First of all, try $n \in N \setminus W$ in (9).
Step 2: If Step 1 was not successful, try the other elements of N.

Now, how many (in principle) correctable errors on a certain codeword cannot be corrected by only using *Step 1*?
For all the following, let C be an $(n, M, 2(k - \mu))$-code constructed from a Ferrero pair as described in Section 2.

Proposition 3 *For any $c \in C$, the number $S_1(C)$ of errors on c, which cannot be corrected in Step 1 of Algorithm 2, is given by*

$$\sum_{i=0}^{k-\mu-2} \binom{n-1}{i}. \tag{9}$$

Proof: Suppose a codeword $\Phi(a) + b$ is sent and W, which is finally coming in at the other end of the line, contains less than $k - \mu - 1$ errors. The only case where *Step 1* is not successful occurs if b belongs to the incoming word W. Thus, $S_1(C)$ is the number of errors on $\Phi(a) + b$, where b is distorted, which obviously equals the sum above. □

If $S_2(C)$ denotes the number of all correctable errors for a fixed $c \in C$, we have (cf. [3])

$$S_2(C) = \sum_{i=0}^{k-\mu-1} \binom{n}{i}. \tag{10}$$

Since there is no closed form for $S_1(\mathcal{C})$ and $S_2(\mathcal{C})$, we must check their relationship for every single code. Table 1 compares the two sums for some examples, where we take the additive group of a finite nearfield $(N, +, \cdot)$ and define

$$\Phi := \{\varphi_a \mid a \in \Phi'\} \tag{11}$$

with $\Phi' \leq (N, \cdot)$ and $\varphi_a : N \to N$, $x \to a \cdot x$. (For a more detailed description see [6].)

Table 1: Comparison of $S_1(\mathcal{C})$ and $S_2(\mathcal{C})$ for some examples

n	k	μ	$S_1(\mathcal{C})$	$S_2(\mathcal{C})$	$\frac{S_1(\mathcal{C})}{S_2(\mathcal{C})}$
8	4	2	1	9	0.11
24	6	2	277	2325	0.12
24	12	6	10903	55455	0.20
15	5	2	15	121	0.12
48	24	12	6930532111	31278197839	0.22
48	8	2	195709	1925357	0.10
48	16	7	75358720	465174935	0.16

We see that in these cases most of the errors can be corrected by only using *Step 1*. Computer test have shown that, by the strategy described above, decoding time for 100 incoming words can be improved up to 30%, especially if the weight of the codewords is not too small (i.e., not too far from $\frac{n}{2}$).

References

[1] T. Beth, D. Jungnickel and H. Lenz, *Design Theory*, Bibl. Inst. Mannheim, 1985.

[2] J. R. Clay, *Nearrings - Geneses and Applications*, Oxford University Press, 1992.

[3] F. J. MacWilliams and N. J. A. Sloane, *The Theory of Error-Correcting Codes*, North-Holland, Amsterdam 1977.

[4] P. R. Fuchs, G. Hofer and G. Pilz, *Codes from Planar Near-Rings*, IEEE-Trans. on Information Theory 36 (1990), 647-651.

[5] P. R. Fuchs, *A decoding method for planar near-ring codes*, Riv. Mat. Univ. Parma (4) **17** (1991), 325-331.

[6] G. G. Wagner, *On Constructing BIB-Designs and Constant Weight Codes from Nearfield-Generated Planar Nearrings*, Diploma Thesis, 1992.